D0930085

Numerical Solution of Elliptic Problems

SIAM Studies in Applied Mathematics

JOHN A. NOHEL, Managing Editor

This series of monographs focuses on mathematics and its applications to problems of current concern to industry, government, and society. These monographs will be of interest to applied mathematicians, numerical analysts, statisticians, engineers, and scientists who have an active need to learn useful methodology for problem solving.

The first five titles in this series are: *Lie-Bäcklund Transformations in Applications*, by Robert L. Anderson and Nail H. Ibragimov; *Methods and Applications of Interval Analysis*, by Ramon E. Moore; *Ill-Posed Problems for Integrodifferential Equations in Mechanics and Electromagnetic Theory*, by Frederick Bloom; *Solitons and the Inverse Scattering Transform*, by Mark J. Ablowitz and Harvey Segur; and *Fourier Analysis of Numerical Approximations of Hyperbolic Equations*, by Robert Vichnevetsky and John B. Bowles.

Garrett Birkhoff and Robert E. Lynch

Numerical Solution of Elliptic Problems

siam *Philadelphia/1984*

This book was set by the second author using TROFF, EQN, TBL, and other UNIX text-processing software, the Purdue University Computer Science Department's VAX 11/780, and the phototypesetting facilities of the University of Illinois, Urbana, Illinois.

Library of Congress Catalog Card Number 84-51823
ISBN: 0-89871-197-5

Contents

Preface

The science of solving elliptic problems has been revolutionized in the last 35 years. Today's large-scale, high-speed computers can solve most two-dimensional boundary value problems at moderate cost *accurately*, by a variety of numerical methods. The aim of this monograph is to provide a reasonably well-rounded and up-to-date survey of these methods.

Like its predecessor (Birkhoff [69]), written at an earlier stage in the Computer Revolution, our book emphasizes problems which have important scientific and/or engineering applications, and which are solvable at moderate cost on current computing machines. Because of this emphasis, it devotes much space to the two-dimensional 'linear source problem'

$$-\nabla \cdot [p(x,y)\nabla u] + q(x,y)u = f(x,y), \qquad p > 0, \quad q \geqslant 0.$$

We make no claim to completeness. Indeed, our main concern is with *linear* boundary value problems, and we say little about eigenproblems. Readers seeking additional information should consult the carefully selected references cited in our many footnotes.

Chapters 1 and 2 are preliminary in nature, and are included to make our book largely self-contained. The first chapter explains the physical origins of several typical, relatively simple elliptic problems, indicating their practical importance. The second chapter reviews some of the most helpful and easily appreciated relevant theorems of classical analysis. It is the properties stated in these theorems (e.g., the smoothness of solutions), that give to elliptic problems their special mathematical flavor. Classical analysis also provides known exact solutions to many 'model problems', which can be used to test the accuracy of numerical methods.

Our analysis of numerical methods begins in Chapter 3. Some of the best known and most successful *difference* approximations to elliptic problems are reviewed, with emphasis on their simplicity and accuracy. Examples are given to show how their accuracy depends on the problem being solved, as well as on roundoff, Richardson extrapolation, etc.

We then devote two chapters to effective algorithms for solving numerically the very large systems of linear algebraic equations (involving

200–5000 or more unknowns) to which such difference approximations give rise. After briefly sketching several direct 'sparse matrix' methods in the first part of Chapter 4, we concentrate on *iterative* and semi-iterative methods. The latter are not only advantageous for treating very large problems and essential for solving *nonlinear* problems, but they seem destined to play a crucial role in solving the *three-dimensional* elliptic problems whose solution will, we hope, become routine during the next decade.

Chapters 6 and 7 return to approximation methods, especially the finite element methods (FEM) that have been adopted so widely during the past two decades. In analyzing these FEM, we emphasize techniques for piecewise polynomial approximations having higher-order accuracy, and simple estimates of their errors. Though our discussion is much less general than that in Philippe Ciarlet's admirable book *The Finite Element Method for Elliptic Problems*, we hope that our sharp error estimates for the most widely used piecewise polynomial approximations will be adequate for many practical purposes.

Chapter 8 gives a brief review of integral equation methods. Although far less versatile and less widely used than difference or FEM methods, these give extremely accurate results with little computation in some important special cases. Moreover, their theoretical analysis is mathematically interesting for its own sake, involving considerations that help to round out and complete Chapter 2.

Our book concludes with a short description of ELLPACK, a powerful new system designed to solve elliptic boundary value problems. The relevant tasks, such as placing a grid on a domain, discretizing the differential equation and the boundary conditions, sequencing the resulting linear algebraic equations, solving the system, and producing printed or plotted output, are all done automatically by ELLPACK. ELLPACK uses the approximation schemes explained in Chapters 3, 6, 7 and 8; it solves the linear system by one of the methods we discuss in Chapters 3, 4, and 5. ELLPACK can also be used to solve nonlinear equations, by methods discussed at the end of Chapter 6. Users' programs are written in a high-level, user-oriented language, which makes it easy to define a problem and to specify a solution method and various kinds of output.

Much of the research whose fruits are summarized here was supported by the Office of Naval Research, to which we are greatly indebted. Purdue University, Carnegie-Mellon University, and the Fairchild Foundation of the California Institute of Technology gave additional support for our work. We also thank William Ames, Donald Anderson, Ronald Boisvert, John Brophy, Wayne Dyksen, Vincent Ervin, Bengt Fornberg, E. C. Gartland, Alan George, Charles Goldstein, Louis Hageman, Elias Houstis, Lois Mansfield, Douglas McCarthy, John Nohel,

Wlodzimierz Proskurowski, John Rice, and Donald Rose for many helpful suggestions, comments, and criticisms. We also thank the Purdue University Computer Science Department and Computing Center and the University of Illinois Computing Center for their cooperation in preparing our computer-produced text.

But above all, we wish to acknowledge the patient and generous advice given us by Richard Varga and David Young. As an expression of our gratitude for this advice, and of our admiration for their many basic contributions to the numerical solution of elliptic problems, we dedicate this monograph to them.

Garrett Birkhoff
Robert E. Lynch

Chapter 1

Typical Elliptic Problems

1. Introduction. The aim of this monograph is twofold: first, to describe a variety of powerful numerical techniques for computing approximate solutions of elliptic boundary value problems and eigenproblems on high speed computers, and second, to explain the reasons why these techniques are effective.

In *boundary value problems*, one is given a partial differential equation (DE), such as the Poisson equation[1] $u_{xx} + u_{yy} = f(x,y)$, to be satisfied in the interior of a region Ω, and also *boundary conditions* to be satisfied by the solution on the boundary $\Gamma = \partial\Omega$ of Ω. Such boundary value problems involving elliptic DE's arise naturally as descriptions of equilibrium states, in many physical and engineering contexts. In contrast, partial DE's of parabolic or of hyperbolic type (like the heat equation $u_t = u_{xx}$ or the wave equation $u_{tt} = u_{xx}$) arise naturally from time-dependent *initial value problems*, in which the DE to be satisfied is supplemented by appropriate *initial conditions*, to be satisfied (for example) at time $t = 0$.

For two independent variables, a general second-order linear differential equation has the form

$$Au_{xx} + 2Bu_{xy} + Cu_{yy} + Du_x + Eu_y + Fu = G, \tag{1.1}$$

where $A = A(x,y), \cdots, G = G(x,y)$. Such a DE is called *elliptic* when $AC > B^2$; this implies that A and C are nonzero and have the same sign. The DE (1.1) is called *self-adjoint* when $B = 0$, $D = A_x$, and $E = C_y$.

We will emphasize the two-dimensional case $n = 2$, because numerical techniques have been most thoroughly tested in this case. Computations involving unknown functions of three or more variables are usually costly. We will also emphasize second-order linear problems; for other problems, see §§7–9. For most *second-order* DE's the following characterization of ellipticity is adequate.

---

[1] A subscripted letter signifies the derivative with respect to the indicated variable.

DEFINITION. A semi-linear second-order DE, of the form[2]

$$\sum_{i=1}^{n} \sum_{j=1}^{n} a_{i,j}(\mathbf{x}) \frac{\partial^2 u}{\partial x_i \partial x_j} = f(\mathbf{x}, u, \nabla u), \tag{1.2}$$

where $\mathbf{x} = (x_1, \cdots, x_n)$ and $\nabla u = \text{grad } u = (\partial u/\partial x_1, \cdots, \partial u/\partial x_n)$, is called *elliptic* when the matrix $A(\mathbf{x}) = ||a_{i,j}(\mathbf{x})||$ is positive definite (or negative definite) identically, for all $\mathbf{x}$ in the domain Ω of interest. This means that, for nonzero $\mathbf{q} = (q_1, \cdots, q_n)$, the quadratic form $\sum a_{i,j}(\mathbf{x}) q_i q_j$ has constant sign. It is *linear* when

$$f(\mathbf{x}, u, \nabla u) = \sum_{j=1}^{n} b_j(\mathbf{x}) \frac{\partial u}{\partial x_j} + c(\mathbf{x}) u + g(\mathbf{x}).$$

Note that, since $\partial^2 u/\partial x_i \partial x_j = \partial^2 u/\partial x_j \partial x_i$, the matrix $A = A(\mathbf{x})$ can be assumed to be *symmetric* without loss of generality. Moreover, off-diagonal terms can be combined as in (1.1), whose (symmetric) matrix is

$$A = \begin{bmatrix} a_{11} & a_{12} \\ a_{21} & a_{22} \end{bmatrix} = \begin{bmatrix} A & B \\ B & C \end{bmatrix},$$

when $n = 2$. Finally, our definition is equivalent to the condition that $\sum a_{i,j}(\mathbf{x}) q_i q_j = 0$ implies $\mathbf{q} = \mathbf{0}$.

Likewise, the fourth-order linear operator

$$\mathbf{L}[u] = \sum_{k=0}^{4} a_k(\mathbf{x}) \frac{\partial^4 u}{\partial x^{4-k} \partial y^k} + \text{lower-order terms}$$

is called *elliptic* at $\mathbf{x}$ when $\sum a_k(\mathbf{x}) \xi^{4-k} \eta^k$ has constant sign for $(\xi, \eta) \neq (0,0)$. The definition of ellipticity in the general case is similar (see Bers-John-Schechter [64, p. 135]).[3]

The remainder of this chapter is devoted to reviewing some physical and engineering problems to which numerical techniques are often applied. We do this for two reasons. First, the most familiar elliptic problems originated in the attempts of nineteenth-century mathematicians like Fourier to develop a science of mathematical physics. Second, scientists and engineers who solve elliptic problems today usually want to describe some specific *physical phenomenon* or *engineering artifact*.

Since these problems are rooted in physics and other sciences, physical intuition often helps one to decide: (a) how to approximate them

---

[2] Differential equations like (1.2), which are linear in the highest derivatives, are called *semi-linear*; if the $a_{i,j}$ depend also on u and ∇u, they are called *quasi-linear*.

[3] Numbers in square brackets abbreviate a year (e.g., [64] for 1964) and refer to an item in the bibliography; letters in square brackets, such as [K], refer to the list of general references given at the beginning of the bibliography.

accurately, (b) which parameters are most important over which ranges, and (c) whether erratic numerical results are due to physical or to numerical instability. For these reasons, we describe the intuitive physical background of some of the most commonly studied elliptic partial DE's of mathematical physics. We include examples which illustrate various specific features of problems that influence the method of numerical solution. Some of these examples may be familiar to the reader, but we hope their inclusion will help to make our mathematical (and numerical) analysis more meaningful. The others are included to indicate the enormous variety of elliptic problems that arise in engineering and physics.

2. Dirichlet and related problems. The most deeply studied elliptic boundary value problem is the *Dirichlet problem*. Mathematically, this consists in finding a function that satisfies the n-dimensional *Laplace equation*

$$\nabla^2 u = \sum_{i=1}^{n} \frac{\partial^2 u}{\partial x_i^2} = 0 \quad \text{in } \Omega, \quad \Omega \subset \mathbb{R}^n, \tag{2.1}$$

in some bounded region $\Omega \subset \mathbb{R}^n$, assumes specified values $g(\mathbf{y})$ for all $\mathbf{y}$ on the *boundary* Γ of Ω, and is continuous in the closed domain $\bar{\Omega} = \Omega \cup \Gamma$. The boundary conditions assumed,

$$u(\mathbf{y}) = g(\mathbf{y}) \quad \text{on } \Gamma, \tag{2.1a}$$

are called *Dirichlet boundary conditions*.

Physically, $n \leqslant 3$, and $u(\mathbf{x}) = u(x,y,z)$ describes the equilibrium temperature in a homogeneous solid occupying Ω, whose boundary Γ is maintained at a temperature $g(\mathbf{y})$. The Laplace equation can be derived by assuming (with Fourier) the Law of Conservation of (thermal) Energy, and that the flow ('flux') of heat energy at any point is proportional to the temperature gradient ∇u there.

The Laplace equation (2.1) arises in a variety of other physical contexts, often in combination with other kinds of boundary conditions. In general, a function which satisfies (2.1) is called *harmonic* (in Ω); the study of harmonic functions (see Chapter 2) is called *potential theory*. Many problems of potential theory are described in Bergman-Schiffer [53], and in Morse-Feshbach [53].

Exterior problem. The Laplace equation (2.1) is satisfied in empty regions of space by gravitational, electrostatic, and magnetostatic potentials (e.g., see [K]). Thus the electrostatic potential due to a charged conductor occupying a closed domain $\bar{\Omega} = \Omega \cup \Gamma$ satisfies (2.1) in the *exterior* Ω' of $\bar{\Omega}$ and, in suitable units,

$$u = 1 \quad \text{on } \Gamma, \quad \text{and} \quad u \to 0 \quad \text{as } r \to \infty, \tag{2.1b}$$

where r is equal to the length of $\mathbf{x}$. The problem of solving (2.1) and (2.1b) is called the *conductor* problem. It can be shown that as r tends to infinity, $u \sim C/r$ (here and elsewhere, $f(x) \sim g(x)$ means that $f(x)/g(x) \to 1$ as $x \to \infty$). In solving this problem, one must also determine the *capacity* C which is the total charge that the conductor can 'hold' when at a unit potential or voltage. The capacity of a sphere of radius a is clearly a, since $u = a/r$ is harmonic and satisfies (2.1b).

Likewise, the irrotational flows of an incompressible fluid studied in classical hydrodynamics (see Lamb [32, Chaps. IV–VI]) have a 'velocity potential' which satisfies (2.1); see §6. For liquids of (nearly) constant density, this remains true under the action of gravity. Moreover, (2.1) is also applicable to some problems of petroleum reservoir mechanics in a homogeneous medium (soil);[4] see §3.

Neumann problem. However, the boundary conditions which are appropriate for hydrodynamical applications are usually quite different from those of (2.1a) or (2.1b). Thus, when $u = \phi$ is the 'velocity potential', they are often of the form[5]

$$\partial\phi/\partial n = h(\mathbf{y}) \qquad \text{on } \Gamma. \tag{2.1c}$$

We will discuss some such applications in §6.

The problem of finding a harmonic function with given normal derivative on the boundary is called the *Neumann problem*; boundary conditions of the form (2.1c) are called *Neumann boundary conditions*. In Neumann problems for heat flow, the normal derivative of u is proportional to the thermal energy flux and $h(\mathbf{y})$ specifies this flux at each point of the boundary.

Mixed boundary conditions. More generally, in the theory of heat conduction, it is often assumed that a solid loses heat to the surrounding air at a rate roughly proportional to its excess surface temperature u (Newton's 'Law of Cooling'). For k the constant of proportionality, this leads one to solve (2.1) for the boundary conditions

$$\partial u/\partial n + ku = g(\mathbf{y}) \qquad \text{on } \Gamma, \quad k > 0, \tag{2.1d}$$

where $g(\mathbf{y})$ is the rate of absorption of radiant energy. Boundary conditions such as (2.1d), of the general form

$$\alpha(\mathbf{y})u + \beta(\mathbf{y})\,\partial u/\partial n = g(\mathbf{y}), \qquad \alpha^2(\mathbf{y}) + \beta^2(\mathbf{y}) \neq 0, \tag{2.1e}$$

[4] See Muskat [37]; also P. Ya. Polubarinova-Kochina, *Advances in Applied Mechanics* 2, Academic Press, 1951, 153–221; A. E. Scheidegger, *Physics of Flow through Porous Media*, Macmillan, 1957; and D. W. Peaceman, *Fundamentals of Numerical Reservoir Simulation*, Elsevier, 1977.

[5] Here and below $\partial/\partial n$ denotes the *exterior* normal derivative.

are called *mixed boundary conditions*. (If the solid is cut out of sheet metal, and so is essentially two-dimensional, the temperature can be assumed to satisfy (2.1a) and the *modified Helmholtz equation* $u_{xx}+u_{yy}=\lambda u$, $\lambda>0$, inside the solid, instead of (2.1).)

3. Membranes; source problems. Potential theory is concerned not only with harmonic functions, but also with solutions of the *Poisson equation*

$$-\nabla^2 u = f(\mathbf{x}), \tag{3.1}$$

in free space and in bounded domains, subject to various boundary conditions such as (2.1a)–(2.1d). In the case $\mathbf{x}=(x_1,x_2)$ of two independent variables, the DE (3.1) is satisfied approximately[6] by the vertical deflection, $z=u(x,y)=u(x_1,x_2)$, of a nearly horizontal *membrane* (or 'drumhead') under uniform lateral tension T, which supports a load $Tf(\mathbf{x})$ per unit area. If such a membrane spans a rigid frame whose height above Γ in the (x_1,x_2)-plane is given by a function $g(\mathbf{y})$, $\mathbf{y}\in\Gamma$, the appropriate Dirichlet boundary condition is

$$u = g(\mathbf{y}) \qquad \text{on } \Gamma. \tag{3.1'}$$

The special case $g(\mathbf{y})\equiv 0$ of (3.1') arises naturally in fluid dynamics. The velocity field $\mathbf{u}(x,y)=(u(x,y),v(x,y))$ of any *plane* flow of an *incompressible* fluid is determined by a *stream function* $\psi(x,y)$. Specifically, we have $u=\partial\psi/\partial y$ and $v=-\partial\psi/\partial x$, and consequently $\operatorname{div}\mathbf{u}=\psi_{xy}-\psi_{yx}=0$; moreover, the *vorticity* $\zeta=\partial v/\partial x-\partial u/\partial y$ satisfies $-\nabla^2\psi=\zeta$, the Poisson equation. If the fluid is in a stationary simply connected container with boundary Γ, then Γ is necessarily a streamline, and so we can assume $\psi=0$ on Γ. Hence the vorticity $\zeta(x,y)$ determines the stream function $\psi(x,y)$ as the solution of the Poisson equation $-\nabla^2\psi=\zeta$ with $\psi\equiv 0$ on Γ.

In our studies of numerical methods in later chapters, we will study repeatedly the following even more special case.

Example 1. The 'Model Problem' [Y, §1.1] defined by the Poisson DE $-\nabla^2 u=f(\mathbf{x})$ in the unit square $S: 0<x,y<1$, with the boundary condition $u\equiv 0$ on $\Gamma=\partial S$, has many physical interpretations.[7]

For example, with $f(x,y)=4$, $u(x,y)$ gives the deflection of a taut elastic membrane held in a square frame, due to a small difference in air pressure on the two sides. It also expresses the velocity profile associated with viscous flow through a square tube parallel to the z-axis. Finally,

[6] In the 'linearized approximation', obtained by replacing $\sin\alpha=\alpha-\alpha^3/3!+\alpha^5/5!-\cdots$ with α.

[7] See Synge [57, p. 130], for fuller discussions of these interpretations.

$u(x,y)+(x^2+y^2)$ represents the 'warping function' of a long straight bar with square cross-section under pure torsion.

For other domains Ω with boundary Γ, the DE $-\nabla^2 u = f(\mathbf{x})$ in Ω with $u \equiv 0$ on Γ has analogous physical interpretations.

When $n=3$, the DE (3.1) with $f(\mathbf{x}) = 4\pi\rho(\mathbf{x})$ is satisfied by the gravitational potential of a continuous distribution of mass with density $\rho(\mathbf{x})$ (mass per unit volume). Likewise, it is satisfied by the electrostatic potential of a continuous charge distribution having this density.

A more general elliptic DE is

$$\text{(3.2)} \quad \mathbf{L}[u] = -\nabla \cdot [p(\mathbf{x})\nabla u] + q(\mathbf{x})u = f(\mathbf{x}), \qquad p > 0 \quad \text{and} \quad q \geqslant 0.$$

Whereas the Laplace operator in (2.1) has *constant* coefficients, the linear differential operator $\mathbf{L}[u]$ in (3.2) has *variable* coefficients. Moreover, the Laplace DE itself leads to second-order linear elliptic problems with variable coefficients when spherical, ellipsoidal, or other coordinate systems are used.

The DE (3.2) is satisfied approximately by the temperature distribution $u(\mathbf{x})$ in a solid having space-dependent thermal conductivity $p(\mathbf{x})$, in which heat is being produced at the rate $f(\mathbf{x})$ (energy per unit volume and time); $q(\mathbf{x})u$ gives the absorption. Since one may think of $f(\mathbf{x})$ as representing a *source* of heat, the DE (3.2) for specified boundary conditions such as (2.1a)–(2.1d) is often said to be a *source problem*. Such source problems arise, typically, in the analysis of diffusion phenomena.

Similar elliptic problems having variable coefficients arise also in the study of electrostatic, magnetostatic, and gravitational potentials, in which the materials involved have physical properties (e.g., dielectric constants or magnetic permeabilities) that depend on position. Moreover, a related elliptic DE also arises from Darcy's Law, in petroleum reservoirs occupying soils (or sands) of variable 'permeability' $k(x,y,z)$. As is explained in Muskat [37, p. 242], the pressure p in such a reservoir satisfies

$$\text{(3.3)} \qquad -\nabla \cdot [k\nabla p] = \rho g\, \partial k/\partial z.$$

In practice, $k(\mathbf{x})$ is known only very roughly, and (like thermal and electrical conductivity) it can vary by orders of magnitude.

4. Two-endpoint problems. The problems described in §§2–3 have counterparts which involve functions of *one* space variable, and hence lead to *ordinary* DE's. Since the boundary of a one-dimensional domain (interval) consists of two points, such problems are often called *two-endpoint problems*. The numerical techniques which are most effective for solving such two-endpoint problems are very different from those used in two or more dimensions. However, we have devoted this section to them because they illustrate so simply various kinds of boundary conditions and other basic ideas.

Example 2. The simplest two-endpoint problem concerns a transversely *loaded string*, in the small deflection or *linear* approximation. (For some nonlinear elliptic problems, see §9.) If the string (assumed nearly horizontal) is under a constant tension T, then the deflection y induced by a load exerting a vertical force $f(x)$ per unit length satisfies the ordinary DE

$$-y''=f(x)/T. \tag{4.1}$$

If the endpoints of the string of length a are fixed, then the deflection also satisfies the two-endpoint conditions

$$y(0)=y(a)=0. \tag{4.1'}$$

The differential operator $\mathbf{L}[u]=-d^2/dx^2$ on the left side of (4.1) is a *linear differential operator* with *constant coefficients*; it is *linear* because for any constants α,β and functions[8] $y,z\in C^2[0,a]$, clearly

$$\mathbf{L}[\alpha y+\beta z]=\alpha\mathbf{L}[y]+\beta\mathbf{L}[z].$$

The differential equation (4.1) is *inhomogeneous* since its right side is nonzero. On the other hand, the boundary conditions (4.1′) are linear and *homogeneous*.

To solve (4.1)–(4.1′), first note that for any continuous $f(x)$, the function

$$g(x)=-\frac{1}{T}\int_0^x(x-t)f(t)\,dt$$

is a solution of (4.1) satisfying the initial conditions $g(0)=g'(0)=0$.[9] The general solution of (4.1) is $g(x)+\alpha+\beta x$, where α and β are arbitrary constants. To construct the solution satisfying (4.1′), set $\alpha=0$ and

$$\beta=\frac{1}{aT}\int_0^a(a-t)f(t)\,dt.$$

The problem of a vertical *loaded spring* is similar. If $p(x)$ is the (variable) stiffness of the spring and $f(x)$ is the load per unit length, then the appropriate DE for vertical (longitudinal) displacement $u(x)$ is

$$-[p(x)u']'=f(x),\qquad p(x)>0. \tag{4.2}$$

[8] The symbol $C^k[0,a]$ denotes the set of functions on the interval $[0,a]$ whose k-th derivative exists and is continuous.

[9] In other words, $G(x;t)=(x-t)$ is the *Green's function* for the operator $\mathbf{L}$ for $x>0$ and the initial data $g(0)=g'(0)=0$ (Birkhoff-Rota [78, Chap. 2, §8]).

Here the linear differential operator $\mathbf{L}[u] = -[p(x)u']'$ has variable coefficients. If the spring is held fixed at $x = 0$ and the other end is free, then one has the boundary conditions

$$u(0) = 0, \qquad u'(a) = 0. \tag{4.2'}$$

If a mass m is attached to the bottom of the spring, then the second condition in (4.2′) is replaced with $u'(a) = mg/p(a)$, where g is the acceleration due to gravity.

Sturm-Liouville systems. As a third example, we consider *Sturm-Liouville systems*. These typically arise from separating out the time variable from simple harmonic solutions of time-dependent problems such as that of a vibrating string $\rho(x)u_{tt} = [p(x)u_x]_x$ with variable density $\rho(x)$ and tension $p(x)$. With $u(x,t) = y(x)\cos kt$, one gets *homogeneous* linear DE's for y of the form $[p(x)y']' + k^2\rho(x)y = 0$, or, more generally,

$$[p(x)y']' + [\lambda\rho(x) - q(x)]y = 0, \qquad p > 0, \quad \rho > 0, \tag{4.3}$$

in which $\lambda = k^2$ is a parameter, and homogeneous linear boundary conditions of the form (4.1′). More generally, Sturm-Liouville systems can involve *separated* boundary conditions of the form

$$\alpha_0 y(0) + \beta_0 y'(0) = \alpha_1 y(a) + \beta_1 y'(a) = 0, \qquad \alpha_i{}^2 + \beta_i{}^2 > 0. \tag{4.3'}$$

Problems like (4.3)–(4.3′) which involve the solution of a *homogeneous* linear (elliptic) DE for *homogeneous* boundary conditions and unknown values of a parameter λ are called *eigenproblems*. The values of the parameter for which nontrivial solutions[10] exist are called *eigenvalues*, and the solutions themselves are called *eigenfunctions*. It is well known that any Sturm-Liouville system (4.3) with separated boundary conditions admits an infinite sequence of real eigenfunctions $\phi_j(x)$ with real eigenvalues $\lambda_1 < \lambda_2 < \lambda_3 < \cdots$, where $\lambda_j \to \infty$.

Other endpoint conditions. Many kinds of 'endpoint conditions' can be prescribed for Sturm-Liouville systems. Thus, for the trigonometric DE $y'' + \lambda y = 0$, the Mathieu equation

$$y'' + (\lambda + \mu \cos x)y = 0, \tag{4.4}$$

and other second-order DE's with periodic coefficients, one typically wants solutions which satisfy *periodic* boundary conditions such as

$$y(0) = y(2\pi), \qquad y'(0) = y'(2\pi). \tag{4.4'}$$

[10] The 'trivial' solution is $y \equiv 0$. For a fuller discussion of Sturm-Liouville problems and the endpoint conditions which are appropriate for them, see Birkhoff-Rota [78, Chap. 10].

For separated boundary conditions of the form (4.3′), the eigenvalues are *distinct*, but for periodic and other nonseparated boundary conditions, one can have *double* eigenvalues. The following examples illustrate this difference.

Example 3. The eigenfunctions of the *trigonometric* DE $y''+\lambda y=0$ for the boundary conditions $y(0)=y(a)=0$ are $\phi_j(x)=\sin(j\pi x/a)$, $j=1,2,3,\cdots$; the corresponding eigenvalues $\lambda_j=j^2\pi^2/a^2$ are distinct.

Example 3′. For the same DE on $(0,2\pi)$, with the periodic boundary conditions (4.4′), the eigenfunctions are

$$1,\ \cos x,\ \sin x,\ \cos 2x,\ \sin 2x,\ \cdots,$$

and the eigenvalues $\lambda_{2j-1}=\lambda_{2j}=j^2$ are *double* for $j>0$.

Example 4. The Bessel DE of order zero,

$$y''+\frac{1}{r}y'+\lambda y=0 \qquad \text{on } 0<r<a,$$

has a *regular singular point* at $r=0$. The only solutions that are analytic on $[0,a]$ and satisfy the boundary condition $y(a)=0$ are multiples of $y=J_0(kr)$, the *Bessel function*[11] of order zero; those satisfying $y(a)=0$ are the $J_0(k_m r)$, where $k_m a=j_{0,m}$ is the m-th zero of $J_0(x)$.

Physical significance. Physically, Sturm-Liouville systems with eigenfunctions $\phi_j(x)$ provide solutions of vibration and diffusion problems. Thus consider the (hyperbolic) linear, constant-coefficient DE for a *vibrating string*,

$$y_{tt}=c^2y_{xx}. \tag{4.5}$$

The general solution of (4.5) is a sum $y=f(x+ct)+g(x-ct)$, where f,g are arbitrary smooth functions. The preceding d'Alembert decomposition represents y as the superposition of two waves moving with the same speed c in opposite directions along the x-axis, without change of form.

For a string of length a, with

$$y(0,t)=y(a,t)=0, \qquad t\geqslant 0, \tag{4.5′}$$

one can use the eigenfunctions $\phi_j(x)=\sin(j\pi x/a)$ for the same boundary conditions to obtain special solutions in 'separable variable' form:

$$u(x,t)=\phi_j(x)\begin{Bmatrix}\sin\\ \cos\end{Bmatrix}(jc\pi t/a). \tag{4.6}$$

For the initial conditions $u(x,0)=f(x)$, $u_t(x,0)=g(x)$, the solution of

[11] See Birkhoff-Rota [78, p. 88]; here again $\lambda=k^2$.

$u_{tt} = c^2 u_{xx}$ has the expansion

$$u = \sum [a_j \cos(\gamma_j t) + b_j \sin(\gamma_j t)]\phi_j(x), \qquad \gamma_j = jc\pi/a. \tag{4.7}$$

Here the a_j and $\gamma_j b_j = \beta_j$ are the coefficients of the expansion of $f(x)$ and $g(x)$, respectively, in sine series.

Likewise, consider the one-dimensional (parabolic) *heat equation*

$$u_t = \alpha u_{xx}, \tag{4.8}$$

for a slab of thickness a with constant 'thermal diffusivity' α whose faces $x = 0$ and $x = a$ are kept in ice water at $0°$C. The problem defined by these conditions has a basis of solutions of the form

$$\exp(-\alpha j^2 \pi^2 t/a^2)\,\phi_j(x), \qquad j = 1, 2, \cdots.$$

Here the general solution of (4.8) subject to the homogeneous boundary conditions is a sum of products of spatial eigenfunctions and exponentially decaying functions of time.

5. Helmholtz equation. The most deeply studied eigenproblems for elliptic DE's have their genesis in the (hyperbolic) DE of a transversely *vibrating membrane*:

$$z_{tt} = c^2(z_{xx} + z_{yy}). \tag{5.1}$$

Here $z(x,y,t)$ gives the transverse deflection of the membrane from its equilibrium position; $c = \sqrt{T/\rho}$ is the wave speed (T the tension and ρ the density of the membrane, both assumed constant). Simple harmonic (in time) oscillations of such a membrane are clearly given by

$$z(x,y,t) = u(x,y)[a \sin(kct) + b \cos(kct)], \tag{5.2}$$

where $u(x,y)$ is a solution of the *Helmholtz* or *reduced wave equation*

$$\nabla^2 u + k^2 u = 0, \qquad \nabla^2 u = \sum_{i=1}^{n} \frac{\partial^2 u}{\partial x_i^2}, \tag{5.3}$$

with $n = 2$ and $x_1 = x$, $x_2 = y$. Hence, to find the possible simple harmonic vibrations of a membrane held in a rigid frame having a given contour Γ, we must solve the Helmholtz equation (5.3) subject to the boundary condition

$$u = 0 \quad \text{on } \Gamma. \tag{5.3$'$}$$

Example 5. Consider the eigenproblem $-\nabla^2 u = \lambda u$ in the unit square $S: 0 < x, y < 1$, for the boundary condition $u \equiv 0$ on ∂S. Its eigenfunctions are the solutions of the Helmholtz equation (5.3) for $n = 2$ which vanish identically on ∂S. These can be found by separating variables and using the eigenfunctions of Example 3; they are

(5.4) $$u = \phi_{j,m} = \sin(j\pi x)\sin(m\pi y), \qquad j, m = 1, 2, \cdots.$$

The eigenvalues are correspondingly

(5.4′) $$k_{j,m}{}^2 = \lambda_{j,m} = (j^2 + m^2)\pi^2, \qquad j, m = 1, 2, \cdots.$$

Similarly, in three-dimensional space, let $u(\mathbf{x}) = u(x,y,z)$ be a solution of the reduced wave equation (5.3) with $n = 3$ in a bounded domain Ω with boundary Γ, and let

(5.5) $$\partial u/\partial n = 0 \qquad \text{on } \Gamma.$$

This is the physically appropriate boundary condition for sound waves, which satisfy $p_{tt} = c^2\nabla^2 p$, with $p = \delta p/p_0$ the relative pressure variation. Then

$$p(x,y,z,t) = u(x,y,z)[a\sin(kct) + b\cos(kct)]$$

describes the pressure variation (from ambient pressure) in a standing sound wave with frequency kc/π in a room (or organ pipe) having the specified (rigid) boundary Γ. For Ω an open organ pipe, (5.5) is replaced by the mixed boundary conditions: $\partial u/\partial n = 0$ on the walls of the pipe and $u = 0$ on its open end.

As with Sturm-Liouville systems (see §4), each of the systems (5.3)–(5.3′) and (5.3)–(5.5) has a sequence of nontrivial solutions called *eigenfunctions* of the system, whose eigenvalues $\lambda_j = k_j^2$ are positive (or zero) and can be arranged in ascending order: $\lambda_1 \leqslant \lambda_2 \leqslant \cdots$, $\lambda_n \uparrow \infty$. Moreover, every square-integrable function can be expanded in a (mean square) convergent series of these (orthogonal) eigenfunctions (completeness property).

However, for n-dimensional eigenproblems, it is usually more natural to consider the eigenfunctions as forming an n-parameter family. Thus, for the reduced wave equation (the Helmholtz equation) in the disk $x^2 + y^2 \leqslant a^2$ and $u(a) = 0$, the eigenfunctions are products

(5.6) $$\phi_{\nu,m} = J_\nu(k_{\nu,m} r)\begin{Bmatrix}\sin\\ \cos\end{Bmatrix}(\nu\theta).$$

Here $\nu = 0, 1, \cdots$, the J_ν are Bessel functions of order ν, and $ak_{\nu,m} = j_{\nu,m}$ is the m-th zero of $J_\nu(x)$.

The possibility of expanding arbitrary solutions in this way has been called by Sommerfeld the 'Ohm-Rayleigh principle'; we shall discuss it in Chapter 2, §6. Various classical examples are worked out in textbooks on sound.[12]

[12] P. M. Morse, *Vibrations and Sound*, McGraw-Hill, 1936.

Maxwell equations.[13] By separating out the spatial variation of simple harmonic (in time) 'standing wave' solutions of Maxwell equations for electromagnetic waves in a homogeneous medium, one is led to still other solutions of the reduced wave equation. However, quantitative results about scattering and transmission by wave guides still usually involve other physical considerations; cf. Chapter 8, §10.[14]

6. Potential flow problems. By definition, a *potential flow* is a flow whose velocity field $\mathbf{u}(\mathbf{x})$ is the gradient $\nabla\phi$ of a harmonic *velocity potential* ϕ, i.e., of a function ϕ with $\nabla^2\phi = 0$. In other words, it is a flow whose velocity field is irrotational (vorticity $\nabla \times \mathbf{u} = 0$) and volume-conserving (divergence $\nabla \cdot \mathbf{u} = 0$). Potential flows have been intensively studied for 150 years (see Lamb [32, Chaps. IV–VI]). The appropriate boundary condition is

$$\partial\phi/\partial n = h(\mathbf{y}) \qquad \text{on } \Gamma; \tag{6.1}$$

i.e., the *normal derivative* is specified. This is because, in order to maintain contact (without cavitation) with a solid, the fluid must have the same normal velocity component on Γ.

The potential of a unit 'dipole' can be taken as

$$\phi = \frac{\partial}{\partial x}\frac{1}{r} = -\frac{x}{r^3} = -\frac{\cos\theta}{r^2}, \tag{6.2}$$

where θ is the colatitude. Hence its gradient

$$\nabla\phi = r^{-5}(2x^2 - y^2 - z^2, 3xy, 3xz) \tag{6.3}$$

is the velocity field of a potential flow. On the surface $r = a$ of a sphere of radius a and center at the origin,

$$\partial\phi/\partial n = \partial\phi/\partial r = U\cos\theta, \qquad U = 2/a^3.$$

Thus the fluid velocity component normal to the surface equals that of a rigid sphere of radius a, moving parallel to the x-axis with speed $U = 2/a^3$. Hence the potential flow induced by a sphere of radius a with its center at the origin, moving with speed U parallel to the x-axis, has for its velocity field the vector field $[a^3U/2r^5](2x^2 - y^2 - z^2, 3xy, 3xz)$ or

$$\frac{a^3U}{2r^5}(2\cos^2\theta - \sin^2\theta, 3\cos\theta\sin\theta\cos\sigma, 3\cos\theta\sin\theta\sin\sigma),$$

where σ denotes the longitude.

---

[13] See Jeans [41].

[14] See R. E. Collin, *Field Theory of Guided Waves*, McGraw-Hill, 1960, Chap. 8; L. Lewin, *Advanced Theory of Wave Guides*, Iliffe, 1951; N. Markuvitz, *Waveguide Handbook*, McGraw-Hill, 1951.

Added mass. The mass of the displaced fluid of density ρ is $m = 4\pi\rho a^3/3$, and the kinetic energy of the surrounding fluid is

$$
\begin{aligned}
T &= \frac{\rho}{2}\iiint_{r>a} (\nabla\phi \cdot \nabla\phi) r^2 \sin\theta \, dr \, d\theta \, d\sigma \\
&= \frac{\pi\rho a^3 U^2}{3} = \frac{1}{2}\frac{m}{2}U^2.
\end{aligned}
\tag{6.4}
$$

The constant $m/2$ is called the *added mass* (or the *induced mass*) of the sphere, due to the inertia of the fluid.[15]

In two dimensions, the analogous 'dipole' potential induced by the forward motion of a circular cylinder of radius a is $-a^2Ux/r^2$, the mass of fluid displaced is $m = \pi\rho a^2$, and the added mass is also m. The sum

$$-Ux - a^2Ux/r^2 = -U(r + a^2/r)\cos\theta$$

is the velocity potential of steady flow relative to the cylinder.

More generally, given the surface $\Gamma = \partial\Omega$ of a solid Ω, there is a unique harmonic 'velocity potential' ϕ which tends to zero as $r \to \infty$ and satisfies the Neumann boundary conditions $\partial\phi/\partial n = U\cos\gamma$ on Γ, where γ is the angle between the outward normal to Γ and the direction of motion of the solid. Hence $\nabla\phi$ is the (ideal) potential flow induced by the motion of the solid in the specified direction with the given speed U.

To find such a function ϕ satisfying

$$
\begin{aligned}
&\nabla^2\phi = 0 \qquad \text{outside } \Omega, \\
&\partial\phi/\partial n = g(\mathbf{y}) \quad \text{on } \Gamma, \quad \text{and} \quad \phi \to 0 \quad \text{as } |\mathbf{x}| \to \infty,
\end{aligned}
\tag{6.5}
$$

is called an *exterior Neumann problem*. Such exterior Neumann problems give the most natural way to define the irrotational[16] flow induced by the motion of a solid through an incompressible, inviscid fluid. In each case, the kinetic energy T is half the density, $\rho = \rho_0$, times the *Dirichlet integral*

$$\mathbf{D}\langle\phi,\phi\rangle = \int_{\Omega'} \nabla\phi \cdot \nabla\phi \, d\Omega, \tag{6.6}$$

where Ω' denotes the exterior of Ω; it is given by $T = \rho\mathbf{D}\langle\phi,\phi\rangle/2$.

More generally, if the solid has the *vector* velocity $\mathbf{U} = (U_1, U_2, U_3)$, then

$$T = \frac{1}{2}\sum T_{i,j}U_iU_j, \qquad T_{i,j} = \rho\mathbf{D}\langle\phi_i,\phi_j\rangle,$$

where $U_i = \partial\phi/\partial x_i$, and $\phi_i(\mathbf{x})$ is the velocity potential for translation of the solid with unit speed in the i-th coordinate direction. The numbers

[15] See G. Birkhoff, *Hydrodynamics*, Princeton University Press, 1960, Chap. VI.

[16] In an 'ideal' incompressible and nonviscous fluid initially at rest, it can be proved that the velocity field will always remain irrotational.

$T_{i,j}$ defined in this way are the components of the *inertial* (or *added mass*) *matrix* for translation.

Convection-diffusion equation. An important elliptic problem that is *not* self-adjoint is the time-independent 'convection-diffusion' equation

$$\mathbf{U}(\mathbf{x}) \cdot \nabla u = \epsilon \nabla^2 u \qquad \text{in } \Omega, \tag{6.7}$$

with Dirichlet boundary conditions $u(\mathbf{y}) = g(\mathbf{y})$ on $\Gamma = \partial\Omega$. This DE is satisfied by the concentration of a chemical, in a solution flowing with a time-independent (steady) velocity field $\mathbf{U}(\mathbf{x})$. Especially interesting mathematically is the case of a uniform velocity field parallel to the x-axis

$$U u_x = \epsilon (u_{xx} + u_{yy}) \tag{6.8}$$

flowing through the rectangular channel $[0,a] \times [0,b]$.

More interesting from an engineering standpoint, as representing diffusion during flow through a long straight pipe, is the general problem

$$U(r) u_x = \epsilon(r)[u_{xx} + (1/r) u_r + u_{rr}], \tag{6.9}$$

where the velocity profile $U(r)$ and the diffusivity $\epsilon(r)$ (both determined by the turbulence level) are allowed to vary with r.

For (6.8), the mean transport time for 'convection' is a/U; that for (longitudinal) 'diffusion' is $O(a^2/\epsilon)$. The *ratio* of these times, aU/ϵ, expresses the relative importance of convection to that of diffusion as a transport mechanism.[17] As we shall see in Chapter 3, §§6 and 11, the comparative effectiveness of methods for solving this problem depends strongly on the size of aU/ϵ.

As $\epsilon \downarrow 0$ and $aU/\epsilon \uparrow \infty$, each of the three DE's (6.7)–(6.9) is a *singular perturbation* of some *first-order hyperbolic* DE, for whose *asymptotic* solution analytical methods are often most informative. (The situation is analogous to that for solving initial value problems for 'stiff' ordinary DE's.)

7. Problems from elasticity. The problems discussed so far have all concerned *second-order* elliptic DE's. In solid mechanics, *fourth-order* elliptic DE's and systems are more prevalent.

The simplest such problems refer to the small deflections of a thin beam or 'rod' by an applied transverse 'load' or force distribution. This problem was solved mathematically by the Bernoullis and Euler, who assumed that the beam or 'elastica' was homogeneous, i.e., had the same physical characteristics in all cross-sections. Around 1700, assuming the linear or 'small deflection' approximation (Hooke's Law), James Bernoulli

[17] This ratio is called the Péclet number; see Jakob [50, p. 491].

showed that the deflection of the centerline of the beam should satisfy

$$u^{iv}(x) = d^4u/dx^4 = f(x), \qquad a \leqslant x \leqslant b \tag{7.1}$$

(see Synge-Griffith [59, p. 90]). Here $f(x)$ is the ratio of the applied transverse load per unit length to the 'stiffness' of the beam, whose undeflected centerline is along the x-axis from $x = a$ to $x = b$. For small deflections, the bending moment is taken to be proportional to u'', the linearized curvature $u''/[1+(u')^2]^{3/2}$, and the stress is proportional to u'''.

The general solution of the DE (7.1) is the sum of any particular solution and some *cubic polynomial*, since the general solution of the ordinary DE $u^{iv} = 0$ is a cubic polynomial. This cubic polynomial is determined by the boundary conditions.

Thin beam problems can involve various sets of homogeneous endpoint conditions, notably the following (cf. Courant-Hilbert [53, pp. 295–296]):

$$u(y) = u''(y) = 0, \qquad y = a \text{ and } b, \quad \text{(simply supported ends)}, \tag{7.2a}$$

$$u(y) = u'(y) = 0, \qquad y = a \text{ and } b, \quad \text{(clamped ends)}, \tag{7.2b}$$

$$u''(y) = u'''(y) = 0, \qquad y = a \text{ and } b, \quad \text{(free ends)}. \tag{7.2c}$$

The two-endpoint problems defined by (7.1) and (7.2a) or (7.2b) have easily determined solutions. Indeed, even for *inhomogeneous* endpoint conditions, such as

$$u(a) = \alpha, \quad u'(a) = \alpha', \quad u(b) = \beta, \quad u'(b) = \beta',$$

one can proceed much as in solving (4.1). Two-endpoint problems defined by (7.1) and the 'free end' conditions (7.2c) do not have solutions in general unless $\int_a^b f(x)\,dx = 0$, since

$$u'''(b) = u'''(a) + \int_a^b f(x)\,dx.$$

Namely, one can first compute a particular solution

$$\phi(x) = \frac{1}{3!}\int_a^x (x-t)^3 f(t)\,dt$$

of (7.1) by performing a quadrature on $(x-t)^3 f(t)$ numerically (e.g., by Simpson's rule). One can then form the sum

$$u(x) = \phi(x) + c_0 + c_1x + c_2x^2 + c_3x^3, \tag{7.3}$$

treating the coefficients c_j as unknowns to be determined from the four endpoint conditions. The case $c_0 = c_1 = c_2 = c_3 = 0$ gives the solution for the homogeneous initial conditions

$$u(a) = u'(a) = u''(a) = u'''(a) = 0.$$

To approximate boundary conditions involving derivatives at b, the formulas

$$\phi^{(j)}(b) = \frac{1}{(3-j)!}\int_a^b (b-t)^{3-j} f(t)\,dt, \qquad j=1,2,3,$$

may be helpful.

Cubic splines. A very useful special case corresponds to 'point-loads' concentrated at some sequence $\pi : a = x_0 < x_1 < \cdots < x_m = b$ of points x_i in $[a,b]$. Since, for any $c < d$,

$$u'''(d) - u'''(c) = \int_c^d u^{iv}(x)\,dx = \int_c^d f(x)\,dx,$$

a total load of w_i concentrated at x_i may be expected to produce a jump of $w_i = u'''(x_i^+) - u'''(x_i^-)$ in the third derivative (shear stress) of the deflection function, whose second derivative is, however, continuous. This shows that the solutions are given by the following class of 'cubic spline' functions.

DEFINITION. A *cubic spline* function on $[a,b]$ with *joints* (or 'knots') at the points x_i, $i = 1, \cdots, m-1$, is a function $u \in C^2[a,b]$ expressible on each segment (x_{i-1}, x_i) by a cubic polynomial

$$p_i(x) = a_{i,0} + a_{i,1}x + a_{i,2}x^2 + a_{i,3}x^3.$$

Splines have been used by naval architects for many years to generate, mechanically, smooth curves which pass through (or 'interpolate' to) preassigned points. The application of more general 'spline functions' for computing accurate numerical solutions of elliptic DE's will be discussed in Chapter 7, §5.

Vibration problems. A homogeneous, freely vibrating thin rod of density ρ and stiffness σ will satisfy $\rho u_{tt} = \sigma u_{xxxx}$. Its normal modes $u = v(x)e^{ikt}$ will therefore satisfy $\sigma v^{iv} + \rho k^2 v = 0$, and so be *eigenfunctions* of the DE $v^{iv} + \lambda v = 0$. These eigenfunctions form a mathematically well-defined family for all three of the sets of boundary conditions (7.2a), (7.2b), and (7.2c).

Plates and shells. The fourth-order ordinary DE (7.1) with the boundary conditions (7.2a)–(7.2c) define one-dimensional analogues of fourth-order elliptic problems. Solid mechanics provides many other challenging fourth-order elliptic problems for mathematicians to solve. One of the simplest of these is provided by Kirchhoff's theory of a transversely loaded flat plate. The transverse deflection, $u = z$, satisfies the deceptively simple looking *biharmonic* equation[18]

[18] Throughout, ∇^2 and ∇^4 denote the Laplacian and the biharmonic operator, respectively; some authors use Δ and Δ^2 for these instead.

(7.4) $$\nabla^4 u = f(x,y), \qquad \nabla^4 u = \frac{\partial^4 u}{\partial x^4} + 2\frac{\partial^4 u}{\partial x^2 \partial y^2} + \frac{\partial^4 u}{\partial y^4}.$$

As in the one-dimensional analogue of the thin beam, the plate can be constrained to satisfy any of a fairly large variety of boundary conditions.

Example 6. The biharmonic equation (7.4) together with the analogue of the simply supported end conditions (7.2a),

(7.5) $$u = 0 \text{ and } \nabla^2 u = 0 \qquad \text{on } \Gamma,$$

gives the deflection u of a transversely loaded simply supported plate. This problem is decomposable into the pair of problems

(7.6) $$\begin{aligned} \nabla^2 u &= v, & \nabla^2 v &= f(x,y), & &\text{on } \Omega, \\ u &= 0, & v &= 0, & &\text{on } \partial\Omega. \end{aligned}$$

For a *rectangular* plate, $\Omega = [0,a] \times [0,b]$, each of these problems is also *separable*. To solve them, expand f in a double Fourier sine series[19]

$$f(x,y) = \sum c_{i,j}\phi_i(x/a)\phi_j(y/b), \qquad \phi_j(z) = \sin j\pi z;$$

u is then given by a series of the same form with coefficients $c_{i,j}/\lambda_{i,j}$, where

$$\lambda_{i,j} = ([i\pi/a]^2 + [j\pi/b]^2)^2.$$

Example 6′. The biharmonic equation and the boundary conditions $u = 0$ and $\partial u/\partial n = 0$ gives the deflection of a loaded clamped plate. This problem is neither decomposable nor separable.

A homogeneous *vertical* plate in a parallel gravity or centrifugal field with potential $V(x,y)$ has stress components σ_x, σ_y, and τ_{xy} most simply expressed in terms of the Airy stress function $\phi(x,y)$:

(7.7) $$\sigma_x = \phi_{yy} + V, \qquad \sigma_y = \phi_{xx} + V, \qquad \tau_{xy} = -\phi_{xy}.$$

The conditions for static equilibrium are given by the compatibility relations

(7.7′) $$\nabla^4\phi + \frac{1-2\nu}{1-\nu}\nabla^2 V = 0,$$

where ν is the 'Poisson ratio'.[20] In a pure gravity field, since $\nabla^2 V = 0$, ϕ is a biharmonic function because it satisfies $\nabla^4\phi = 0$.

Curvilinear elastic shells satisfy much more complicated but analogous systems of linear elliptic equations with variable coefficients.

---

[19] In the sense of *mean square convergence* see Chapter 2, §5.

[20] Timoshenko-Woinowsky-Krieger [59].

8. Multigroup diffusion. Another important area of application for numerical methods is provided by the steady state multigroup diffusion equations of nuclear reactor theory. These constitute a *system* of DE's. For an idealized 'thermal' reactor, they are often assumed to be of the following cyclic form (see Glasstone-Edlund [52, Chap. 8]):

$$(8.1) \qquad -\nabla\cdot[D_1(\mathbf{x})\nabla\phi_1]+\sigma_{a,1}(\mathbf{x})\phi_1=\nu\sigma_n(\mathbf{x})\phi_n,$$

$$(8.1') \qquad -\nabla\cdot[D_i(\mathbf{x})\nabla\phi_i]+\sigma_{a,i}(\mathbf{x})\phi_i=\sigma_{i-1}(\mathbf{x})\phi_{i-1}, \qquad i=2,\cdots,n,$$

and the coefficients D_i, $\sigma_{a,i}$ and $\sigma_i \leqslant \sigma_{a,i}$ to be piecewise constant. For example, if $n=3$, the system (8.1)–(8.1′) can be displayed in vector notation as

$$\begin{bmatrix} L_1[\phi_1] \\ L_2[\phi_2] \\ L_3[\phi_3] \end{bmatrix} = \begin{bmatrix} 0 & 0 & \nu\sigma_3(\mathbf{x}) \\ \sigma_1(\mathbf{x}) & 0 & 0 \\ 0 & \sigma_2(\mathbf{x}) & 0 \end{bmatrix} \begin{bmatrix} \phi_1 \\ \phi_2 \\ \phi_3 \end{bmatrix}, \qquad L_i = -\nabla\cdot D_i(\mathbf{x})\nabla + \sigma_{a,i}(\mathbf{x})I.$$

The cyclic pattern of the nonzero entries in the matrix corresponds to the fact that 'fast' neutrons slow down to 'intermediate' to 'thermal', while the last produce 'fast' neutrons by being captured and causing fission.

The DE's (8.1)–(8.1′) are to hold in the 'reactor domain' Ω; on the boundary Γ, it is assumed that

$$(8.1'') \qquad \beta_i\,\partial\phi_i/\partial n+\phi_i=0, \qquad i=1,\cdots,n, \quad \beta_i>0.$$

The dependent variable $\phi_i(\mathbf{x})=v_iN_i(\mathbf{x})$ stands for the 'flux' at $\mathbf{x}$ of neutrons of the i-th velocity group, where v_i is their (nominal) mean velocity and $N_i(\mathbf{x})$ is their expected density (population per unit volume) in the vicinity of $\mathbf{x}$. The velocities satisfy $v_1>v_2>\cdots>v_n$; ϕ_1 is the flux of 'fast' (high energy) neutrons produced by fission and ϕ_n is the flux of 'slow' or 'thermal' neutrons. The D_i are the mean 'diffusivities' of neutrons of the i-th velocity group. The (macroscopic) absorption and down-scattering cross-sections are $\sigma_{a,i}$ and σ_i; ν is the mean neutron yield per fission.

The term on the right side of (8.1) is the *source term* for the fast neutron flux and the term on the right side of (8.1′) is the source term for the flux of the i-th velocity group. Thus the system is a *cyclic* system of (elliptic) *source problems*.

In addition, it can be regarded as an 'eigensystem' with $\nu>0$ considered as a parameter. Of greatest practical interest is the smallest eigenvalue ν_0 (the critical yield per fission). In general, complex eigenvalues can occur because the system (8.1)–(8.1′) is not 'self-adjoint' (cf. Chapter 2, §5). However, the smallest (in magnitude) eigenvalue ν_0

is always *positive* (and simple if Ω is connected); moreover, the associated eigenfunction (the *critical flux distribution*) is also positive.

9. Some nonlinear problems. The problems described in §§2–8 involve *linear* DE's (and boundary conditions). Many important elliptic problems are *non*linear; here we describe only a few examples of such problems. For second-order problems in two variables, these are of the form

$$au_{xx} + 2bu_{xy} + cu_{yy} = f(x, y, u, u_x, u_y). \tag{9.1}$$

When a, b, and c depend only on x and y, then the left side can be expressed in terms of the linear operator $\mathbf{L}[u] = au_{xx} + 2bu_{xy} + cu_{yy}$ and the equation is called *semi-linear*. If the coefficients a, b, c depend also on u, u_x, u_y, then the equation is called *quasi-linear*.

Probably the simplest nonlinear elliptic problem is that of a loaded string or *cable* having a large deflection. If we use the exact expression for the curvature $\kappa = y''/[1+(y')^2]^{3/2}$ instead of the approximation $\kappa \approx y''$ as in Example 2, the DE of a loaded string under a horizontal tension T and vertical load $Tf(x)$ per unit length is

$$y'' = f(x)\sqrt{1+(y')^2}. \tag{9.2}$$

If $f(x)$ is constant, then y is a catenary. Only slightly less simple is the nonlinear thin beam problem, whose DE is (in terms of arc-length s)[21]

$$d^2\theta/ds^2 + K\sin\theta = 0, \qquad \tan\theta = dy/dx.$$

Plateau problem. An interesting nonlinear elliptic problem whose solution is a function of two independent variables is the Plateau problem (Courant [50, Chap. 3]). In its simplest form, the problem is to minimize the *area*

$$A = \iint \sqrt{1 + {z_x}^2 + {z_y}^2}\, dx\, dy \tag{9.3}$$

of a variable surface $z(x, y)$ spanning a fixed simple closed curve $\gamma: x = x(\theta), y = y(\theta), z = z(\theta)$. Physically, this surface can be realized by a thin *soap-film* spanning a wire loop tracing out the curve γ (a special-purpose 'analog computer').

The associated Euler-Lagrange variational equation is

$$(1 + {z_y}^2)z_{xx} - 2z_x z_y z_{xy} + (1 + {z_x}^2)z_{yy} = 0, \tag{9.4}$$

which clearly reduces to the Laplace DE for a nearly flat surface, with

[21] Its solutions are described detail in A. E. H. Love, "A Treatise on the Mathematical Theory of Elasticity," 4th ed., Dover, 1944, §262.

$|z_x| << 1, |z_y| << 1$. Equation (9.4) is also the DE of a surface with *mean curvature zero*.

A related nonlinear problem is that of determining (e.g., computing) the surface or surfaces spanning γ which have given *constant mean curvature* $(\kappa_1+\kappa_2)/2 = M$. Such a problem arises when a soap bubble is constrained to enclose a fixed volume.

Note that since z is the solution of two different problems, (9.3) and (9.4), one has a *choice* as to which problem one solves numerically.

Nonlinear heat conduction. In reality, conductivity and specific heat are temperature-dependent, while heat transfer rates in fluids depend on the temperature gradient as well as the temperature. Therefore, more exact mathematical descriptions of heat conduction also lead to nonlinear DE's.

Of these, $\nabla^2 u + e^u = 0$ has been a favorite among mathematicians because of its simplicity, but it is by no means typical. Some idea of the complexity of real heat transfer problems can be obtained by skimming through Jakob [57, Chap. 26].

Semiconductor problems. An interesting system of nonlinear elliptic equations arises in the study of *semiconductors*. With u, v, w denoting the electrostatic potential and the density of electrons and holes, respectively, the system is

$$-\nabla^2 u + e^{u-v} - e^{w-u} = k(x,y),$$
$$-\nabla \cdot (\mu_n e^{u-v} \nabla v) = 0,$$
$$-\nabla \cdot (\mu_p e^{w-u} \nabla w) = 0.$$

The functions μ_n, μ_p, and the 'doping' profile k are given. For a discussion of methods for solving discretizations of this system, see W. Fichtner and D. J. Rose, in Schultz [81, pp. 277–284].

Chapter 2

Classical Analysis

1. Separation of variables. Nineteenth-century mathematical physicists displayed enormous ingenuity in solving *linear* partial DE's with constant coefficients, in terms of series and integrals involving tabulated special functions of one variable. Their most versatile method for obtaining such series (or integrals) consisted in 'separating variables', as in Examples 3–4 below.

This chapter will recall some of the infinite series and integral formulas obtained in this way. It will also recall some of the most essential and commonly used results of classical analysis. So as to get to *numerical algorithms* with a minimum of preliminaries, we will state most theorems only in a simple and easily derived form, and omit most proofs. Sharper, deeper, and more general formulations of many of these theorems will be presented in Chapters 6–8.

Besides solving many linear elliptic problems with constant coefficients, classical formulas perennially provide significant new solutions of important nonlinear problems.[1] Moreover, they play an essential role in scientific computing, because classical asymptotic and perturbation methods are generally much more effective than numerical methods for treating DE's whose coefficients differ by large factors, some very small ($\epsilon \ll 1$) and some very large ($\lambda \gg 1$). In addition, many algorithms for solving (elliptic) problems numerically have been inspired by classical variational principles, conservation laws, series expansions, and asymptotic formulas from classical analysis and mechanics. As von Mises predicted in 1952, writing about difference ("network") methods:[2]

> most or all of the results obtained in analytic studies of differential equations will have to be utilized . . . in supplementing the computational procedure[s] based on the network idea.

---

[1] See, for example, P. Neményi, *Advances in Applied Mechanics*, vol. 2, Academic Press, 1951, pp. 123–151; R. Berker, *Handbuch der Physik*, VIII/2, pp. 1–384; W. F. Ames, *Nonlinear Partial Differential Equations in Engineering*, Academic Press, 1965.

[2] R. von Mises in Beckenbach [52, pp. 1–5].

For these reasons, we will devote this chapter to a systematic review of relevant results from classical analysis, most of which will be applied later.

Particular solutions of linear partial DE's and ΔE's (difference equations) with constant coefficients also play a special role in contemporary numerical analysis, by providing *model problems* whose exact analytical solutions can be compared with those computed by approximate numerical methods (e.g., difference schemes). Among such model problems, the following are especially noteworthy.

Example 1. The Dirichlet problem for $\nabla^2 u = 0$ in the unit disk can be solved by expanding its (periodic) boundary values $u(1,\theta) = g(\theta)$ in Fourier series:

$$g(\theta) = \frac{1}{2}a_0 + \sum_{j=1}^{\infty} (a_j \cos j\theta + b_j \sin j\theta). \tag{1.1}$$

Its solution is then given by the infinite series

$$u(r,\theta) = \frac{1}{2}a_0 + \sum_{j=1}^{\infty} r^j (a_j \cos j\theta + b_j \sin j\theta), \tag{1.1'}$$

where

$$a_j = \frac{1}{\pi}\int_0^{2\pi} g(\theta)\cos j\theta \, d\theta, \qquad b_j = \frac{1}{\pi}\int_0^{2\pi} g(\theta)\sin j\theta \, d\theta. \tag{1.1''}$$

Each term of the series (1.1) is a solution $r^j \cos j\theta$ or $r^j \sin j\theta$ of $\nabla^2 u = 0$ having 'separated variables', i.e., each term is a product $\phi_j(r)\psi_k(\theta)$ of functions of one variable.

Example 1a. In particular, if $g(\theta) = 1$ on $(0,\pi)$ and $g(\theta) = -1$ on $(\pi,2\pi)$,

$$\begin{aligned} u(r,\theta) &= \frac{4}{\pi}[r\sin\theta + \frac{1}{3}r^3 \sin 3\theta + \frac{1}{5}r^5 \sin 5\theta + \cdots], \\ &= \frac{4}{\pi}\sum_{j=1}^{\infty} \frac{r^{2j-1}}{2j-1}\sin([2j-1]\theta). \end{aligned}$$

The series for $u(1,\theta)$ is zero for $\theta = 0$ and $\theta = \pi$, and converges pointwise to $g(\theta)$ for $\theta \in (0,\pi) \cup (\pi,2\pi)$. The function $u(r,\theta)$ is *analytic* for $|r| < 1$.

Example 1b. The (bounded) solution of the Laplace equation for the boundary values (1.1) in the *exterior* of the unit disk is given by

$$u(r,\theta) = \frac{1}{2}a_0 + \sum_{j=1}^{\infty} r^{-j}(a_j \cos j\theta + b_j \sin j\theta). \tag{1.2}$$

Example 2. It is easy to construct algebraically a basis of $2j+1$ homogeneous harmonic polynomials of degree j in three variables. Thus for

$j = 0, 1, 2$, such polynomial bases are:

$$1; \quad x, y, z; \quad \text{and} \quad xy, yz, zx, x^2 - y^2, x^2 + y^2 - 2z^2.$$

If θ is colatitude and ϕ is longitude, so that $z = r\cos\theta$, $x = r\sin\theta\cos\phi$, and $y = r\sin\theta\sin\phi$, then evidently

$$2xy = r^2\sin^2\theta\sin 2\phi, \qquad 2yz = r^2\sin 2\theta\sin\phi,$$
$$x^2 + y^2 - 2z^2 = -2r^2 P_2(\cos\theta),$$

where $P_2(u) = (3u^2 - 1)/2$ is the Legendre polynomial of degree 2.

The 'spherical harmonics'

$$r^n P_n^m(\cos\theta)\cos m\phi \quad \text{and} \quad r^n P_n^m(\cos\theta)\sin m\phi,$$

obtained similarly from harmonic polynomials of different degrees, are *orthogonal* on any sphere whose center is the origin.[3] Using them, one can solve the Dirichlet problem for the *interior* or *exterior* of any such sphere $r = a$, by series analogous to those displayed in Examples 1 and 1b.

Example 3. The solution of the Poisson equation $-\nabla^2 u = f(x,y)$ in the unit square S, with $u \equiv 0$ on ∂S, is easily expressed as a double sine series in terms of the eigenfunctions of Example 5 of Chapter 1, §5. If the 'source term' is

$$f(x,y) = \sum_{j=1}^{\infty}\sum_{k=1}^{\infty} c_{j,k}\sin j\pi x\sin k\pi y, \tag{1.3}$$

then the solution is

$$u(x,y) = \sum_{j=1}^{\infty}\sum_{k=1}^{\infty} a_{j,k}\sin j\pi x\sin k\pi y, \tag{1.4}$$

where

$$a_{j,k} = c_{j,k}/\lambda_{j,k} = c_{j,k}/(j^2 + k^2)\pi^2. \tag{1.4'}$$

For instance, setting $f \equiv 4$ in Example 3, one obtains the solutions of the Model Problem of Example 1 in Chapter 1, §3, in the form (1.4) with

$$a_{j,k} = \begin{cases} 64/jk(j^2 + k^2)\pi^4 & \text{if } j \text{ and } k \text{ are odd,} \\ 0 & \text{otherwise.} \end{cases} \tag{1.4''}$$

Likewise, from this $u(x,y)$, the solution of the torsion problem of Chapter 1, §3, is obtained as $v(x,y) = u(x,y) - x^2 - y^2$.

[3] See Courant-Hilbert [53, Chap. VII, §5].

Many other more sophisticated separations of variables (e.g., into Lamé functions [K, p. 205]) have been worked out by mathematical physicists. However, it has gradually become apparent that this tool, though versatile, is not truly general, and that its possibilities have been almost completely exhausted by the ingenuity of classical analysts.[4] Indeed, the main advantage of the *numerical methods* described in this monograph consists in their much greater *generality*.

Example 4. To solve the Dirichlet problem in the unit square S: $[0,1]\times[0,1]$, we use products like

$$\psi_k(x,y) = \sin k\pi x \sinh k\pi(1-y).$$

These are solutions of the Laplace equation which for integer k are zero on three sides. Their sum can be made equal to $g(x)=u(x,0)$ for $0<x<1$ on the fourth side, by setting

$$u(x,y) = \sum_{k=1}^{\infty} c_k \sin k\pi x \sinh k\pi(1-y), \tag{1.5}$$

with $c_k \sinh k\pi$ the Fourier sine coefficient b_k of $g(x)$. Harmonic functions which assume the given values on the other three sides can be constructed similarly using $\psi_k(x,1-y)$, $\psi_k(y,x)$, and $\psi_k(1-y,x)$.

By combining Examples 3 and 4, we can solve the Poisson equation with Dirichlet boundary conditions on the unit square S. We can construct similarly solutions satisfying Neumann boundary conditions on one or more sides, or periodic boundary conditions.

This method extends to the Poisson equation in any n-dimensional 'box' $\mathrm{B}=[0,a_1]\times\cdots\times[0,a_n]$. The Laplace operator has a basis of eigenfunctions of the form

$$\phi_{\mathbf{k}} = \prod_{j=1}^{n} \sin(k_j\pi x_j/a_j) \quad \text{with eigenvalues} \quad \pi^2 \sum_{j=1}^{n} k_j^2/a_j^2,$$

for the boundary condition $u\equiv 0$ on $\partial\mathrm{B}$. These can be used to solve the Poisson equation for this boundary condition. Likewise, one can construct harmonic functions which are zero on all but one 'face' of B; for example,

$$\sinh(k_1\pi x_1/a_1)\prod_{j=2}^{n}\sin(k_j\pi[a_j-x_j]/a_j), \qquad (k_1/a_1)^2 = \sum_{j=2}^{n}(k_j/a_j)^2.$$

By taking linear combinations of the functions described above, one can

[4] See L. P. Eisenhart, Annals of Math. 35 (1934), 284–305; Willard Miller, Jr., *Symmetry and Separation of Variables*, Addison-Wesley, 1977.

solve the Poisson equation in B for general Dirichlet-type boundary conditions.

2. Complex variable techniques. In §1, we described some elliptic problems which are solvable in terms of tabulated functions of one *real* variable. Other elliptic problems, especially those involving *harmonic* functions of two real variables (i.e., functions satisfying $u_{xx}+u_{yy}=0$), are most easily solved in terms of functions of one *complex* variable.

This is because the real and imaginary parts of any complex analytic function $w=f(z)$ of a complex variable $z=x+iy$ are conjugate harmonic functions of the two real variables x and y. That is, writing $w=u(x,y)+iv(x,y)$, we have the Cauchy-Riemann equations $u_x=v_y$ and $v_x=-u_y$, whence $\nabla^2 u=\nabla^2 v=0$. Conversely, if u is any harmonic function of two variables, then $v(x,y)=\int(u_x\,dy-u_y\,dy)$ is also, and $w=u+iv$ is a complex analytic function of $z=x+iy$.

Example 5. The translational and rotational added masses (Chapter 1, §6) of the square S: $[-1,1]\times[-1,1]$, are the Dirichlet integrals of the harmonic functions (velocity potentials) ϕ_1, ϕ_2 which satisfy, respectively, the Neumann boundary conditions

$$\left.\begin{aligned}\frac{\partial\phi_1}{\partial x}(\pm1,y)&=1\\ \frac{\partial\phi_2}{\partial x}(\pm1,y)&=y\end{aligned}\right\}\ \text{for } -1\leqslant y\leqslant 1,\qquad \left.\begin{aligned}\frac{\partial\phi_1}{\partial y}(x,\pm1)&=0\\ \frac{\partial\phi_2}{\partial y}(x,\pm1)&=-x\end{aligned}\right\}\ \text{for } -1\leqslant x\leqslant 1,$$

and $\phi_i\to 0$ as $x^2+y^2\to\infty$. Since the Dirichlet integrals of conjugate harmonic functions are equal, these are equal to the Dirichlet integrals of the conjugate harmonic functions ψ_1, ψ_2 (stream functions). They satisfy $\nabla^2\psi_i=0$ with the Dirichlet boundary conditions

$$\psi_1=y,\quad \psi_2=c+(x^2+y^2)/2\qquad \text{for } (x,y)\in\partial S$$

and $\psi_i\to 0$ as $x^2+y^2\to\infty$. The constant c must also be determined.

Remark. For any positive integer n, n-dimensional potential theory (which is the study of harmonic functions) can be viewed as generalizing some aspects of complex analysis from functions of two real variables to functions of n variables. (The study of harmonic functions of one real variable is trivial: they are all linear, of the form $u=ax+b$.)

Harmonic functions of two variables include the *harmonic polynomials* $1, x, y, x^2-y^2, xy$, etc. Thus, since

$$z^k=r^k\cos k\theta+ir^k\sin k\theta=(x+iy)^k,$$

the terms in (1.1′) can also be written as superpositions of harmonic polynomials like

$$r^4\cos 4\theta=x^4-6x^2y^2+y^4\quad\text{and}\quad r^4\sin 4\theta=4xy(x^2-y^2).$$

Conformal mapping. Clearly, the set of harmonic functions $h(x,y)$ is invariant under one-one *conformal mappings* of the form $t=f(z)$, $z=x+iy$, f any complex analytic function. Also, since $dt=f'(z)\,dz$ and $|dw/dt|=|dw/dz|/|dt/dz|$, normals go into normals, and normal derivatives are divided by $|f'(z)|$ under such a conformal mapping. This makes it easy to reduce boundary value problems in a given region to similar problems in any conformally equivalent region, *if* one knows the appropriate conformal transformation.

Example 6. The exterior E : $r>1$ of the unit circle is mapped conformally onto the exterior R of the unit slit S : $|x|\leqslant 1$ by

$$w=\frac{1}{2}(z+z^{-1})=\frac{1}{2}[(r+r^{-1})\cos\theta+i(r-r^{-1})\sin\theta]. \tag{2.1}$$

Hence one can combine Example 1 with (2.1) to solve the (exterior) Dirichlet problem for R. Moreover, since (2.1) maps each circle $r=a$ $(a>1)$ onto an ellipse, one can solve the exterior Dirichlet problem for each such ellipse similarly.

Example 7. The upper half-plane H : $y>0$ can be mapped onto the interior of a rectangle of any shape or size by a conformal transformation of the form

$$w(z)=\beta\int_0^z\frac{dz}{\sqrt{(1-z^2)(1-k^2z^2)}},\qquad \beta\text{ real, }0<k^2<1. \tag{2.2}$$

This follows because $dw=\beta\,dz/\sqrt{(1-z^2)(1-k^2z^2)}$ is *real* for z real except when $1\leqslant|x|\leqslant 1/k$, while on these intervals $(1-x^2)(1-k^2x^2)$ is negative and so dw is pure imaginary. The imaginary axis is an axis of symmetry for the rectangle (and the transformation).

More generally, the upper half-plane H can be mapped conformally onto the interior of any polygon with n sides by a so-called Schwarz-Christoffel transformation

$$w=\alpha+\beta\int\frac{dz}{(z-a_1)^{\mu_1}\cdots(z-a_n)^{\mu_n}}, \tag{2.3}$$

where all a_i and μ_i are real, and $\sum_i\mu_i=2$; the $\pi(1-\mu_i)$ are the interior angles, measured in radians, and the $\pi\mu_i$ are the angles of bend.

Riemann mapping theorem. Indeed, the fundamental theorem of conformal mapping is the following, much more general result (Ahlfors [79, p. 229]); see also Chapter 8, §6.

THEOREM 1. *Every simply connected bounded plane domain can be mapped one-one and conformally onto the unit disk.*

Since $z=(1+it)/(1-it)$ maps the upper half t-plane onto the unit disk, and since every simple closed curve can be approximated arbitrarily closely by a polygon, formula (2.3) makes the theorem plausible. We shall study its effective implementation in Chapter 8, §§6–7.

The method of conformal mapping is much less fruitful in $n > 2$ dimensions, because when $n > 2$ all conformal transformations of $\mathbb{R}^n$ are generated by inversions, and carry spheres into spheres or planes; this is a classic theorem of Liouville. In §3, we will discuss analytical methods which *are* applicable to $\mathbb{R}^n$ for $n > 2$, emphasizing the case $n = 3$.

3. Integral formulas. The divergence theorem states that, if a vector-valued function $\mathbf{v}(\mathbf{x})$ is continuously differentiable in the closure $\bar{\Omega} = \Omega \cup \Gamma$ of a bounded domain Ω (in symbols, for any $\mathbf{v} \in C^1(\bar{\Omega})$), then

$$(3.1) \qquad \int_\Omega \nabla \cdot \mathbf{v}\, dR = \int_\Gamma v_n(\mathbf{y})\, dS.$$

Here and below we use dR for the differential of 'volume' ($dx\, dy$ in the plane, $dx\, dy\, dz$ in space and dS for the differential of 'surface area' (arc length in the plane). In (3.1), $v_n = \boldsymbol{\gamma} \cdot \mathbf{v}$ denotes the component of $\mathbf{v}$ in the direction of the *outward* unit *normal* vector $\boldsymbol{\gamma}(\mathbf{y})$ at each $\mathbf{y} \in \Gamma$. In words, the integral of the divergence of $\mathbf{v}(\mathbf{x})$ over any domain is the outward *flux* of $\mathbf{v}$ through its boundary. In particular, if $\mathbf{v} = \nabla u$ is the *gradient* of a scalar function $u \in C^2(\bar{\Omega})$, then

$$(3.2) \qquad \int_\Omega \nabla^2 u\, dR = \int_\Gamma \frac{\partial u}{\partial n}\, dS.$$

Arithmetic mean. We next prove Gauss' theorem of the arithmetic mean: the value of a harmonic function at any point is the average (or 'arithmetic mean') of its values over any sphere in $\mathbb{R}^n$ having that point as center.

THEOREM 2 (Gauss' theorem of the arithmetic mean). *If u is harmonic in a closed sphere of radius a and center c, then*

$$(3.3) \qquad u(\mathbf{c}) = \frac{1}{4\pi a^2} \int_\Gamma u\, dS.$$

That is, the value of u at the center of any sphere is the average of its values on the surface.

Proof. Without loss of generality, we can assume that the center of the sphere is the origin $\mathbf{0}$. By (3.2), $\int_{S(r)} \partial u / \partial r\, dS = 0$, where $S(r)$ is the sphere of radius r with center $\mathbf{0}$, $dS = r^2\, d\omega$ is the differential of surface area, and $d\omega$ is the differential of solid angle. Therefore, we have

$$\int_{S(r)} \frac{\partial u}{\partial r}\, d\omega = \frac{\partial}{\partial r} \int_{S(r)} u\, d\omega = 0,$$

and so $\int u(r, \omega)\, d\omega$ is a constant, independent of r. Letting $r \downarrow 0$, we see by continuity that this constant is $4\pi u(\mathbf{0})$:

$$u(\mathbf{0}) = \frac{1}{4\pi}\int u(a,\omega)\,d\omega = \frac{1}{4\pi a^2}\int_{S(a)} u(a,\omega)\,dS.$$

Similarly, in the plane, it follows from (3.2) that

$$u(a,b) = \frac{1}{2\pi}\int_0^{2\pi} u(a+c\cos\theta, b+c\sin\theta)\,d\theta. \tag{3.4}$$

In words, the value of a harmonic function at the center of any circle is also the average of its values on the circumference; see Ahlfors [79, p. 165]. Analogous formulas hold in $\mathbb{R}^n$ for any $n > 2$.

Gauss' integral formula. In $\mathbb{R}^3$, the potential of a charge e at a point $\boldsymbol{\xi} \in \Omega$ is $u = e/r$. It is geometrically obvious that the flux of the field $\nabla(e/r) = -(\cos\gamma)e/r^2$ through an infinitesimal boundary element dS of $\Gamma = \partial\Omega$ is $e\,d\omega$, where $d\omega$ is the solid angle subtended at $\boldsymbol{\xi}$ by the cone through dS (whose projected area is $r^2\,d\omega$). By combining these observations, we get Gauss' integral formula [K, p. 43].

THEOREM 3. *Let Γ be the boundary of a domain Ω in $\mathbb{R}^3$. Then the flux through Γ of any distribution of charge over $\mathbb{R}^3$ is $-4\pi e(\Omega)$, where $e(\Omega)$ is the total charge inside Ω.*

In particular, if the charge density $\rho(\boldsymbol{\xi})$ is *continuous*, and if its potential $u \in C^2(\overline{\Omega})$,[5] then u satisfies the *Poisson DE*

$$-\nabla^2 u = f(\mathbf{x}) \quad \text{in } \Omega, \qquad f(\mathbf{x}) = 4\pi\rho(\mathbf{x}). \tag{3.5}$$

When $\Omega = \mathbb{R}^3$, this formula gives an *integral formula* for solving the Poisson DE (3.5) in empty space (i.e., in $\mathbb{R}^3$):

$$u(\mathbf{x}) = \int_{\mathbb{R}^3} G(\mathbf{x};\boldsymbol{\xi}) f(\boldsymbol{\xi})\,dR(\boldsymbol{\xi}) = \mathbf{G}[f], \tag{3.6}$$

with 'kernel' $G(\mathbf{x};\boldsymbol{\xi}) = 1/4\pi|\mathbf{x}-\boldsymbol{\xi}|$.

Green's functions. Note that, setting $\mathbf{L} = -\nabla^2$ in free space, (3.5) can also be written symbolically in operator notation as $\mathbf{L}[\mathbf{G}[f]] = f$. This formula asserts that the *integral* operator $\mathbf{G}$ is a *right-inverse* of the *differential* operator $\mathbf{L}$. The 'kernels' of integral operators with this property are called *Green's functions* (sometimes 'fundamental solutions'). Note also that $G(\mathbf{x};\boldsymbol{\xi}) = G(\boldsymbol{\xi};\mathbf{x})$; this *symmetry* is typical of the Green's functions of self-adjoint operators.

One can construct analogous 'Green's functions' for the Laplace operator $-\nabla^2$ in any *bounded* domain $\Omega \subset \mathbb{R}^3$, by setting

$$G(\mathbf{x};\boldsymbol{\xi}) = \frac{1}{4\pi|\mathbf{x}-\boldsymbol{\xi}|} + \Phi(\mathbf{x};\boldsymbol{\xi}),$$

---

[5] For a careful discussion of these assumptions, see G. Birkhoff and L. J. Burton, Canadian J. Math. 1 (1949), 199–208.

where for each $\boldsymbol{\xi} \in \Omega$, Φ is harmonic in $\mathbf{x}$ and satisfies

$$\Phi(\mathbf{y};\boldsymbol{\xi}) = -\frac{1}{4\pi|\mathbf{y}-\boldsymbol{\xi}|}$$

on $\Gamma = \partial\Omega$. It follows that G vanishes identically on Γ, and is asymptotic to $1/4\pi|\mathbf{x}-\boldsymbol{\xi}|$ near $\boldsymbol{\xi}$. Physically, G represents the potential induced by a unit charge at $\boldsymbol{\xi}$ inside a cavity bounded by a conductor fitted to Γ.

Kelvin transformation. Geometrical *inversion* in any sphere S is accomplished by the following 'transformation of reciprocal radii'. Set up spherical coordinates having the center of S as origin, and map (r,θ,ϕ) into $(a^2/r,\theta,\phi)$, where a is the radius of S. This transformation is *conformal*, and carries spheres into spheres (or planes). Kelvin proved further that if $u(r,\theta,\phi)$ is any function harmonic *inside* S, then the function

$$v(r,\theta,\phi) = \frac{a}{r}u(a^2/r,\theta,\phi) \tag{3.7}$$

is harmonic *outside* S, and vice versa.

Green's functions for spheres and disks. The Green's function of the Laplace operator $-\nabla^2$ in a sphere can be constructed by the method of images. A charge e at a point a distance c from the center of a sphere of radius a and a charge $-ea/c$ at a point inverse to the first in the sphere, produce together a potential which is 0 on the surface of the sphere [K, p. 235]. In symbols,

$$G(\mathbf{x};\boldsymbol{\xi}) = \frac{1}{4\pi}\left[\frac{1}{|\mathbf{x}-\boldsymbol{\xi}|} - \frac{k}{|\mathbf{x}-k^2\boldsymbol{\xi}|}\right], \qquad k = \frac{a}{|\boldsymbol{\xi}|}, \tag{3.8}$$

where $k^2\boldsymbol{\xi}$ is the inverse, with respect to the sphere, of $\boldsymbol{\xi}$. If γ is the angle between $|\mathbf{x}|$ and $|\boldsymbol{\xi}|$, then, because

$$\frac{1}{k^2}|\mathbf{x}-k^2\boldsymbol{\xi}|^2 = \frac{1}{a^2}|\mathbf{x}|^2|\boldsymbol{\xi}|^2 + a^2 - 2|\mathbf{x}||\boldsymbol{\xi}|\cos\gamma,$$

the Green's function is *symmetric*: $G(\mathbf{x};\boldsymbol{\xi}) = G(\boldsymbol{\xi};\mathbf{x})$.

Similarly, in the plane, the inversion with respect to a circle of radius a, $(r,\theta) \mapsto (a^2/r,\theta)$, maps a harmonic function u to a harmonic function U:

$$U(r,\theta) = u(a^2/r,\theta),$$

and the Green's function for the disk is

$$G(\mathbf{x};\boldsymbol{\xi}) = \frac{1}{2\pi}\log\left[\frac{|\mathbf{x}-k^2\boldsymbol{\xi}|}{k|\mathbf{x}-\boldsymbol{\xi}|}\right]. \tag{3.8$'$}$$

Poisson's integral.[6] It is easy to derive a formula which expresses the solution to the Dirichlet problem on the unit disk, $\Delta: x^2+y^2 \leqslant 1$, as a weighted average of its boundary values $g(\theta) = u(1,\theta)$. Expanding $g(\theta)$ in Fourier series, we have, as in (1.1),

$$g(\theta) = \frac{1}{2} a_0 + \sum (a_k \cos k\theta + b_k \sin k\theta),$$

where the a_k and b_k are given by (1.1″). Replacing θ by ϕ in (1.1″), and substituting into (1.1′), elementary trigonometric manipulations give

$$u(r,\theta) = \frac{1}{2\pi} \oint [1 + 2 \sum_{k=1}^{\infty} r^k \cos k(\theta - \phi)]\, g(\phi)\, d\phi.$$

Denoting the 'Poisson integral kernel' in square brackets by $P(r,\theta,\phi)$, we have

$$u(r,\theta) = \frac{1}{2\pi} \oint P(r,\theta,\phi) g(\phi)\, d\phi \tag{3.9}$$

where, by definition, setting $t = r e^{i(\theta-\phi)}$, $t^* = r e^{-i(\theta-\phi)}$,

$$\begin{aligned} P(r,\theta,\phi) &= \mathrm{Re}\{1 + 2 \sum_{k=1}^{\infty} t^k\} = \mathrm{Re}\{(1+t)/(1-t)\} \\ &= \mathrm{Re}\{[1 + (t - t^*) - tt^*]/[(1-t)(1-t^*)]\} \\ &= (1 - tt^*)/|1-t|^2, \end{aligned} \tag{3.9′}$$

since $t - t^*$ is always a pure imaginary. This formula has a simple geometric interpretation. The denominator, $\rho^2 = 1 - 2r\cos(\theta - \phi) + r^2$, is the square of the distance between the radius vector to (r,θ) and the vector to the point $(1,\phi)$ on the boundary, and the numerator is $1 - r^2$. Hence the positive number $P(r,\theta,\phi) = (1 - r^2)/\rho^2$ expresses the *weighting factor* by which values on the boundary must be multiplied, to express $u(r,\theta)$ as a *weighted mean* of boundary values.

4. Green's identities. By discovering the divergence theorem and applying it systematically, the British mathematician George Green derived in 1828 a large number of identities involving the Laplace operator and integrals over a general region Ω and its boundary $\Gamma = \partial\Omega$. He tacitly assumed all functions to be twice differentiable (smooth) in the *closure* $\bar{\Omega} = \Omega \cup \Gamma$ of Ω, an assumption that is often unnecessary.[7]

[6] Cf. Courant-Hilbert [53, pp. 513–514]. For the n-dimensional case, see Gilbarg-Trudinger [77, §2.5].

[7] Green's identities hold more generally; see [K, Chaps. IV and VIII] for the case $n = 3$; Gilbarg-Trudinger [77, §2.4] for general n.

Nowadays, it seems most natural to relate Green's identities to the properties of the *Dirichlet inner product* $\mathbf{D}\langle u,v\rangle = \mathbf{D}_\Omega\langle u,v\rangle$, defined by the formula

$$\mathbf{D}\langle u,v\rangle = \int_\Omega \nabla u \cdot \nabla v \, dR. \tag{4.1}$$

Clearly the functional[8] $\mathbf{D}\langle u,v\rangle$ is *symmetric* in the sense that $\mathbf{D}\langle u,v\rangle = \mathbf{D}\langle v,u\rangle$, *bilinear* because

$$\mathbf{D}\langle u, \alpha v + \beta w\rangle = \alpha \mathbf{D}\langle u,v\rangle + \beta \mathbf{D}\langle u,w\rangle$$

for any real scalars α and β, and *nonnegative* in the sense that $\mathbf{D}\langle u,u\rangle \geqslant 0$ for all[9] $u \in D^1(\overline{\Omega})$. It is *almost* positive: $\mathbf{D}\langle u,u\rangle > 0$ unless $u(\mathbf{x})$ is constant. On the subspace of functions on Ω for which $u(\mathbf{y}) = 0$ on Γ, $\mathbf{D}\langle u,v\rangle$ *is* an inner product (cf. §5); i.e., $\mathbf{D}\langle u,u\rangle = 0$ implies $u \equiv 0$. Hence unless u is a constant, $\mathbf{D}$ is an *inner product* (Birkhoff-Mac Lane [77, Chap. 7]); it makes the space of functions which are in $C^2(\overline{\Omega})$, and satisfy $u(\mathbf{y}) = 0$ on Γ, a *Euclidean vector space*, with all the associated geometrical properties.

Green's first identity. The inner product $\mathbf{D}\langle u,v\rangle$ is especially convenient for analyzing harmonic functions. Thus since

$$\nabla u \cdot \nabla v = \nabla \cdot [v \nabla u] - v \nabla^2 u,$$

evidently

$$\mathbf{D}\langle u,v\rangle = \int_\Gamma v \frac{\partial u}{\partial n} \, dS - \int_\Omega v \nabla^2 u \, dR. \tag{4.2}$$

Equation (4.2) is Green's first identity.

If u is *harmonic* in Ω, so that $\nabla^2 u \equiv 0$ there, then for all $v \in C^1(\overline{\Omega})$

$$\mathbf{D}\langle u,v\rangle = \int_\Gamma v \frac{\partial u}{\partial n} \, dS \qquad \text{if } \nabla^2 u = 0. \tag{4.3}$$

In particular, if also $v \equiv 0$ on Γ, then $\mathbf{D}\langle u,v\rangle = 0$.

Dirichlet principle. Now let $u(\mathbf{x})$ satisfy $\nabla^2 u = 0$ in a closed domain $\overline{\Omega}$, and let u be given everywhere on Γ. Then for any $u + \delta u$ satisfying the same boundary conditions, δu vanishes identically on $\Gamma = \partial\Omega$. Set $v = \delta u$; it then follows from (4.3) that $\mathbf{D}\langle u, \delta u\rangle = 0$. Consequently,

$$\mathbf{D}\langle u + \delta u, u + \delta u\rangle = \mathbf{D}\langle u,u\rangle + \mathbf{D}\langle \delta u, \delta u\rangle.$$

But $\mathbf{D}\langle \delta u, \delta u\rangle \geqslant 0$, with equality holding if and only if δu is constant;

---

[8] We recall that a function*al* is a function that assigns to each element of its domain (here the pair of functions u and v) a number.

[9] The set D^k is defined to consist of those functions that are of class C^{k-1}, and of class C^k except for finite jumps in $(k-1)$-st derivatives across a finite set of hypersurfaces.

moreover, since $\delta u \equiv 0$ on Γ, this constant must be 0. This proves the following theorem, which will be generalized in Chapter 8.

THEOREM 4 (Dirichlet principle). *For given Dirichlet boundary conditions, a function $u \in C^2(\bar{\Omega})$ minimizes the Dirichlet integral, if and only if u is harmonic.*

Furthermore, setting $u = v$ in (4.3) we obtain

$$\mathbf{D}\langle u,u\rangle = \int_\Gamma u \frac{\partial u}{\partial n} \, dS \qquad \text{if } \nabla^2 u = 0. \tag{4.4}$$

This leads to the following uniqueness theorem for mixed boundary value problems [K, p. 214].

THEOREM 5. *At most one harmonic function $u \in C^1(\bar{\Omega})$ can satisfy the mixed boundary condition*

$$\partial u/\partial n + h(\mathbf{y})u = g(\mathbf{y}) \qquad \text{on } \Gamma, \quad h(\mathbf{y}) \geqslant 0, \tag{4.5}$$

unless $h(\mathbf{y}) \equiv 0$. If $h(\mathbf{y}) \equiv 0$, then two solutions must differ by a constant.

Proof. For any other solution $v(\mathbf{x})$ of the same problem, the difference $w = v - u$ is harmonic and satisfies (4.5) with $g \equiv 0$. Therefore, by (4.1) and (4.4),

$$\mathbf{D}\langle w,w\rangle = \int_\Omega \nabla w \cdot \nabla w \, dR = \int_\Gamma w \frac{\partial w}{\partial n} \, dS = -\int_\Gamma h(\mathbf{y}) w^2 \, dS \leqslant 0.$$

Hence $\nabla w \equiv 0$ in Ω, so w is constant in $\bar{\Omega} = \Omega \cup \Gamma$; moreover, $\partial w/\partial n = 0$ on Γ, so $h(\mathbf{y})w \equiv 0$ there. Unless $h(\mathbf{y}) \equiv 0$, this implies $w \equiv 0$ in $\bar{\Omega}$, which completes the proof.

Since any constant function $u = C$ is a solution of the Neumann problem, the exception $h(\mathbf{y}) \equiv 0$ on Γ is of course genuine.

Setting $v = 1$ in (4.1), so that $\nabla v \equiv 0$ in Ω and $\mathbf{D}\langle u,v\rangle = 0$, we also have from (4.3) that

$$\int_\Gamma \frac{\partial u}{\partial n} \, dS = 0 \qquad \text{for any harmonic } u. \tag{4.6}$$

(In words: the flux through a bounding surface Γ of any function harmonic inside and on Γ is zero.)

Green's second identity. Now suppose that both of the functions u and v are in $C^2(\bar{\Omega})$. Subtracting from (4.2) the identity obtained from it by interchanging u and v, we get *Green's second identity*:

$$\int_\Omega (u\nabla^2 v - v\nabla^2 u) \, dR = \int_\Gamma \left(u \frac{\partial v}{\partial n} - v \frac{\partial u}{\partial n}\right) dS. \tag{4.7}$$

Since $\mathbf{D}\langle u,v\rangle = \mathbf{D}\langle v,u\rangle$, this has the following immediate consequence.

THEOREM 6. *If u and v are both harmonic in $\bar{\Omega}$, then*

$$\int_\Gamma u \frac{\partial v}{\partial n} \, dS = \int_\Gamma v \frac{\partial u}{\partial n} \, dS. \tag{4.8}$$

The same formulas hold for the region Ω exterior to a given boundary Γ, provided the integral defining **D** exists, as in the added mass problems of Chapter 1, §6. To derive them, one must appeal to the result that any function *harmonic* and bounded outside a sphere $x^2+y^2+z^2=R$ can be expanded there in a series of terms of the form

$$P_n^m(\cos\theta)\cos m\phi/r^{n+1} \quad \text{and} \quad P_n^m(\cos\theta)\sin m\phi/r^{n+1},$$

where the P_n^m are associated Legendre functions (Courant-Hilbert [53, p. 505]). This generalizes the theorem that a complex analytic function $f(z)$ which is bounded outside a circle $|z|>R$ can be expanded in a convergent series of negative powers of z.

Using Green's identities, one can show the outward normal derivative[10] $\partial G/\partial\nu$ (with respect to ξ) of the Green's function (3.8) is a *Poisson kernel* in the following sense. If $u(\mathbf{x})$ is the solution of the Dirichlet problem for the boundary values $u(\mathbf{x})=g(\mathbf{y})$ on Γ, then we have

$$u(\mathbf{x})=\int_\Gamma \frac{\partial G}{\partial\nu}(\mathbf{x};\xi)g(\xi)\,dS(\xi). \tag{4.9}$$

Analogous formulas hold in two dimensions (i.e., the plane); the role of $1/4\pi r$ is replaced by $(\log r)/2\pi$, however.

Green's third identity. We now apply Green's second identity (4.7) to the region Δ which is *interior* to a bounding surface Γ and is *exterior* to a small sphere γ of radius ρ and center $\mathbf{x}$ inside γ. Continuing to let r denote the distance of ξ from $\mathbf{x}$, and noting that $\nabla^2(1/r)\equiv 0$ for $r>0$, we obtain

$$\int_{\Gamma^*}\frac{1}{r}\frac{\partial u}{\partial\nu}\,dS=\int_{\Gamma^*}u\frac{\partial}{\partial\nu}\frac{1}{r}\,dS+\int_\Delta \frac{1}{r}\nabla^2u\,dR, \tag{4.10}$$

where $\Gamma^*=\Gamma\cup\gamma$ and $\partial/\partial\nu$ denotes the normal derivative[10] at a point ξ of Γ^*. As $\rho\downarrow 0$, $\int_\gamma(1/r)(\partial u/\partial n)\,dS=O(\rho)$ tends to zero,[11] while (cf. [K, p. 219])

$$-\int_\gamma u\frac{\partial}{\partial\nu}\frac{1}{r}\,dS=\int_\gamma u\cos\theta\,d\theta\,d\phi\longrightarrow 4\pi u(\mathbf{x}). \tag{4.11}$$

Substituting into (4.10), passing to the limit as $\rho\downarrow 0$, and transposing $4\pi u(\mathbf{x})$, we get Green's third identity in the following form.

THEOREM 7. *Let* $u\in C^2(\overline{\Omega})$. *Then*

$$4\pi u(\mathbf{x})=\int_\Gamma \frac{1}{r}\frac{\partial u}{\partial\nu}\,dS-\int_\Gamma u\frac{\partial}{\partial\nu}\frac{1}{r}\,dS-\int_\Omega\frac{1}{r}\nabla^2u\,dR. \tag{4.12}$$

[10] Here and below, we use $\partial/\partial n$ and $\partial/\partial\nu$ to denote normal derivatives with respect to $\mathbf{x}$ and ξ, respectively.

[11] A function $f(\rho)$ is said to be "order $g(\rho)$ as $\rho\downarrow 0$" if the ratio $f(\rho)/g(\rho)$ is bounded when ρ tends (downward) to zero; in symbols: $f(\rho)=O(g(\rho))$.

Hence if u *is harmonic in* $\overline{\Omega}$, *then*

$$u(\mathbf{x}) = \frac{1}{4\pi}\int_\Gamma \frac{1}{r}\frac{\partial u}{\partial \nu}\, dS - \frac{1}{4\pi}\int_\Gamma u \frac{\partial}{\partial \nu}\frac{1}{r}\, dS. \tag{4.12'}$$

Formula (4.12′) expresses any function harmonic in Ω as the potential of a distribution of charges and dipoles on the boundary $\Gamma = \partial\Omega$, the latter oriented normal to Γ (see [K, pp. 55, 68]). This result will be applied to solve elliptic problems by integral equation methods in Chapter 8, §§4, 8, and 9.

In the plane, where $1/4\pi r$ is replaced with $(\log r)/2\pi$, (4.12′) is replaced with

$$u(\mathbf{x}) = \frac{1}{2\pi}\int_\Gamma u \frac{\partial}{\partial \nu}\log r\, ds - \frac{1}{2\pi}\int_\Gamma \log r \frac{\partial u}{\partial \nu}\, ds. \tag{4.13}$$

5. Inner products; norms. An 'inner product' $\langle u,v\rangle$ on a real vector space V is, by definition, a functional (on $V \times V$) which satisfies

$$\begin{aligned} &\langle u,u\rangle > 0 && \text{unless } u = 0,\\ &\langle u,v\rangle = \langle v,u\rangle && u, v \in V,\\ &\langle \alpha u + \beta v, w\rangle = \alpha\langle u,w\rangle + \beta\langle v,w\rangle && \alpha,\beta \in \mathbb{R},\ u, v, w \in V. \end{aligned} \tag{5.1}$$

For example, $\langle u,v\rangle = \int uv\, dR$ is an inner product and, for $\mathbf{x}, \mathbf{y} \in \mathbb{R}^n$, $\mathbf{x}^T\mathbf{y}$ is the usual algebraic 'inner product'.

An 'inner product' $\mathbf{P}\langle u,v\rangle$ analogous to $\mathbf{D}\langle u,v\rangle$ in (4.1) can be constructed for any 'source problem with absorption' defined as in Chapter 1, (3.2). Inserting the absorption term $q(\mathbf{x})u$ on the left side, we get the DE

$$\mathbf{L}[u] = -\nabla \cdot [p(\mathbf{x})\nabla u] + q(\mathbf{x})u = f(\mathbf{x}), \qquad p > 0, \quad q \geqslant 0. \tag{5.2}$$

Moreover, solutions of such general linear source problems share many other properties with solutions of the Laplace and Poisson equations.

Thus, for given p, q, the functional

$$\mathbf{P}\langle u,v\rangle = \int_\Omega [p(\mathbf{x})\nabla u \cdot \nabla v + q(\mathbf{x})uv]\, dR \tag{5.3}$$

is obviously symmetric (i.e., $\mathbf{P}\langle u,v\rangle = \mathbf{P}\langle v,u\rangle$) and bilinear. Then, if $p > 0$ and $q \geqslant 0$, $\mathbf{P}\langle u,u\rangle > 0$ unless either $u \equiv 0$, or u is constant and $q \equiv 0$. This proves the following result.

THEOREM 8. *With* $p > 0$ *and* $q \geqslant 0$, *the functional* $\mathbf{P}\langle u,v\rangle$ *defined by* (5.3) *is an inner product on any subspace of* $C^1(\overline{\Omega})$, *unless* $q \equiv 0$. *Even if* $q \equiv 0$, *it is an inner product unless the subspace contains a nonzero constant function.*

We can integrate the first term of (5.3) by parts, much as we did in proving Green's identities. Since at every point $\mathbf{x} \in \Omega$,

$$\nabla\cdot[pv\nabla u]=p\nabla v\cdot\nabla u+v\nabla\cdot[p\nabla u],$$

the divergence theorem (3.1) implies the following analogue of (4.2):

$$\mathbf{P}\langle u,v\rangle=\int_\Gamma pv\frac{\partial u}{\partial n}\,dS-\int_\Omega[p\nabla v\cdot\nabla u+v\nabla\cdot p\nabla u]\,dR. \tag{5.4}$$

If u is a *solution* of (5.2), then

$$\mathbf{P}\langle u,v\rangle=\int_\Gamma pv\frac{\partial u}{\partial n}\,dS+\int_\Omega vf(\mathbf{x})\,dR. \tag{5.5}$$

Among other things, this formula allows us to prove the following generalization of the Dirichlet principle.

THEOREM 9. *If u is any solution of the homogeneous linear DE $\mathbf{L}[u]=0$ obtained by setting $f=0$ in (5.2), then $\mathbf{P}\langle u,u\rangle<\mathbf{P}\langle w,w\rangle$ for any other smooth function w assuming the same boundary values on Γ.*

Proof. For any such function w, set $v=w-u$. Then $v\equiv 0$ on Γ and by (5.5) with $f=0$, $\mathbf{P}\langle u,v\rangle=0$. Thus

$$\mathbf{P}\langle w,w\rangle=\mathbf{P}\langle u+v,u+v\rangle=\mathbf{P}\langle u,u\rangle+\mathbf{P}\langle v,v\rangle>\mathbf{P}\langle u,u\rangle$$

because $\mathbf{P}\langle v,v\rangle>0$ by Theorem 8.

Self-adjoint operators. Second-order linear operators $\mathbf{L}[u]$ are called *self-adjoint* when, like $-\nabla\cdot[p(x)\nabla u]$, they satisfy an identity of the form

$$v\mathbf{L}[u]-u\mathbf{L}[v]=\nabla\cdot[p(\mathbf{x})(v\nabla u-u\nabla v)]$$

whose right side is the *divergence* of a bilinear expression in u, v, ∇u, and ∇v. An analogue of Theorem 9 holds for self-adjoint linear operators in general; see §6.

Norms. A real vector space in which an inner product satisfying (5.1) is defined, such as $\int uv\,dR$ or $\mathbf{P}\langle u,v\rangle$, is called a *Euclidean* vector space.[12] In any Euclidean vector space, the functional $\|u\|=[\langle u,u\rangle]^{1/2}$ is a *norm*. That is, it possesses the key defining properties

$$\begin{aligned}
&\|u\|>0 &&\text{unless } u=0,\\
&\|\lambda u\|=|\lambda|\cdot\|u\| &&\text{for every } \lambda\in\mathbb{R},\\
&\|u+v\|\leqslant\|u\|+\|v\| &&\text{for all } u,v.
\end{aligned} \tag{5.6}$$

In general, any real-valued functional defined on a vector space and having properties (5.6) is called a *norm*. If $\|u\|$ has the last two properties in (5.6), and $\|u\|\geqslant 0$ for all u, then it is a *semi-norm* ($\|u\|=0$ does not necessarily imply $u=0$).

With respect to *any* norm, any (real) vector space is a metric space with the distance between u and v defined as $\|u-v\|$.

[12] Some authors say "inner product space".

In addition to quadratic norms such as $\mathbf{P}\langle u,u\rangle^{1/2}$ and $\mathbf{D}\langle u,u\rangle^{1/2}$, we will also use the L_1, L_∞, and L_2 norms:

$$(5.7)\qquad \begin{aligned} &\|f\|_1=\int |f(\mathbf{x})|\,dx, \qquad \|f\|_\infty=\sup\{|f(\mathbf{x})|\},\\ &\text{and}\quad \|f\|_2=\langle f,f\rangle^{1/2},\quad \text{where}\quad \langle f,f\rangle=\int f(\mathbf{x})^2\,dx,\end{aligned}$$

for continuous f on compact domains. In discretizations approximating solutions u by finite-dimensional vectors $\mathbf{U}=(U_1,\cdots,U_n)$, we will also use the l_2 norm $\|\mathbf{U}\|_2=\langle \mathbf{U},\mathbf{U}\rangle^{1/2}=[\mathbf{U}^T\mathbf{U}]^{1/2}$, as well as

$$(5.7')\qquad \|\mathbf{U}\|_1=\sum |U_j|, \qquad \|\mathbf{U}\|_\infty=\max\{|U_j|\}.$$

Hilbert spaces. A (real) 'Hilbert space' is a Euclidean vector space which is *complete* when considered as a metric space with respect to the distance function $\|u-v\|$. By this it is meant that "every Cauchy sequence has a limit", i.e., that $\lim_{m,n\to\infty}\|u_m-u_n\|=0$ implies that a 'limit point' u exists in the space such that $\|u_n-u\|\to 0$.

Mean square convergence. Let $\bar{\Omega}$ be the (compact) closure of any bounded domain in $\mathbb{R}^n$, and let $q(\mathbf{x})$ be any *positive* weight function, continuous in $\bar{\Omega}$. Then

$$\mathbf{Q}\langle u,v\rangle=\int_\Omega q(\mathbf{x})u(\mathbf{x})v(\mathbf{x})\,dR$$

is obviously an inner product. Moreover, *metric* convergence in the resulting Hilbert space $\mathbf{H}$, defined by

$$\lim_{n\to\infty}\mathbf{Q}\langle u_n-u,u_n-u\rangle=0,$$

is independent of q, and is called *mean square convergence*.

Orthonormal bases. Every Hilbert space $\mathbf{H}$ has an *orthonormal basis*. By this is meant a set of vectors $\phi_1,\phi_2,\phi_3,\cdots$, satisfying

$$(5.8)\qquad \mathbf{P}\langle\phi_i,\phi_j\rangle=\delta_{i,j},\qquad \delta_{i,i}=1 \text{ and } \delta_{i,j}=0\quad \text{if } i\neq j,$$

in terms of which every vector $u\in\mathbf{H}$ can be expressed as a (possibly infinite) unique linear combination $u=\Sigma a_j\phi_j$. Relative to any such orthonormal basis, the inner product has a very simple form. Since

$$(5.9)\qquad \mathbf{P}\langle\Sigma a_j\phi_j,\Sigma b_k\phi_k\rangle=\sum\sum a_jb_k\delta_{j,k}=\sum a_jb_j,$$

we have the following.

THEOREM 10. *Relative to any orthonormal basis, the inner product in a Hilbert space assumes the explicit form* (5.9).

COROLLARY. *Any two Hilbert spaces having orthonormal bases of the same cardinality are isomorphic.*

It is also easy to prove that the elements of any Hilbert space with orthonormal basis $\Phi=\{\phi_j\}$ are the sums $\Sigma a_j\phi_j$ with *square-summable* coefficients, satisfying $\sum a_j^2 < +\infty$. Actually, all orthonormal bases of any Hilbert space have the same cardinality: the *dimension* of $\mathbf{H}$. We shall consider in this book only *separable* Hilbert spaces, in which the ϕ_j can be arranged in an ordinary (finite or infinite) sequence.

Direct sums. Let $\mathbf{H}$ and $\mathbf{H}'$ be any two (real, separable) Hilbert spaces, with orthonormal bases $\Phi = \{\phi_j\}$ and $\Psi = \{\psi_k\}$. By the 'direct sum' $\mathbf{H} \oplus \mathbf{H}'$ of $\mathbf{H}$ and $\mathbf{H}'$ is meant the Hilbert space having as orthonormal basis the 'disjoint union' $\Phi \sqcup \Psi$, consisting of all the ϕ_j and ψ_k.[13]

Example 8. Relative to the inner product

$$\langle u,v \rangle = \int_{-1}^{1} u(x)v(x)\,dx, \tag{5.10}$$

the Hilbert space of all square-integrable functions on $[-1,1]$ is the direct sum $\mathbf{H} \oplus \mathbf{H}'$ of the subspace $\mathbf{H}$ of all *even* such functions, and the subspace $\mathbf{H}'$ of all *odd* such functions. One pair of bases can be obtained from the normalized Legendre polynomials, $p_n(x) = \sqrt{n+1/2}P_n(x)$: $\phi_j = p_{2j-2}(x)$, $j = 1,2,\cdots$, and $\psi_k = p_{2k-1}(x)$, $k = 1,2,\cdots$. Alternatively, as a basis for $\mathbf{H}$, one can take $\phi_0 = 1/\sqrt{2}$, $\phi_j = \cos j\pi x$, $j = 1,2,\cdots$, and $\psi_k(x) = \sin k\pi x$, $k = 1,2,\cdots$ as a basis for $\mathbf{H}'$.

Example 9. Let $u(\mathbf{x})$ be any function defined on Ω with finite $\mathbf{P}\langle u,u \rangle$. Let $w(\mathbf{x})$ be the solution of the DE $\mathbf{L}[w] = 0$ in Ω for the boundary values $u(\mathbf{y})$ on Γ, and let $v(\mathbf{x})$ be the solution of $\mathbf{L}[v] = f$ in Ω for the boundary conditions $v \equiv 0$ on Γ. Then since $\mathbf{L}[v+w] = f$ in Ω, while $v(\mathbf{y}) + w(\mathbf{y}) = u(\mathbf{y})$ on Γ, $v + w = u$.

Furthermore, by Green's identities

$$\begin{aligned}\mathbf{P}\langle v,w \rangle &= \int_\Omega [p(\mathbf{x})\nabla v \cdot \nabla w + q(\mathbf{x})vw]\,dR \\ &= \int_\Gamma pv\frac{\partial w}{\partial n}\,dS + \int_\Omega v\mathbf{L}[w]\,dR = 0,\end{aligned}$$

since $v \equiv 0$ on Γ and $\mathbf{L}[w] \equiv 0$ in Ω. This decomposes the Hilbert space (Sobolev space $\mathbf{H} = H^1(\Omega)$) of all functions in Ω with finite $\mathbf{P}\langle u,u \rangle$ into two *orthogonal subspaces*, $\mathbf{V}$ and $\mathbf{W}$, so that $\mathbf{H} = \mathbf{V} \oplus \mathbf{W}$. Moreover, the two mappings $\phi : u \mapsto v$ and $\psi : u \mapsto w$ are *orthogonal projections* of $\mathbf{H}$ onto $\mathbf{V}$ and $\mathbf{W}$, respectively.

Homogenation. However, a more practical (if less elegant) procedure consists in first simply constructing *an* interpolant $w(\mathbf{x})$ to the given boundary values, and then *defining* $v(\mathbf{x})$ as $u(\mathbf{x}) - w(\mathbf{x})$. Then v is the solution of $\mathbf{L}[v] = f(\mathbf{x}) - \mathbf{L}[w]$ for the homogeneous boundary condition $v \equiv 0$ on Γ. Techniques for constructing such interpolants $w(\mathbf{x})$ will be described in Chapter 7, §9.

---

[13] For the notion of 'disjoint union', see either Birkhoff-Bartee [70, p. 59] or Mac Lane-Birkhoff, *Algebra*, 2nd. ed., Macmillan, 1979, p. 152.

For example, consider the problem of solving $\nabla^2 u = 0$ in the square $S: [-1,1] \times [-1,1]$, for the boundary condition $u \equiv x^2 + y^2$ on $\Gamma = \partial S$. Then, by setting $w = x^2 + y^2$, we trivially replace the problem with that of solving $\nabla^2 v = -4$ for $v \equiv 0$ on Γ.

Direct products. Example 3 of §1 illustrates an important general concept of algebraic analysis. Namely, since the basis functions are the products $\phi_j \psi_k$ of the members of two orthonormal bases of functions of one variable, $\phi_j(x) = \sqrt{2} \sin j\pi x$ and $\psi_k(y) = \sqrt{2} \sin k\pi y$, we have the Hilbert space decomposition $L_2(S) = L_2([0,1]) \otimes L_2([0,1])$ corresponding to the Fubini theorem. Moreover, since in Example 3 the Laplace operator $\nabla^2 = \partial^2/\partial x^2 + \partial^2/\partial y^2$ is *separable*, we have the tensor product *operator* decomposition

$$\nabla^2[\phi_j \psi_k] = A\,[\phi_j]\psi_k + \phi_j B\,[\psi_k], \tag{5.11}$$

where $A = \partial^2/\partial x^2$ and $B = \partial^2/\partial y^2$. That is, in operator notation, we have

$$\nabla^2 = (A \otimes I) + (I \otimes B), \qquad I \text{ the identity}.$$

In general, by the *direct product* (or 'tensor product') $\mathbf{H} \otimes \mathbf{H}'$ of two Hilbert spaces with orthonormal bases $\Phi = \{\phi_j\}$ and $\Psi = \{\psi_k\}$, is meant the Hilbert space having as basis all symbolic products $\phi_j \psi_k$ (i.e., having as orthonormal basis the Cartesian product $\Phi \times \Psi$). By the Fubini theorem, if $\mathbf{H}$ is the space of all square-integrable functions on $[-1,1]$ with inner product (5.10), and $\mathbf{H}'$ that of all square integrable (periodic) functions on $(-\pi, \pi]$, then $\mathbf{H} \otimes \mathbf{H}'$ is the space of all square-integrable functions on the finite *cylinder* $[-1,1] \times (-\pi, \pi]$, with inner product

$$\langle u,v \rangle = \iint u(x,\theta)\, v(x,\theta)\, dx\, d\theta.$$

6. Variational principles. The solutions of many boundary value problems of mathematical physics minimize appropriate functionals. In particular, every configuration of static equilibrium in classical (Lagrangian) mechanics minimizes a suitable energy function. In Chapter 6, we will exploit systematically *variational principles* expressed by statements like the above; here we will merely give a few examples.

First, we show that the solution of the linear source problem (5.2), $-\nabla \cdot (p(\mathbf{x}) \nabla u) + q(\mathbf{x}) u = f(\mathbf{x})$, for Dirichlet *or* Neumann boundary conditions, minimizes the functional

$$\begin{aligned} \mathbf{J}[u] &= \mathbf{P}\langle u,u \rangle - 2\int_\Omega f(\mathbf{x})\, u\, dR \\ &= \int_\Omega [p(\mathbf{x}) \nabla u \cdot \nabla u + q(\mathbf{x}) u^2 - 2f(\mathbf{x}) u]\, dR, \qquad p > 0,\ q \geqslant 0, \end{aligned} \tag{6.1}$$

in the class of 'smooth' functions satisfying these boundary conditions.[14]

[14] More precise versions of this principle will be stated in Chapter 6.

In (6.1), $dR = dx_1 \cdots dx_n$ is the volume differential in $\mathbb{R}^n$. In particular, the first variation of $\mathbf{J}$ is zero:

$$\delta \mathbf{J}[u] = \frac{d}{d\epsilon} \mathbf{J}[u + \epsilon v] \Big|_{\epsilon = 0} = 0, \tag{6.2}$$

for all v satisfying $v \equiv 0$ or $\partial v / \partial n \equiv 0$ on the boundary. In the special case $p = 1$ and $q = f = 0$, this reduces to the classical Dirichlet principle of §4, Theorem 4, which states that the solutions of the Laplace equation (i.e., harmonic functions) minimize the Dirichlet integral $\mathbf{D}\langle u, u \rangle$.

Moreover, for any $p(\mathbf{x}) > 0$ and $q(\mathbf{x}) \geqslant 0$, the minimization of the *quadratic* functional $\mathbf{J}$ in (6.1) with respect to *linear* homogeneous boundary conditions has a simple geometric interpretation in terms of the inner product $\mathbf{P}\langle u, v \rangle$. Namely, in the *affine subspace* of all functions satisfying $\mathbf{L}[u] = f(\mathbf{x})$, the function *minimizing* (6.1) is the point *nearest* to the origin in the *distance* function or 'norm' $\langle u - v, u - v \rangle^{1/2} = \| u - v \|$ associated with (6.1). Equivalently, this point is the foot of the perpendicular from the origin to that affine subspace.

The following informally stated theorem generalizes the preceding result. For a general smooth function $F(x, y, u, u_x, u_y)$ and $\delta u = 0$ on the boundary,

$$\delta \iint F(x, y, u, u_x, u_y) \, dx \, dy = 0 \tag{6.3}$$

if and only if u satisfies the *Euler-Lagrange equation*

$$\frac{\partial F}{\partial u} = \frac{\partial}{\partial x} \frac{\partial F}{\partial u_x} + \frac{\partial}{\partial y} \frac{\partial F}{\partial u_y}. \tag{6.4}$$

A more sophisticated variational principle refers to a simply supported plate with Poisson ratio ν, load density $\rho(x, y)$, and boundary on the plane $z = 0$. Here the equilibrium condition is $\delta \mathbf{J} = 0$, for

$$\mathbf{J}[u] = \int_\Omega \left[(\nabla^2 u)^2 - 2(1 - \nu)(u_{xx} u_{yy} - u_{xy}{}^2) - 2\rho u \right] dx \, dy. \tag{6.5}$$

Surprisingly, the Euler-Lagrange DE for the minimum of (6.5) is $\nabla^4 u = \rho(x, y)$ regardless of ν; we will explain the reason for this in Chapter 6, §8.

A third example is furnished by the Plateau problem of Chapter 1, §9: to find a 'minimal surface' (surface of least area) spanning a given contour γ. To prove the *existence* of solutions of the Plateau problem for a reasonably general class of curves taxed the skill of mathematicians for many years. A readable discussion of its (partial) solution around 1935 is given in Courant [50].

Rayleigh quotients. As Rayleigh first observed, the mean *kinetic energy* of a sinusoidally vibrating string or membrane, whose displacement from equilibrium is $\sin(ck_j t)\phi_j(\mathbf{x})$, is proportional to $\mathbf{D}\langle \phi_j, \phi_j \rangle$, and its *potential energy* is proportional to $\langle \phi_j, \phi_j \rangle = \int \phi_j^2 \, dR$. More precisely, the energy is $\rho \mathbf{D}\langle \phi_j, \phi_j \rangle + T\langle \phi_j, \phi_j \rangle$, where ρ is density and T is tension. He observed that the averages of these taken over any period are the same. Moreover, the eigenfunctions ϕ_j themselves satisfy a remarkable variational principle. The critical or *stationary* points of the functional $\mathbf{R}[u] = \mathbf{D}\langle u,u \rangle / \langle u,u \rangle$ are the *eigenfunctions* of the Helmholtz DE $\nabla^2\phi + k^2\phi = 0$, already discussed in Chapter 1, §5. In other words, $\delta\mathbf{R}[u] = 0$ if and only if u is an eigenfunction ϕ_j. The functional $\mathbf{R}[u]$, where u measures the displacement from equilibrium, is called the *Rayleigh quotient* to honor Rayleigh's observation.

More generally, the eigenfunctions of the generalized Helmholtz equation

$$\nabla \cdot [p(\mathbf{x}) \nabla \phi] - q(\mathbf{x})\phi + k^2 \rho(\mathbf{x})\phi = 0 \qquad \text{in } \Omega, \tag{6.6}$$

subject to the boundary condition $u = 0$ on $\partial\Omega$, are the functions whose Rayleigh quotient

$$\mathbf{R}[u] = \frac{\int_\Omega [p(\mathbf{x}) \nabla u \cdot \nabla u + q(\mathbf{x}) u^2] \, dR}{\int_\Omega \rho(\mathbf{x}) u^2 \, dR} = \frac{\mathbf{P}\langle u,u \rangle}{\mathbf{Q}\langle u,u \rangle} \tag{6.7}$$

has a stationary value, i.e., satisfies $\delta\mathbf{R} = 0$. Moreover, if $\lambda_0 = k_0{}^2$ is the smallest eigenvalue of (6.6), we have the Poincaré inequality

$$\int [p(\mathbf{x}) \nabla u \cdot \nabla u + q(\mathbf{x}) u^2] \, dR \geqslant \lambda_0 \int \rho(\mathbf{x}) u^2 \, dR, \tag{6.8}$$

valid for all functions satisfying $u = 0$ on $\partial\Omega$.

When the eigenvalues of (6.6) are all distinct, the preceding Rayleigh principle has as a corollary the following *minimax* characterization of the n-th eigenvalue $\lambda_n = k_n{}^2$ of (6.6). For each n-dimensional subspace S of (smooth) functions (with 0 deleted), let $\lambda(S) = \max_{u \in S} \mathbf{R}[u]$. Then

$$\lambda_n = \min_S \lambda(S) = \min_S \{ \max_{u \in S} \mathbf{R}[u] \}.$$

Moreover, the functions ϕ_n where this 'minimax' value is achieved are the eigenfunctions of the DE (see Courant-Hilbert [53, Chap. 6, §§1–3]).

Ohm-Rayleigh principle. On physical grounds, Rayleigh asserted much more: that in any bounded domain Ω with compact closure $\bar{\Omega}$, the DE (6.6) has a *basis* of eigenfunctions. (It is easily shown that any two eigenfunctions having distinct eigenvalues are orthogonal *both* with respect to $\mathbf{P}\langle u,u \rangle$ *and* with respect to $\mathbf{Q}\langle u,u \rangle$.) Sommerfeld has named this assertion, which was first proved in appropriate generality after 1900 by Hilbert, Weyl, and others, the 'Ohm-Rayleigh principle'.[15]

---

[15] See A. Sommerfeld, *Partial Differential Equations in Physics*, Academic Press, 1949, p. 179.

7. Maximum principle. We now turn our attention to general linear elliptic differential operators with *variable* coefficients,

$$\mathbf{L}[u] = -\sum_{i,j=1}^{n} a_{i,j}(\mathbf{x})\frac{\partial^2 u}{\partial x_i \partial x_j} + \sum_{i=1}^{n} b_i(\mathbf{x})\frac{\partial u}{\partial x_i} \tag{7.1}$$

having continuous coefficient-functions in a compact domain $\bar{\Omega}$. The operator in (5.2) for the linear source problem has this form with $a_{i,j}(\mathbf{x}) = p(\mathbf{x})\delta_{i,j}$ and $b_i(\mathbf{x}) = \partial p/\partial x_i$ ($\delta_{i,i} = 1$, $\delta_{i,j} = 0$ if $i \neq j$). So does the convection-diffusion operator with $a_{i,j}(\mathbf{x}) = \nu\delta_{i,j}$ and $b_i(\mathbf{x}) = U_i(\mathbf{x})$. In the plane, $\mathbf{L}[u] = f(\mathbf{x})u + g(\mathbf{x})$ has the form

$$Au_{xx} + 2Bu_{xy} + Cu_{yy} + Du_x + Eu_y + Fu = G, \tag{7.2}$$

displayed in Chapter 1, (1.1).

At any local interior *extremum* (maximum or minimum) of u, all $\partial u/\partial x_i = 0$ and so $\mathbf{L}[u] \geqslant 0$. From this, one can prove the following results.[16]

THEOREM 11. *Let the coefficients in* (7.1) *be continuous, and let* $\mathbf{L}[u] \geqslant 0$ *for all* $\mathbf{x} \in \Omega$. *Then* u *cannot have a local interior maximum unless* u *is a constant and* $\mathbf{L}[u] \equiv 0$.

COROLLARY 1. *If* $\mathbf{L}[u] \equiv 0$, *then the maximum and minimum values of* u *must be assumed on the boundary.*

COROLLARY 2. *If* $\mathbf{L}[u] + q(\mathbf{x})u \geqslant 0$ *and* $q(\mathbf{x}) \geqslant 0$ *in* Ω, *and* $u \geqslant 0$ *on* Γ, *then* $u \geqslant 0$ *on* $\bar{\Omega}$. *For given* $g(\mathbf{y})$ *on* Γ *and* $f(\mathbf{x})$ *on* Ω, *there is at most one solution of* $\mathbf{L}[u] = f(\mathbf{x})$ *in* Ω *satisfying* $u(\mathbf{y}) = g(\mathbf{y})$ *on* Γ.

The second result of Corollary 2 follows from the first, since $\mathbf{L}[u] = 0$ and $u(\mathbf{y}) = 0$ imply $u \equiv 0$.

This maximum principle has been greatly generalized, and it has many important applications. For example, it not only implies the uniqueness of solutions of the Dirichlet problem for given boundary values, it also implies the continuous dependence of solutions on their boundary values (in the uniform norm).

COROLLARY 3. *Let* $u, h_1, h_2, \cdots, h_n$ *be harmonic on* $\bar{\Omega}$ *and let* $U = \sum \alpha_i h_i$. *Then*

$$\max_{\mathbf{x} \in \Omega} |u(\mathbf{x}) - U(\mathbf{x})| \leqslant \max_{\mathbf{y} \in \Gamma} |u(\mathbf{y}) - U(\mathbf{y})|.$$

Consequently, one gets an accurate approximation on Ω if one can choose coefficients α_i that make the error on the *boundary* small (see Chapter 8, §3).

---

[16] The result was originally due to E. Hopf, Sitzungber. Preuss. Akad. Wiss. 19 (1927), 147–152. Proofs are also given in Bers-John-Schechter [64, pp. 150–152], Protter-Weinberger [67, Chap. 2, § 3], and Gilbarg-Trudinger [77, Chaps. 3 and 9].

Clamped plate. The special case of a clamped plate has been treated by Kondrat'ev and by Coffman and Duffin.[17] In particular, these authors show that the biharmonic Green's function of a plane region whose boundary contains a corner cannot be of one sign.

Monotone operators. An operator $\mathbf{L}$ such that $\mathbf{L}[u] \geqslant 0$ implies $u \geqslant 0$ is said to be *monotone* or of 'monotone type' (Collatz [60, p. 43]). Thus $\mathbf{L}$ in (5.2) and (7.1) are monotone on the space of twice differentiable functions which are zero on Γ. Differential operators of monotone type have positive Green's functions and conversely. Likewise, the inverse of a matrix of monotone type is nonnegative.

Error bounds. We now consider an important application due to Collatz.[18] Suppose one has an approximation U to a solution u of $\mathbf{L}[u]=f(\mathbf{x})$ on Ω assuming given values on Γ. If one also knows a function v such that $\mathbf{L}[v]>0$ on Ω and $v>0$ on $\bar{\Omega}$, then for appropriate constants m and M, we have

$$\begin{aligned} m\mathbf{L}[v] \leqslant \mathbf{L}[U-u] &\leqslant M\mathbf{L}[v] \qquad \text{on } \Omega, \\ mv \leqslant \quad U-u \quad &\leqslant Mv \qquad\qquad \text{on } \Gamma. \end{aligned} \tag{7.3}$$

Since $\mathbf{L}$ is linear, this implies

$$\begin{aligned} \mathbf{L}[(U-u)-mv] \geqslant 0, \quad \text{and} \quad \mathbf{L}[Mv-(U-u)] \geqslant 0 \qquad \text{on } \Omega, \\ (U-u)-mv \geqslant 0, \quad \text{and} \quad Mv-(U-u) \geqslant 0 \qquad \text{on } \Gamma, \end{aligned}$$

and so gives upper and lower bounds on the error:

$$mv \leqslant U-u \leqslant Mv \qquad \text{on } \bar{\Omega}. \tag{7.3'}$$

8. Smoothness. A characteristic property of elliptic equations, not shared by partial DE's of other types, is the fact that their solutions are as smooth as their coefficient-functions: if these coefficients and the right side (source term) of an elliptic DE are all of class C^n, then so is any solution at all interior points. Likewise, if its coefficients and right side are all analytic functions, then any solution is analytic at all interior points (Bers-John-Schechter [64, p. 136]). This generalizes the analyticity property of harmonic functions, to be proved in Chapter 8.

This smoothness makes it much easier to compute accurate solutions of elliptic problems by numerical methods, using either difference or finite element approximations, than accurate solutions of hyperbolic initial value problems.

[17] V. A. Kondrat'ev, Trans. Moscow Math. Soc. 16 (1967), 227–313; C. V. Coffman and R. J. Duffin, Adv. Appl. Math. 1 (1980), 373-389.

[18] See Collatz [60, p. 42] and L. Collatz, *Functional Analysis and Numerical Mathematics*, Academic Press, 1966, Chap. III.

Moreover, solutions of analytic elliptic problems are also analytic on the boundary wherever the boundary and boundary data are analytic. Such solutions can therefore be continued analytically across the boundary, by the Cauchy-Kowalewski theorem, into an open region containing the boundary in its interior. For harmonic functions, we have also in special cases various reflection principles. Of these, the following *Schwarz reflection principle* is the most famous and familiar (cf. Ahlfors [79, p. 171]).

THEOREM 12. *Let* $w = f(z)$ *be an analytic function of the complex variable* z *in a domain* Ω *of the upper half-plane, whose boundary* Γ *includes a segment* S *of the real axis. Let* w *be continuous in* $\bar{\Omega}$ *and real on* S. *Then* f *can be continued analytically into the mirror image* Ω' *of* Ω *in the lower half-plane by setting* $w = [f(z^*)]^*$ *there, where* z^* *designates the complex conjugate of* z.

COROLLARY 1. *Let* $u(x,y) \in C(\bar{\Omega})$ *be harmonic in* Ω, *continuous in* $\bar{\Omega}$, *and let* $u(x,0) \equiv 0$ *on* S. *Then its continuation by the formula* $u(x,-y) = -u(x,y)$ *is harmonic in the union of* $\bar{\Omega}$ *and* $\bar{\Omega}'$.

COROLLARY 2. *If, in Corollary* 1, *we replace* $u \in C(\bar{\Omega})$ *by* $v \in C^1(\bar{\Omega})$, *and the condition* $u \equiv 0$ *by* $v_y \equiv 0$, *then the formula* $v(x,-y) = v(x,y)$ *continues* v *to a function harmonic on* $\bar{\Omega} \cup \bar{\Omega}'$.

Singularities at corners. Although solutions of elliptic DE's with analytic coefficients are analytic at interior points, and solutions of elliptic boundary value problems with analytic coefficients *and* boundary data are analytic where the boundary is analytic, such solutions generally have singularities at corners, edges, etc.

Example 10. Consider the Poisson DE $-\nabla^2 u = 1$ in the unit square, $S = (0,1) \times (0,1)$, with $u \equiv 0$ on $\Gamma = \partial S$. Here the solution u has a logarithmic singularity at each corner.[19] The same is true of the deflection of a uniformly loaded clamped plate: the solution of $\nabla^4 u = 1$ in S with $u = \partial u/\partial n = 0$ on Γ.

Example 10 shows that *a priori* truncation error estimates which involve differentiability assumptions are often inapplicable at corners of surfaces, edges and corners of solids, etc., where boundaries are nonanalytic (cf. Chapter 3, §7, and Chapter 7, §10).

Using the Schwarz reflection principle, H. Lewy and his students have found the most general form of such singularities for a fairly wide family of conditions. One of the most powerful results is the following, due to Lehman.[20]

---

[19] See G. Birkhoff in Aziz [69, p. 229].

[20] R. S. Lehman, J. Math. Mech. 8 (1959), 727–760.

THEOREM 13. *Let* $u(x,y)$ *be a solution of*

$$\nabla^2 u + a(x,y)u_x + b(x,y)u_y + c(x,y)u = f(x,y),$$

with analytic coefficients and right side, in an analytic corner subtending an interior angle π/β, *and analytic values on sides (except at the vertex). Then* $u(x,y)$ *is asymptotic to an analytic function of* $z, z^\beta, z^*, z^{*\beta}$ *if* β *is irrational, and of these variables and* $z^q \log z, z^{*q} \log z^*$ *if* $\beta = q/p$ *is a rational fraction (in lowest terms).*

This result has been extended by Wigley to the case of sufficiently smooth (i.e., highly differentiable) mixed boundary conditions and by Eisenstat to more general operators.[21]

For early work on the numerical treatment of singularities of elliptic equations, see Fox [62, Chap. 24], and the references given there.

Interfaces. In various applications, including nuclear reactors, petroleum reservoir mechanics, and so on, physical properties abruptly change at interfaces separating different materials. In other words, the p in (5.2) is often discontinuous. Conservation principles then require that the solution of (5.2) and its flux be continuous across an interface. For instance, if the interface is the plane $x = x_1$, this gives the *interface conditions*

$$\begin{aligned} u(x_1^+, y, z) &= u(x_1^-, y, z), \\ p_1 u_x(x_1^+, y, z) &= p_2 u_x(x_1^-, y, z), \end{aligned} \tag{8.1}$$

where the p_i are suitable *diffusivities*. Note that this results in a discontinuity in the normal derivative $\partial u/\partial n$ at an interface;[22] the tangential derivative, however, is obviously continuous.

The behavior of solutions near corners of interfaces will be discussed in Chapter 7, §10.

[21] N. Wigley, J. Math. Mech. 13 (1964), 549–576; Pacific J. Math. 15 (1965), 1435–1461; Math. Comp. 23 (1969), 395–401; Math Z. 115 (1970), 33–52; and S. C. Eisenstat, SIAM J. Numer. Anal. 11 (1974), 654–682.

[22] See R. B. Kellogg, SYNSPADE 1970, Academic Press, 1971, pp. 351–400; in Aziz [72, pp. 589–602]; Applicable Analysis 4 (1975), 101–129. See also G. Birkhoff, J. Approx. Theory 6 (1972), 215–230.

Chapter 3

Difference Approximations

1. Arithmetizing analysis. In this chapter, we take up our main theme: the idea that *elliptic problems can be solved to any desired precision by the systematic application of arithmetic methods*, and that the functions defined by *exact solutions* should be regarded simply as *limits of these approximate solutions*, in much the same way that $e = 1 + 1 + 1/2 + 1/6 + \cdots + 1/n! + \cdots$ is the limit of a sequence of ordinary fractions.

This idea, of approximating a *differential* equation by a system of linear *algebraic* equations is just one aspect of the so-called *Arithmetization of Analysis*. This represented the culmination of 19th century mathematical analysis of problems like the Dirichlet problem and the heat and wave equations. Much as the infinitesimal Calculus of Newton and Leibniz, with its suggestive symbolism for describing the infinitely small, precipitated one mathematical revolution, the germ of a second revolution was the recognition by leading 19th century mathematicians, from Cauchy to Riemann and from Weierstrass to Poincaré, that all Analysis could be based logically on a combination of ordinary arithmetic and passage to the limit.

In this second revolution, an important role was played by Fredholm's theory of integral equations, whose kernels $K(x,y)$ can be treated as limits of matrices. We will give numerical applications of this approach in Chapter 8. Also important were direct variational methods, pioneered by Rayleigh (1873) and Ritz (1906); they will be the main theme of Chapters 6 and 7. But most directly applicable are *difference methods*, and we shall devote this chapter to them.

Example 1. We first consider a sample Dirichlet problem: solving the Laplace equation $\nabla^2 u = 0$ in the L-shaped region Ω of Fig. 1a: $([0,1] \times [0,2]) \cup ([1,2] \times [0,1])$ of the (x,y)-plane, subject to the boundary conditions $u = x^2 + y^2$ on $\Gamma = \partial\Omega$.

To construct a difference approximation to a given elliptic problem, one must first select a suitable subdivision or *mesh* Ω_h, subdividing Ω. The

L-shaped domain of Example 1 has the convenient property that its boundary falls on lines $x = x_i$ and $y = y_j$ of a square *mesh*. This is true also of many other domains bounded by vertical and horizontal lines, and other (straight) lines making 45° angle with the vertical or horizontal. We shall call such domains *regular polygonal domains*.

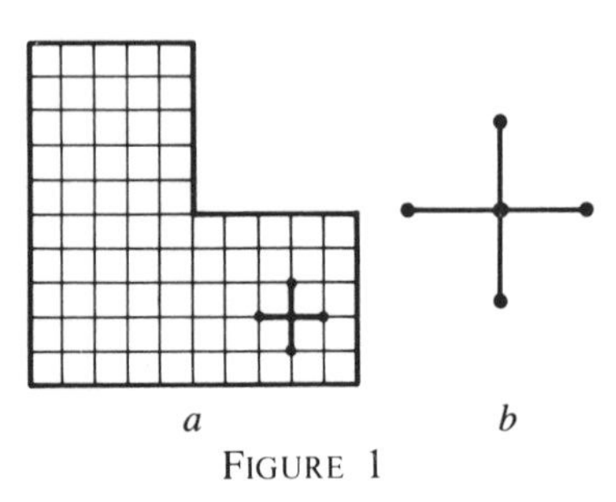

FIGURE 1

Thus, for the regular polygonal L-shaped domain of Example 1, for any positive integer n, and for the corresponding mesh length $h = 1/n$, consider the values $u_{i,j} = u(x_i, y_j)$ of u at each of the $3n^2 + 4n + 1$ interior and boundary *mesh-points* $(x_i, y_j) = (ih, jh)$ in $\bar{\Omega} = \Omega \cup \Gamma$. At the $8n$ *boundary* mesh-points on $\Gamma \cap \Omega_h = \Gamma_h$, $u_{i,j} = (i^2 + j^2)h^2$ is *known*; on the set Ω_h of $N = 3n^2 - 4n + 1$ *interior* mesh-points, $u_{i,j}$ is unknown. At the interior mesh-points, we can approximate *derivatives* by *difference quotients*. Thus, for any function $u(x)$ with continuous u_{xxxx}, expanding each term in Taylor's series with remainder, then cancelling and dividing by h^2, we get

$$(u_{i-1,j} - 2u_{i,j} + u_{i+1,j})/h^2 = u_{xx}(x_i, y_j) + Kh^2/12,$$

where $|K| \leqslant \max|u_{xxxx}(x, y_j)|$ for $x \in [x_i - h, x_i + h]$. Adding a similar expression involving u_{yy}, we see that $\nabla^2 u = u_{xx} + u_{yy} = 0$ implies

$$(1.1) \quad -\nabla_h^2 u_{i,j} \equiv (4u_{i,j} - u_{i-1,j} - u_{i,j-1} - u_{i+1,j} - u_{i,j+1})/h^2 = \epsilon_{i,j}.$$

In (1.1), $\epsilon_{i,j} = K_{i,j}h^2$, the $K_{i,j}$ being *uniformly bounded* on any closed subset of $\bar{\Omega}$ which excludes its corners, for any harmonic function u.[1]

Already in 1908, Runge[2] proposed neglecting the $\epsilon_{i,j}$ in (1.1), and replacing $\nabla^2 u$ by the 5-point central difference quotient $\nabla_h^2 u$. This replaces the Laplace equation in Example 1 by a system of $N = 3n^2 - 4n + 1$ simultaneous *linear algebraic* equations in N unknowns:

$$(1.2) \quad U_{i,j} - (U_{i-1,j} + U_{i+1,j} + U_{i,j-1} + U_{i,j+1})/4 = 0, \quad (x_i, y_j) \in \Omega_h,$$

$$(1.2') \quad U_{i,j} = u_{i,j}, \quad (x_i, y_j) \in \Gamma_h.$$

Equation (1.2) relates the unknowns at five points; see Fig. 1b.

The fact that[3] $\epsilon_{i,j} = O(h^2)$ in (1.1) suggests that the solutions $U_{i,j}$ of (1.1)–(1.2) should differ from the corresponding $u(x_i, y_j)$ by $O(h^2)$. In

[1] See Chapter 2, §8.

[2] C. Runge, Zeits. Math. Phys. 56 (1908), 225–232.

[3] For an error $\epsilon = \epsilon(x, y; h)$ depending on h, one says that ϵ is "order h^{ν}" and writes $\epsilon = O(h^{\nu})$ when $|\epsilon(x, y; h)| \leqslant Mh^{\nu}$ for some constant M and all h sufficiently small.

other words, this fact suggests that at any mesh-point (x_i, y_j), $u(x_i, y_j) = U(x_i, y_j) + O(h^2)$ as $h \downarrow 0$.

Weaker but much more general results were proved rigorously by R. G. D. Richardson (1917),[4] Phillips-Wiener [23], and Courant-Friedrichs-Lewy [28]. Their papers considered the Dirichlet problem in general bounded 'regular' regions of $\mathbb{R}^n$. They concluded that convergence of the nearest $U_{i,j}$ to $u(x,y)$ was *uniform*, and used this to prove a constructive *existence theorem* for the Dirichlet problem. Their results corroborated the idea of arithmetizing analysis very convincingly, in this special case.

Summarizing, we may say that results like the preceding showed that one could 'arithmetize analysis' in *principle*. But to implement the idea *practically*, in the sense of getting approximations whose errors are (say) only 0.1% or even 5%, required a third revolution. This is now in full swing, and is based on the availability of large-scale, high-speed computers at moderate cost. In the remainder of our monograph, we will explain some of the key algorithms, mathematical ideas, and computational methods that, together with modern computers, have made possible enormous progress in the numerical solution of elliptic problems over the past 30 years.[5] Naturally, our objective is not only to make these methods more readily available; it is to pave the way for continuing progress in the coming decades, based on the availability of still more powerful computers (such as vector machines), and further advances in computational methods.

To 'arithmetize' the solution of elliptic problems on general domains, many technical problems must be solved. The first of these concerns the treatment of boundary conditions on a square mesh, when the boundary does *not* fall on mesh lines, i.e., when the domain is *not* a regular polygonal one.

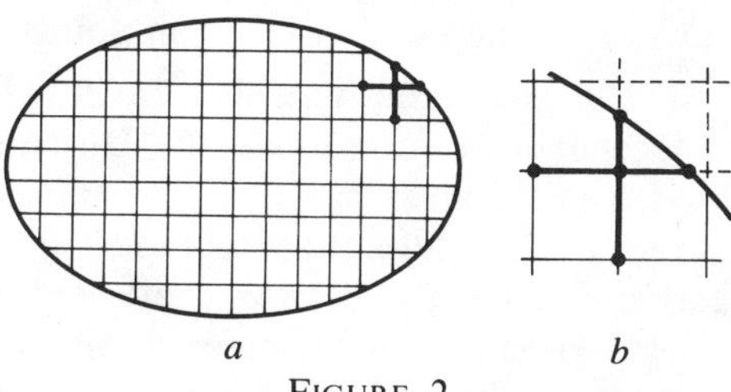

FIGURE 2

Example 2. Consider the Dirichlet problem of Example 1 in a general bounded *convex plane region* Ω, for given continuous boundary values $g(\mathbf{y})$ on Γ (see Fig. 2a).[6] In this case, every mesh line that intersects Ω at all will do so in a closed interval with endpoints (ih, c_i) and (ih, d_i), $c_i \leqslant d_i$, or (a_j, jh) and (b_j, jh), $a_j \leqslant b_j$. Let Γ_h denote the set of all the 'boundary' mesh-points

[4] R. G. D. Richardson, Trans. Am. Math. Soc. 18 (1917), 489–518.

[5] See G. Birkhoff, in Schultz [81, pp. 17–38] for a survey of advances in the solution of elliptic problems between 1930 and 1980.

[6] The techniques described here also suffice for many non-convex regions. Thus they apply to any region that is 'regular' in the sense of Kellogg [K, Chap. IV].

terminating such intervals; only exceptionally will any of them be one of the (ih, jh). However, at all *interior* mesh-points, defined as mesh-points (ih, jh) in Ω but not on Γ, we can approximate the DE $u_{xx} + u_{yy} = 0$.

To explain how, we first note that each such interior mesh-point $(x_i, y_j) = (ih, jh) \in \Omega_h$, at which the unknown value of u can again be designated by $u_{i,j}$, will be adjacent to four other mesh-points in $\overline{\Omega}_h$, one or more of which may be in Γ_h. Hence it will be the central point of a *stencil* of 5 points (see Fig. 2b), and we can still designate the values of u at these points as $u_{i,j-1}, u_{i-1,j}$, etc., with the proviso that one or more of them may be *known* if (x_i, y_j) is adjacent to a boundary mesh-point.

Discrete harmonic functions. The difference equation $\nabla_h^2 u = 0$ on a square mesh is evidently equivalent to the condition that the value of u at each mesh-point is the *arithmetic mean* of its values at the four adjacent mesh-points. It discretizes (is an 'arithmetization' of) the arithmetic mean property of Gauss

$$u(x_i, y_j) = \frac{1}{2\pi} \int_0^{2\pi} u(x_i + h\cos\theta, y_j + h\sin\theta)\, d\theta \tag{1.3}$$

described in Chapter 2, §3. Indeed, it approximates this by trapezoidal quadrature, using the four evaluation points $\theta_k = k\pi/2$, $k = 0, 1, 2, 3$.

A function which satisfies $\nabla_h^2 u = 0$ is called a 'discrete harmonic function'. Discrete harmonic functions of n variables can be defined similarly on any discrete 'cubic' mesh Ω_h filling a domain Ω in $\mathbb{R}^n$. Discrete harmonic functions have been extensively studied by Duffin and others.[7] Since the number of equations is the number of *interior* mesh-points, the dimension of the vector 'space' of discrete harmonic functions on any such Ω_h is the number of mesh-points on the *boundary* Γ_h of Ω_h.

2. Poisson equation; ELLPACK. Difference methods, like the one proposed in §1 for solving the Dirichlet problem in an L-shaped region, have a great advantage over the analytical methods of Chapter 2. As we stated in §1, difference methods are applicable in principle to general partial DE's with *variable* coefficients on *general domains* (but see §7). This is because one can approximate *any* DE arbitrarily closely by a suitable difference equation (ΔE). This chapter will be devoted to methods for constructing such approximating ΔE's and appraising their errors, while the next two will describe methods for solving the resulting systems of algebraic equations. As was explained in the preface, the resulting 'difference methods' were used almost exclusively to solve elliptic problems numerically until 1960 or so.

We begin by approximating the Laplacian. We saw in §1 that, at the mesh-points $(x_i, y_j) = (ih, jh)$ of a square mesh, the Laplacian $\nabla^2 u$ is

---

[7] H. A. Heilbronn, Proc. Camb. Philos. Soc. 45 (1949), 194–206; Duffin [56]; R. J. Duffin and E. P. Shelly, Duke Math. J. 25 (1958), 209–238.

approximated with $O(h^2)$ accuracy by using central difference quotients

$$\nabla_h^2 u(x_i, y_j) \equiv (u_{i+1,j} + u_{i-1,j} + u_{i,j+1} + u_{i,j-1} - 4u_{i,j})/h^2, \tag{2.1}$$

where $u_{i,j} = u(ih, jh)$. To estimate the *truncation error* $\nabla_h^2 u - \nabla^2 u$, we again assume $u \in C^4(\Omega)$ and expand in Taylor series about (x_i, y_j) through terms in h^4. We find, after dividing by h^2, that

$$\nabla_h^2 u - \nabla^2 u = \frac{h^2}{12}\left[\frac{\partial^4}{\partial x^4} u(x_i + \theta_1, y_j) + \frac{\partial^4}{\partial y^4} u(x_i, y_j + \theta_2)\right], \tag{2.2}$$

where $|\theta_k| < h$. The right side is the *truncation error*; it is $O(h^2)$ for any $u \in C^4(\Omega)$.

Two-dimensional linear difference operators on uniform grids can be displayed schematically in *stencil* form. In this notation, the discrete Laplacian of (2.1) appears as

$$\nabla_h^2 u(x_i, y_j) = \frac{1}{h^2}\begin{bmatrix} & 1 & \\ 1 & -4 & 1 \\ & 1 & \end{bmatrix} u(x_i, y_j). \tag{2.3}$$

Note that this stencil establishes a natural correspondence between *nodal values* $u_{i,j}$ and *equations*.

Example 3. The preceding formulas suggest the idea of obtaining an *approximate* solution of the Dirichlet problem for the Poisson equation $-\nabla^2 u = f$, say in the L-shaped region of Fig. 1a, by replacing the Poisson equation at each of the 56 interior mesh-points shown there with the linear (algebraic) difference equation $\nabla_h^2 U(x_i, y_j) = f(x_i, y_j)$. One sets $U(x_i, y_j) = u(x_i, y_j)$ at the 40 mesh-points on the boundary where u is known and one then has 56 linear equations in 56 unknown values $U(x_i, y_j) \approx u(x_i, y_j)$ at interior mesh-points. If h is replaced by $h/2$, there result 261 equations in 261 unknowns, and so on.

The above technique applies more generally if terms $q_{i,j} U_{i,j}$ are inserted to approximate the DE $-\nabla^2 u + q(x,y)u = f(x,y)$.

There are many ways of setting up meshes in plane domains. In Chapters 3–5, we will deal mostly with *rectangular* meshes, which (by definition) subdivide a given region by mesh lines $x = x_i$ and $y = y_j$ into interior rectangles and boundary 'cells' (or elements); see Fig. 3.

Irregular stencils. If the boundary does not intersect mesh lines at mesh-points, as in Figs. 2a and 3b, then, for mesh-points adjacent to (or on) it,[8] equations ('irregular stencils') must be derived by special methods, to be described in §3 and §4.

[8] For Neumann boundary conditions, and 'mixed' boundary conditions of the form $\partial u/\partial n + h(\mathbf{y})u = g(\mathbf{y})$, equations are associated also with boundary mesh-points.

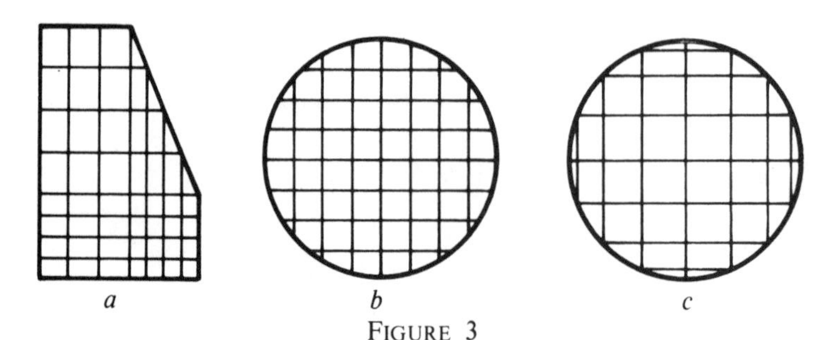

a *b* *c*

FIGURE 3

The approximation (2.1) again 'reduces' the analytical problem defined by the Poisson DE and any given Dirichlet boundary condition $u(x,y) = g(x,y)$ on $\partial\Omega = \Gamma$, to the *algebraic* problem of solving a system of N simultaneous linear equations in N unknowns, where N is the number of interior mesh-points. In vector notation, this algebraic problem consists in solving a *vector* equation of the form $A\mathbf{U} = \mathbf{b}$. Here $\mathbf{U}$ is the vector whose components U_k are the approximations $U_{i,j}$ to the unknown values $u_{i,j} = u(x_i, y_j)$ of the exact solution at interior mesh-points; the matrix A and the vector $\mathbf{b}$ of source and boundary values, $f_{i,j}$, $g_{i,j}$, are known.

In practice, systems of simultaneous linear equations arising from elliptic problems can be very large, consisting of 10^4 or more equations with as many unknowns. We will consider their solution by some simple 'direct' methods in §5 of this chapter, progressing in the next two chapters to explanations of their solution by more sophisticated and more powerful procedures.

ELLPACK. It is a formidable task to write and debug a program that will discretize an elliptic problem, and then solve all the resulting algebraic equations! Fortunately, there exists a package of portable Fortran subroutines especially designed to help perform these tasks. This package, ELLPACK, is described in Chapter 9. At computing centers where ELLPACK is available, many elliptic problems can be solved quite accurately by programs less than 30 lines long, written in a high level language developed by John R. Rice of Purdue University.[9]

[9] For surveys of software for solving elliptic problems, see R. A. Sweet, in *Information Linkage between Applied Mathematics and Industry*, Schoenstadt et al. (eds.), Academic Press, 1979; and R. F. Boisvert and R. A. Sweet, in Cowell [84]. Many special purpose packages exist for solving elliptic problems arising in structural mechanics. NASTRAN is the oldest; mention should also be made of ADINA, designed to solve deformation problems. See K.-J. Bathe (ed.), *Nonlinear Finite Element Analysis and ADINA*, Pergamon Press, 1981.

For example, suppose one wishes to solve

$$-u_{xx} - u_{yy} = f(x,y), \qquad f(x,y) = 4(r^2 - 1),$$

on the unit disk $\Omega : x^2 + y^2 < 1$ with boundary conditions $u = y^2$ on the upper half circle and $u = 0$ on the lower half. An ELLPACK program to solve the difference equation problem $-\nabla_h^2 U_{i,j} = f_{i,j}$ at interior points of a square mesh with $h = 1/4$ (and (3.3′), below, next to the boundary) is given in Table 1.

TABLE 1. *Sample* ELLPACK *program.*

```
EQUATION.          - UXX - UYY = 4.*( X**2 + Y**2 - 1. )
BOUNDARY.
 U = Y**2 ON X = COS(PI*THETA), Y = SIN(PI*THETA) FOR THETA = 0. TO 1.
 U = 0    ON X = COS(PI*THETA), Y = SIN(PI*THETA) FOR THETA = 1. TO 2.
GRID.              9 X POINTS  $  9 Y POINTS
DISCRETIZATION.    5 POINT STAR
INDEXING.          AS IS
SOLUTION.          LINPACK BAND
OUTPUT.            TABLE(U)  $  PLOT(U)
END.
```

The 'GRID' segment specifies that the interval $-1 \leqslant x \leqslant 1$ (known from the 'BOUNDARY' segment) be partitioned with 9 equal spaced mesh lines, $x_i = i/4$, $i = -4, \cdots, 4$, and similarly for $-1 \leqslant y \leqslant 1$. The 'INDEXING' and 'SOLUTION' segments specify that the 'natural' (AS IS) ordering of unknowns and equations be used and that the LINPACK band elimination be used to solve the linear system (see §5). The output is a table of values at mesh-points and a contour plot. After running this program, the mesh can be refined by replacing the pair of 9's in the 'GRID' segment by a pair of, say, 17's and then a pair of 33's. The three solutions on the meshes with $h = 1/4, 1/8$, and $1/16$ can be compared to appraise the accuracy: the differences between the values at common mesh-points provide estimates of the orders of magnitudes of the error.

Triangulations. For many applications, it is most convenient to *triangulate* Ω, i.e., to cut it up into interior and boundary triangles, each of the latter usually having one curvilinear side. For example, one can subdivide each of mesh squares in Fig. 1a into two or four triangles, as indicated in Fig. 4. The approximation of DE's on triangular nets will be discussed in Chapters 6 and 7.

Example 4.[10] Consider the eigenproblem $-\nabla^2 u = \lambda u$ in Ω, a regular hexagon, with the condition $u \equiv 0$ on its boundary Γ. This region can be subdivided to obtain a set of mesh-points Ω_h on a regular triangular mesh as in Fig. 5; the set of boundary mesh-points Γ_h is a subset of Γ. On Ω_h,

[10] Collatz [60, pp. 392–395] also treats this problem; he uses the μ_k to estimate the λ_k.

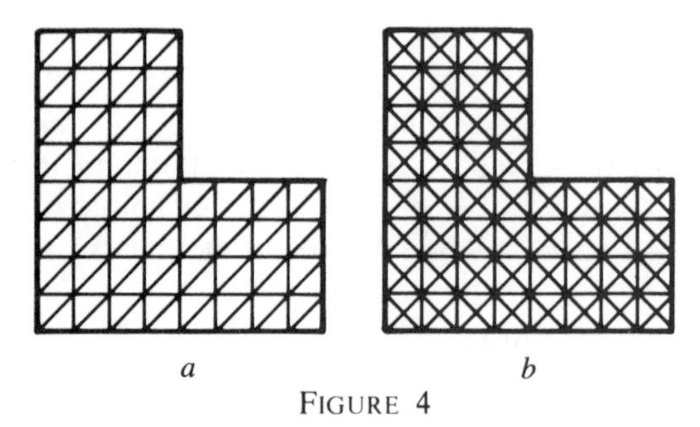

a *b*

FIGURE 4

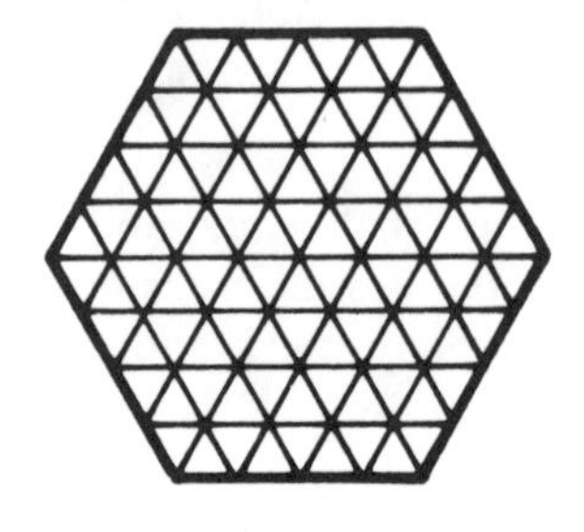

FIGURE 5

a convenient approximation to $-\nabla^2$ is given by

$$\mathbf{L}_h[U]_{i,j} = 2(6U_{i,j} - \textstyle\sum_k U_k)/3h^2,$$

where the summation is over the six mesh-points nearest to (x_i, y_j); this gives another discretization of (1.3). Let ϕ be the (positive) eigenfunction of $\mathbf{L}$ (for the boundary condition $\phi = 0$ on Γ) having the smallest eigenvalue, λ. Then

$$\mathbf{L}_h[\phi] + \nabla^2\phi = -\frac{h^2}{16}\nabla^4\phi + O(h^4) = -\frac{h^2\lambda^2}{16}\phi + O(h^4).$$

For fixed h, consider the eigenvector Φ_h of

$$\mathbf{L}_h[\Phi]_{i,j} = \mu_h \Phi_{i,j} = (\lambda_h - h^2\lambda_h^2/16)\Phi_{i,j},$$

satisfying $\Phi_{i,j} = 0$ on Γ_h, having the smallest eigenvalue μ_h. Solving this algebraic eigenproblem, one presumably gets an approximation

$$\lambda_h = (8/h^2)[1 - \sqrt{1 - h^2\mu_h/4}] \approx \lambda.$$

See Chapter 6, §7, for experimental results.

3. Discretizing boundary conditions. We next describe three different methods for approximating *Dirichlet-type* boundary conditions, i.e., for constructing ΔE's approximately satisfied by solutions of linear source problems that assume specified values on the boundary Γ of a domain Ω.

Method A. Generalizing the procedure proposed in §1, one can supplement the mesh-points (x_i, y_j) in $\bar{\Omega}$ by the set of intersections (x, y_j) and (x_i, y) of the mesh lines of the rectangular mesh M with Γ. The resulting set Γ_h of boundary mesh-points will be contained in Γ. This will ordinarily lead to 'irregular stencils' on the boundary, and can give rise to boundary pentagons, quadrangles with a curvilinear side, and triangles; see Figs. 2 and 3b.

Irregular stencils. To approximate $\nabla^2 u(x_i, y_j)$ on the irregular stencils that arise next to the boundary, we first consider u_{xx}. For given $u_0 = u(x_i - \theta h, y_j)$, $u_1 = u(x_i, y_j)$, and $u_2(x_i + h, y_j)$, we expand in

Taylor series about x_i to get

$$u_0 - (1+\theta)u_1 + \theta u_2 = (1+\theta)\theta h^2 u_{xx}(x_i, y_j)/2 + (1-\theta^2)\theta h^3 u_{xxx}(x_i, y_j)/6 + Kh^4,$$

where $K = |u^{(iv)}(\xi, y_j)|/12$ for some $\xi \in (x_i - \theta h, x_i + h)$. Dividing by $(1+\theta)\theta h^2/2$, we obtain

$$u_{xx}(x_i, y_j) = 2[u_0 - (1+\theta)u_1 + \theta u_2]/(1+\theta)\theta h^2 + (1-\theta)hu_{xxx}(x_i, y_j)/3 + K'h^2. \tag{3.1}$$

Combining this with the analogous formula for $u_{yy}(x_i, y_j)$, we get

$$\nabla^2 u_{ij} = l_{ij}u_{i-1,j} + b_{ij}u_{i,j-1} + r_{ij}u_{i+1,j} + t_{ij}u_{i,j+1} - d_{ij}u_{ij} + \epsilon_{ij}, \tag{3.2}$$

where $d_{i,j} = 2/h^2$, $l_{i,j} = 2/h^2\theta(1+\theta), \cdots$, and $\epsilon_{i,j} = O(h)$.

For approximating solutions of $\nabla^2 u = 0$, this suggests setting[11]

$$U_{i,j} = (l_{i,j}U_{i-1,j} + b_{i,j}U_{i,j-1} + r_{i,j}U_{i+1,j} + t_{i,j}U_{i,j+1})/d_{i,j} \tag{3.3}$$

(which, for a square mesh, reduces to (1.2) away from the boundary) at each interior mesh-point, and replacing the $U_{r,s}$ in (3.3) which *are* on Γ_h by the given (known) boundary values of u.[12] This results in a very large system of linear algebraic equations when h is small. R. G. D. Richardson and the other authors cited in §1 essentially proved that if this is done, the approximations $U_{i,j}(h)$ to the $u(x_i, y_j)$, obtained by solving these equations, will form a *Cauchy sequence*, giving in the limit a solution to the stated Dirichlet problem.

Actually, if $u \in C^4(\bar{\Omega})$, the solution has $O(h^2)$ accuracy even though the difference approximation (3.2) only has $O(h)$ accuracy. The same is true of the following alternative boundary approximations.

Similarly, to approximate the Poisson equation, $-\nabla^2 u = f(x,y)$, near the boundary, one can use

$$d_{ij}U_{ij} - l_{ij}U_{i-1,j} - b_{ij}U_{i,j-1} - r_{ij}U_{i+1,j} - t_{ij}U_{i,j+1} = f_{ij}. \tag{3.3'}$$

Method B. Given any rectangular mesh M with mesh lines $x = x_i$ and $y = y_j$, we can choose the subset $\bar{\Omega}$ of mesh *cells* (rectangles and border triangles) which seems to best *approximate* $\bar{\Omega}$, and let its boundary mesh-points constitute Γ_h. For each $(x_i, y_j) \in \Gamma_h$, set $U_{i,j}$ equal (for Dirichlet-type boundary conditions) to the value obtained by interpolation and/or extrapolation from the given boundary values. (Approximating normal derivatives is harder, by all methods!) For example, one can let this

---

[11] This is the approximation used by Shortley-Weller [38].

[12] Alternatively, one can replace Γ_h by the boundary of the largest *rectangular polygon* in $\bar{\Omega}$. This gives simpler formulas, but is less accurate.

subset be the largest connected component of the union of the mesh rectangles contained in $\bar{\Omega}$, and assign to each mesh-point (x_i, y_j) on its boundary the value $u_{i,j} = g(\mathbf{y})$ specified at some nearby $\mathbf{y} \in \Gamma$.

Method C. If Ω is convex and has an *axis of symmetry*[13] (say the y-axis), then any pair of mesh lines $x = \pm x_i$ symmetric about this axis will intersect Γ in four points $(\pm x_i, c_i)$ and $(\pm x_i, d_i)$, except in the case of *tangent* mesh lines touching Γ at $(\pm a_i, y_j)$. This modification of Method B can be thought of as approximating Ω by an *inscribed polygon*, subdivided into rectangles and boundary triangles. Note that if a coordinate axis is an axis of symmetry, one can use the edges of *inscribed rectangles* as mesh lines; see Fig. 3c. This method also avoids irregular stencils.

Neumann conditions. For solving the Poisson equation with Neumann boundary conditions, $\partial u/\partial n = g(\mathbf{y})$, $y \in \Gamma$, Fox[14] proposed the following scheme.

For mesh-points, take the points of a square mesh which are in $\bar{\Omega}$ as well as those outside and adjacent to $\bar{\Omega}$. A configuration near Γ is illustrated in Fig. 6. At each mesh-point in $\bar{\Omega}$, such as B, C, E, F in Fig. 6, use the usual 5-point approximation $-\nabla_h^2 U = f$. The difference equation for the stencil centered at C includes U_A and U_D at external mesh-points. Here U_A can be eliminated from the ΔE by using the divided central difference approximation to the normal derivative:

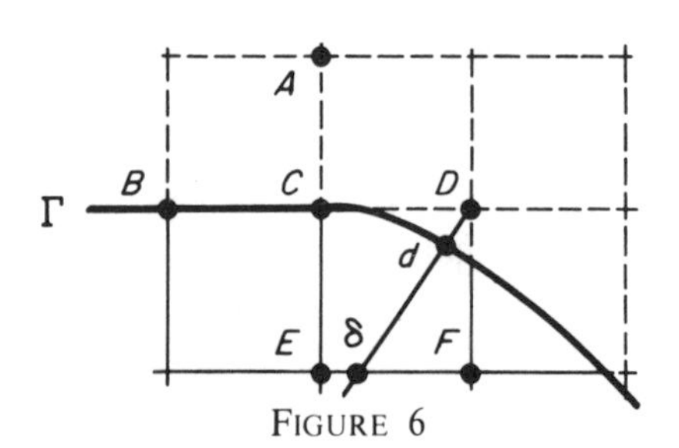

FIGURE 6

$$(-U_E + U_A)/2h = g(\mathbf{y}_C).$$

To eliminate U_D, we can drop a perpendicular to Γ from D; let d and δ denote the points of its intersections with Γ and the first interior mesh line. A divided difference approximation to $\partial u/\partial n$ is obtained in terms of U_D and U_δ; then U_δ is expressed as the linear interpolant to U_E and U_F:

$$(-U_\delta + U_D)/k = g(\mathbf{y}_d), \qquad U_\delta = \alpha U_E + \beta U_F,$$

where $k = \overline{\delta D}$ is the length of the line segment between δ and D, and $\alpha = \overline{\delta F}/h$. $\beta = \overline{E\delta}/h$, The final formula is

$$[4U_C - U_B - (2+\alpha)U_E - \beta U_F]/h^2 = f_C + 2g(\mathbf{y}_C)/h + k\,g(\mathbf{y}_d)/h^2.$$

[13] Otherwise, difficulties can arise; see G. Birkhoff in Schoenberg [69, pp. 202–203].

[14] L. Fox, Quart. Appl. Math. 2 (1944), 251–257; see also Fox [50]. For other schemes, see [FW, §20.10] and the references cited; [V, §6.3] treats mixed boundary conditions for the linear source problem.

Mixed boundary conditions. The ELLPACK module 5 POINT STAR uses $O(h)$ approximations to 'mixed' boundary conditions of the form

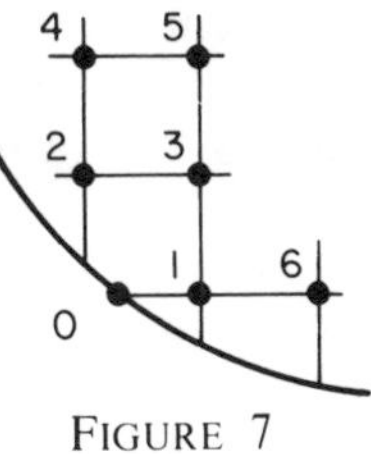

FIGURE 7

$$(3.4) \qquad a(\mathbf{y})u + b(\mathbf{y})u_x + c(\mathbf{y})u_y = g(\mathbf{y}),$$

for $\mathbf{y} \in \Gamma$ (a, b, c not all zero). To approximate u_x at $\mathbf{y} = \mathbf{0}$ in Fig. 7, it uses the $O(h^2)$ approximation:

$$u_x(\mathbf{0}) \approx -\frac{2\alpha+1}{(\alpha^2+\alpha)h}u_0 + \frac{\alpha+1}{\alpha h}u_1 - \frac{\alpha}{(\alpha+1)h}u_6,$$

The right side is the derivative at $\mathbf{y} = \mathbf{0}$ of the quadratic interpolant to u at the points $x_0 = \mathbf{0}$, $x_1 = \alpha h$, $x_6 = (\alpha+1)h$. Linear interpolation to u_2 and u_3 is used to obtain $O(h^2)$ approximation to $u(0,h)$, and likewise, $u(0,2h)$ is approximated with u_4 and u_5; then

$$u_y(\mathbf{0}) \approx \frac{1}{2h}[-3u_0 + \alpha(4u_2 + u_4) + (1-\alpha)(4u_3 + u_5)],$$

with $O(h)$ error.

Remark. By using a different procedure, one can approximate $u_y(0,0)$ with $O(h^2)$ accuracy. Setting $\beta = \alpha - 1$, we have the following Taylor's series expansions for $u_1, \cdots, u_5$:

$$\begin{aligned} u_1 - u_0 &= \alpha h u_x + \alpha^2 h^2 u_{xx}/2 + O(h^3), \\ u_2 - u_0 &= (-\beta u_x + u_y)h + (\beta^2 u_{xx} - 2\beta u_{xy} + u_{yy})h^2/2 + O(h^3), \\ &\vdots \\ u_5 - u_0 &= (\alpha u_x + 2u_y)h + (\alpha^2 u_{xx} + 4\alpha u_{xy} + 4u_{yy})h^2/2 + O(h^3). \end{aligned}$$

After eliminating u_x, u_{xx}, u_{xy}, and u_{yy}, we solve for u_y. Now substitute these $O(h^2)$ approximations of $u_x(0,0)$ and $u_y(0,0)$ into (3.4), to get

$$\begin{aligned} a(\mathbf{0})U_0 + b(\mathbf{0})&\left[-\frac{2\alpha+1}{(\alpha^2+\alpha)h}U_0 + \frac{\alpha+1}{\alpha h}U_1 - \frac{\alpha}{(\alpha+1)h}U_6\right] \\ + c(\mathbf{0})&\left[-\frac{3}{2h}U_1 - \frac{\alpha}{h}U_2 + \frac{2+\alpha}{h}U_3 + \frac{\alpha}{h}U_4 - \frac{1+2\alpha}{2h}U_5\right] = g(\mathbf{0}). \end{aligned}$$

The preceding approximations cover the case that the boundary intersects a horizontal mesh line of a square mesh as in Fig. 7, with point 0 at (0,0), and point 1 at $(\alpha h, 0)$. Other configurations of interior and boundary mesh points can be treated similarly; see Rice-Boisvert [84].

Numerical experiment. The *rate of convergence* of the $U_{i,j}$ to the $u(x_i, y_j)$ can be easily tested numerically. For example, consider the solution U of the difference approximation $-\nabla_h^2 U = 4(r^2-1)$ to the Poisson DE $-\nabla^2 u = 4(r^2-1)$, in the unit disk $x^2 + y^2 \leqslant 1$, satisfying the boundary condition $u \equiv 0.75$ on $r = 1$. The exact solution is $u = r^2 - r^4/4$, and its local *truncation error* is $\nabla^2 u - \nabla_h^2 u = h^2$ at all interior mesh-points not adjacent to the boundary. Next to the boundary $U_{i,j}$ satisfies (3.3′); see Fig. 3b.

Since the formula

$$V_{i,j} = u_{i,j} + h^2(r_{i,j}^2 - 1)/4$$

defines a function V on Ω_h which satisfies $-\nabla_h^2 V_{i,j} = 4(r_{i,j}^2 - 1)$ at the interior grid points specified, the *discretization error* (or 'global' truncation error), $e(x_i, y_j) = U_{i,j} - u(x_i, y_j)$, is nearly

$$e(x_i, y_j) \approx h^2(r_{i,j}^2 - 1)/4. \tag{3.5}$$

Solving for the $U_{i,j}$ numerically on a square mesh with $h = 2/n$, we obtain the results displayed in Table 2. These show that, in this example, the pointwise error is indeed approximately proportional to h^2.

TABLE 2. *Poisson equation in disk.*

$n = 2/h$	4	8	16	32	64
e_{max}	.049	.0129	.00357	.000926	.000268
e_{max}/h^2	.196	.206	.228	.237	.274

Exploiting reflection symmetry. For linear differential equations with constant coefficients, some boundary conditions permit exploitation of reflection symmetries. Thus, for the loaded *thin beam* problem $u^{iv} = f(x)$ of Chapter 1, §7, the endpoint condition (7.1a) of *simple support*, $u(1) = u''(1) = 0$, is equivalent to a continuation of the load by anti-symmetry: $f(1+t) = -f(1-t)$, $u(1+t) = -u(1-t)$. This substitution replaces the ΔE

$$U_{j-2} - 4U_{j-1} + 6U_j - 4U_{j+1} + U_{j+2} = h^4 f_j$$

at the mesh-point x_{J-1} adjacent to the boundary by

$$U_{J-3} - 4U_{J-2} + 5U_{J-1} = h^4 f_{J-1}. \tag{3.6}$$

Likewise, the condition $u'(1) = 0$ that the beam be *clamped* (Chapter 1, (7.1b)) is equivalent to $f(1+t) = f(1-t)$, $u(1+t) = u(1-t)$. Hence, with $u(1) = 0$, the ΔE can be replaced with

$$U_{J-3} - 4U_{J-2} + 7U_{J-1} = h^4 f_{J-1}. \tag{3.6′}$$

Analogous, but more complicated reflection symmetries can be used for *square plates*; see §5 for symmetry of the linear source problem.

4. Linear source problem. We now consider the standard five-point difference approximation of the general linear *source problem* in a bounded plane region Ω (Chapter 1, (3.3)):

$$L[u] = -\nabla \cdot [p(x,y)\nabla u] + q(x,y)u = f(x,y), \qquad p > 0, \quad q \geq 0, \tag{4.1}$$

on a general rectangular mesh. This approximation is

$$D_{i,j}U_{i,j} - L_{i,j}U_{i-1,j} - R_{i,j}U_{i+1,j} - T_{i,j}U_{i,j+1} - B_{i,j}U_{i,j-1} = F_{i,j}. \tag{4.2}$$

In (4.2), the 'left', 'right', 'bottom', and 'top' coefficients are defined by

$$\begin{aligned} L_{i,j} &= p_{i-1/2,j}\frac{y_{j+1}-y_{j-1}}{2(x_i-x_{i-1})}, \qquad & R_{i,j} &= p_{i+1/2,j}\frac{y_{j+1}-y_{j-1}}{2(x_{i+1}-x_i)}, \\ B_{i,j} &= p_{i,j-1/2}\frac{x_{i+1}-x_{i-1}}{2(y_j-y_{j-1})}, \qquad & T_{i,j} &= p_{i,j+1/2}\frac{x_{i+1}-x_{i-1}}{2(y_{j+1}-y_j)}, \end{aligned} \tag{4.2a}$$

where, as in [V, (6.39)],

$$p_{i-1/2,j} = p([x_{i-1}+x_i]/2, y_j), \qquad p_{i,j-1/2} = p(x_i, [y_{j-1}+y_j]/2).$$

The 'diagonal' and 'source' coefficients are defined by

$$D_{i,j} = L_{i,j} + R_{i,j} + B_{i,j} + T_{i,j} + q(x_i,y_j)\alpha_{i,j}, \quad F_{i,j} = f(x_i,y_j)\alpha_{i,j}, \tag{4.2b}$$

where the $\alpha_{i,j}$ are areas ('weights'),

$$\alpha_{i,j} = (x_{i+1}-x_{i-1})(y_{j+1}-y_{j-1})/4. \tag{4.2c}$$

Note that, when all mesh-points are intersections of mesh lines, $L_{i,j} = R_{i-1,j}$ and $B_{i,j} = T_{i,j-1}$. Hence, if each unknown approximate nodal value $U_{i,j}$ is made to correspond to the equation whose stencil is centered at that nodal value, as in (2.3), the coefficients of (4.2) defined by (4.2a) and (4.2b) lead to a *symmetric* matrix A. We will describe some other properties of these matrices in §5.

Because of this symmetry, on the average it suffices to store *four* coefficients per mesh-point. For Dirichlet boundary conditions, moreover, some of the $U_{i,j}$ are given as boundary values $u(x_i,y_j)$ and the difference approximation (4.2) can be used at any interior mesh-point, such as those in Figs. 1a, 3a, and 3c.

Caution. The irregular stencils at the boundaries in Figs. 2a and 3b make A *non*symmetric. Note that Figs. 19 and 20 of [V, p. 183] avoid irregular stencils by making mesh lines intersect the boundary only at points of a rectangular grid.

Central differences. We first consider a *square* mesh with mesh length h. In this case, the *central* difference operator δ_x is defined as usual by

$$\delta_x[u] = u(x+h/2) - u(x-h/2);$$

the subscript x signifies the independent variable being varied. Iterating, we get (for fixed h)

$$\delta_x^2[u] = \delta_x[\delta_x[u]] = u(x+h) - 2u(x) + u(x-h).$$

Clearly, the standard 5-point stencil (or 'star') (2.3) for approximating the Laplacian on a square mesh is the central difference *quotient* approximation

$$\nabla_h^2 u(x,y) = (\delta_x^2[u] + \delta_y^2[u])/h^2.$$

Likewise, in the approximation of (4.2)–(4.2a), one has

$$(\delta_x[p\delta_x[u]])_{i,j} = p_{i+1/2,j}(u_{i+1,j} - u_{i,j}) - p_{i-1/2,j}(u_{i,j} - u_{i-1,j}),$$

and the right side of this involves values of u only at mesh-points and values of p at midpoints between adjacent mesh-points. Since for any function $v \in C^5$, expansion in Taylor series with remainder gives

$$\delta_x[v(x,y)] = hv_x(x,y) + v_{xxx}(x,y)h^3/24 + O(h^5),$$

we have for any $u \in C^6$ and $p \in C^5$ that

$$\frac{1}{h^2}\delta_x[p\delta_x[u]] = (pu_x)_x + \frac{h^2}{24}[(pu_{xxx})_x + (pu_x)_{xxx}] + O(h^4). \tag{4.3}$$

Therefore, on a *square* mesh, the difference quotient operator

$$\mathbf{L}_h[u] = -(\delta_x[p\delta_x[u]] + \delta_y[p\delta_y[u]])/h^2 + qu \tag{4.4}$$

yields an $O(h^2)$ approximation to $\mathbf{L}$ in (4.1).

Formula (4.4) can be extended to any *uniform rectangular* mesh with constant spacing h in the x-direction and $k = \theta h$ in the y-direction: δ_x is as above and $\delta_y[u] = u(x, y+k/2) - u(x, y-k/2)$; that is, one has

$$\mathbf{L}_h[u] = -\delta_x[p\delta_x[u]]/h^2 - \delta_y[p\delta_y[u]]/k^2 + qu. \tag{4.4'}$$

THEOREM 1. *For any uniform rectangular mesh with spacing* $\Delta x = h$, $\Delta y = k = \theta h$, *and* $u \in C^4$, $p \in C^3$, *with* $\mathbf{L}$ *and* $\mathbf{L}_h$ *as in* (4.1) *and* (4.4'),

$$\mathbf{L}_h[u] = \mathbf{L}[u] + O(h^2). \tag{4.5}$$

Consider the linear source problem in a polygon Ω having horizontal, vertical, and 45° sides that has been subdivided by a *square mesh*. That is, let u satisfy (4.1) in Ω and Dirichlet boundary conditions:

$$\mathbf{L}[u] = -\nabla \cdot [p(x,y)\nabla u] + q(x,y)u = f(x,y), \qquad p > 0, \quad q \geqslant 0 \tag{4.6}$$

$$u = g(x,y) \qquad \text{on } \partial\Omega. \tag{4.6'}$$

As usual, Ω_h denotes the set of mesh-points (x_i, y_j) in the interior of Ω, and $\partial\Omega_h = \Gamma_h$ the set of boundary mesh-points. For instance, Ω might be a rectangle with corners (0,0), $(Ih,0)$, (Ih,Jh), $(0,Jh)$ for some positive integers I and J. In this example, the discretized problem (4.6)–(4.6′) consists in solving $(I-1)(J-1)$ linear algebraic equations; there is one equation for each of the unknown values $U_{i,j} = U(ih,jh)$ used to approximate $u(x_i, y_j) = u(ih,jh)$.

Normalization. Note that on a square mesh, the *standard* 5-point difference approximation (4.2) is

$$h^2 \mathbf{L}_h[U_{i,j}] = h^2 f_{i,j}.$$

It differs from (4.4′) by the *normalizing* factor $h^2 = \alpha_{i,j}$. Consequently, if an exact solution u is substituted into (4.2), the two sides differ by only $O(h^4)$, even though the *relative* error of the solution is $O(h^2)$.

Divided differences. The approximation (4.2)–(4.2a) can be derived on *any* rectangular mesh by using 'divided differences'. Such divided difference approximations to partial differential operators can be obtained by iterating difference approximations to ordinary derivatives. Thus, for derivatives with respect to x, we have the first and second divided difference formulas[15]

$$u[x_0,x_1] = \frac{u_1 - u_0}{x_1 - x_0}, \qquad u[x_0,x_1,x_2] = \frac{u[x_1,x_2] - u[x_0,x_1]}{x_2 - x_0},$$

and, in general, if $x_0 \neq x_n$,

$$u[x_0,x_1,\cdots,x_n] = \frac{u[x_1,x_2,\cdots,x_n] - u[x_0,x_1,\cdots,x_{n-1}]}{x_n - x_0}.$$

The n-th divided difference approximates $n!$ times the n-th derivative. For any x, $x_0 \leqslant x \leqslant x_n$:

$$d^n u(x)/dx^n = u^{(n)}(x) = n!\,u[x_0,\cdots,x_n] + O(h), \qquad h = x_n - x_0.$$

Moreover, some special choices of points, x_0, $\cdots$, x_n, and x, give approximations to $u^{(n)}(x)$ with higher order accuracy. Thus with a uniform mesh, where $x_j = jh$, $u''(x_1) = 2u[x_0,x_1,x_2] + O(h^2)$; likewise, $u^{iv}(x_2) = 24\,u[x_0,x_1,x_2,x_3,x_4] + O(h^2)$.

Following this idea, for any $x_{i-1} < x_i < x_{i+1}$ and corresponding $h_i = x_i - x_{i-1}$, $h_{i+1} = x_{i+1} - x_i$, one can first approximate u_x at the midpoint $x_{i-1/2} = (x_{i-1} + x_i)/2$ by the divided difference $u[x_{i-1}, x_i]$, multiply this by $p_{i-1/2} = p(x_{i-1/2})$, and then form the first divided difference of the result. This gives

$$d[p(x)\,du/dx]/dx = l_i u_{i-1} - c_i u_i + r_i u_{i+1} + \epsilon_i,$$

[15] See [KK, p. 165], Isaacson-Keller [66, Chap. 6, §1], or Conte-de Boor [80, Chap. 2].

where

$$l_i = \frac{2p_{i-1/2}}{h_i(h_i+h_{i+1})}, \qquad c_i = l_i + r_i, \qquad r_i = \frac{2p_{i+1/2}}{(h_i+h_{i+1})h_{i+1}}.$$

Adding the corresponding formula for derivatives with respect to y and multiplying by the area $\alpha_{i,j}$ in (4.2c) as a *normalizing factor* (after neglecting the ϵ's), we get (4.2)–(4.2a). In three dimensions, multiply divided differences similarly by the *volume*

$$(x_{i+1}-x_{i-1})(y_{j+1}-y_{j-1})(z_{k+1}-z_{k-1})/8.$$

Inner products. In Chapter 2, §5 (see especially (5.2) and Example 9), the linear source problem was related to the inner product

$$\mathbf{P}\langle u,v\rangle = \int_\Omega \{p[u_x v_x + u_y v_y] + quv\}\, dR \tag{4.7}$$

on the space $C_0^1(\Omega)$ of functions continuously differentiable in Ω which vanish on $\Gamma = \partial\Omega$. For *twice*-differentiable $v \in C_0^2(\Omega)$, integration by parts gives also

$$\mathbf{P}\langle u,v\rangle = \int_\Omega u\mathbf{L}[v]\, dR = \int_\Omega v\mathbf{L}[u]\, dR. \tag{4.7$'$}$$

Formula (4.7′) shows that $\mathbf{L}$ is *self-adjoint*.

An analogous relation holds for the discrete linear source problem. Namely, by definition, a bilinear form $\langle \mathbf{u}, A\mathbf{u}\rangle$ on the space of real $\mathbf{u} = (u_1, \cdots, u_N)^T$ is an inner product if and only if A is symmetric and positive definite. When $A = \mathbf{L}_h$ is the difference approximation on the left side of (4.2) on a rectangular mesh, this bilinear form is

$$\begin{aligned}\mathbf{P}_h\langle \mathbf{u},\mathbf{v}\rangle = \sum_{i,j} [&p_{i-1/2,j}\Delta_x u_{i,j}\Delta_x v_{i,j} + p_{i,j-1/2}\Delta_x u_{i,j}\Delta_y v_{i,j} \\ &+ q_{i,j}u_{i,j}v_{i,j}\Delta x_i \Delta y_j],\end{aligned} \tag{4.8}$$

for $u, v = 0$ on Γ_h, and where

$$\Delta_x u_{i,j} = u(ih+h, jh) - u(ih, jh), \qquad \Delta x_i = x_{i+1} - x_i,$$
$$\Delta_y u_{i,j} = u(ih, jh+h) - u(ih, jh), \qquad \Delta y_j = y_{j+1} - y_j.$$

In the 'bilinear form' of (4.8), the matrix P of coefficients of the terms involving the (first) forward differences Δ_x and Δ_y is clearly *symmetric* (this is the discrete analogue of being self-adjoint), while the matrix Q of the coefficients of $u_{i,j}v_{i,j}$ is diagonal. Moreover, $\mathbf{P}_h\langle \mathbf{u},\mathbf{u}\rangle > 0$ if $\mathbf{u} \neq \mathbf{0}$, since it then reduces to a sum of positive multiples of squares; hence the matrix of coefficients of (4.8) is *symmetric and positive definite*.

Three-dimensional problems. Almost all two-dimensional elliptic problems have natural three-dimensional (3D) analogues. Indeed, only the cost of solving them has prevented 3D problems from being the central concern of physicists and engineers. It seems likely that, over the next 20 years, more powerful computing machines and improved computational methods will make it possible to solve many such problems at moderate cost. We will, therefore, discuss them from time to time in this book.

Among three-dimensional elliptic problems, the following is one of the simplest.

Example 5. Let $\overline{\Omega}$ be a (closed, bounded) connected *rectangular polytope* whose boundary Γ is a finite union of *rectangles* having *edges* parallel to the coordinate axes. Then $\overline{\Omega}$ can be subdivided by planes $x = x_i$, $y = y_j$, and $z = z_k$ into a finite number of *boxes*

$$\mathrm{B} = [x_{i-1}, x_i] \times [y_{j-1}, y_j] \times [z_{k-1}, z_k],$$

whose corners (x_i, y_j, z_k) are the *nodes* of a *rectangular mesh* (possibly uniform).

Consider the linear source problem defined in Ω by the DE

$$-\nabla \cdot [p(x,y,z)\nabla u] + q(x,y,z)u = f(x,y,z), \tag{4.9}$$

and Dirichlet boundary conditions. Taking as unknowns the nodal values $U_{i,j,k} \approx u(x_i, y_j, z_k)$, we can approximate (4.9) at each interior mesh-point by a 7-point difference approximation similar to the 5-point approximation (4.2)–(4.2a), replacing the area $\alpha_{i,j}$ in (4.2c) by volume. More precisely, writing out

$$\nabla \cdot [p\nabla u] = (pu_x)_x + (pu_y)_y + (pu_z)_z,$$

we can approximate the additional term $(pu_z)_z$ as in (4.3), with x replaced with z, $p_{i-1/2,j}$ replaced with $p_{i,j,k-1/2}$, and so on.

Truncation and discretization errors. As in §2, one can expand $\nabla_h^2 u$ in terms of derivatives of u. For any $u \in C^6(\Omega)$, we have

$$\nabla_h^2 u = \nabla^2 u + (u_{xxxx} + u_{yyyy} + u_{zzzz})h^2/12 + \cdots.$$

The error in approximating ∇^2 by ∇_h^2 for $u \in C^6(\Omega)$ is thus found to be

$$\nabla_h^2 u - \nabla^2 u = (u_{xxxx} + u_{yyyy} + u_{zzzz})h^2/12 + O(h^4)$$

and we will call this the *truncation error*. Following [FW, p. 54] and Ames [77, p. 24], we will call the difference $e_{i,j} = U_{i,j} - u_{i,j}$ between the exact solutions of $-\nabla^2 u = f$ and $-\nabla_h^2 U_{i,j} = f_{i,j}$ at mesh-points the *discretization error*.

5. Band elimination. In this section, we will consider the problem of *solving* the set of N simultaneous linear equations in N unknowns arising from the ΔE (4.2). The choice of methods for solving this problem depends on many properties of the symmetric, positive definite matrices A produced by this ΔE. For simplicity, we will consider only Dirichlet boundary conditions of the form $u = g$ on Γ. We will also ignore the complications that arise when boundary grid points lie on only *one* mesh line, i.e., we consider only regular polygonal domains.

In this case, the approximations $U_{i,j} = \mathrm{U}_s$ to the exact $u(x_i, y_j)$ at interior mesh-points form a convenient basis of unknowns. However, even in this simple case, we have not specified the *order* in which these interior points are to be listed. For any such *ordering*, we can write the equations to be satisfied by the discrete approximation as

$$\text{(5.1)} \qquad \mathrm{E}_r: \quad \sum_{s=1}^{N} a_{r,s}\mathrm{U}_s = b_s, \qquad r = 1, \cdots, N,$$

or in matrix notation, with $\mathbf{U} = (\mathrm{U}_1, \cdots, \mathrm{U}_N)^T$, and $\mathbf{b} = (b_1, \cdots, b_N)^T$, as

$$\text{(5.1}') \qquad A\mathrm{U} = \mathbf{b}.$$

In (5.1′), the components U_s of $\mathbf{U}$ are the mesh values $U_{i,j}$ on Ω_h, while the components of $\mathbf{b}$ involve the $f_{i,j}$ and the boundary values $g_{i,j}$.

Sparsity. Many properties of the matrix A are independent of the ordering of nodes. Thus, for any ordering of equations and unknowns, when a 5-point approximation like (4.2) is used, each equation has at most 5 nonzero coefficients; hence so does each row of A. Thus A is a *sparse matrix*, with 'density' at most $5/N$. (In general, the 'density' δ of a matrix is defined as the fraction of nonzero entries; its 'sparsity' is $1-\delta$.)

Ordering schemes. In (4.2), the unknowns $U_{i,j}$ are *nodal values* at the *nodes* or mesh-points (x_i, y_j) of a rectangular mesh, and the equations refer to *stencils* centered at these same points. Therefore any sequential ordering (or indexing) of these mesh-points also orders the equations and unknowns by a *bijection*:

$$\text{(5.2)} \qquad \beta: \quad \left\{ \begin{array}{l} (i,j) \leftrightarrow r = \beta(i,j) \\ (i,j) \leftrightarrow s = \beta(i,j) \end{array} \right\} \quad r, s = 1, \cdots, N,$$

which orders the equations E_r and the unknowns U_s alike. The next result follows because $\mathbf{P}\langle \mathbf{u}, \mathbf{v} \rangle$ in (4.8) is an inner product.

THEOREM 2. *For any ordering of the nodes, the standard 5-point discretization (4.2) with Dirichlet conditions on a rectangular mesh gives rise through (5.2) to a symmetric, positive definite coefficient-matrix* $A = \|a_{r,s}\|$.

Note that the matrix A of Theorem 2 depends only on the mesh, the difference approximation (4.2), and the ordering; it is independent of the source term and the (Dirichlet) boundary values.

The natural order. On rectangular meshes, the *natural* order of mesh-points (x_i, y_j) (by mesh rows) is most commonly used. In this, (i,j) precedes (i', j') if and only if either $j < j'$, or $j = j'$ and $i < i'$, this holding alike for equations and unknowns as in Fig. 8a (other indexing schemes will be discussed in Chapter 4).

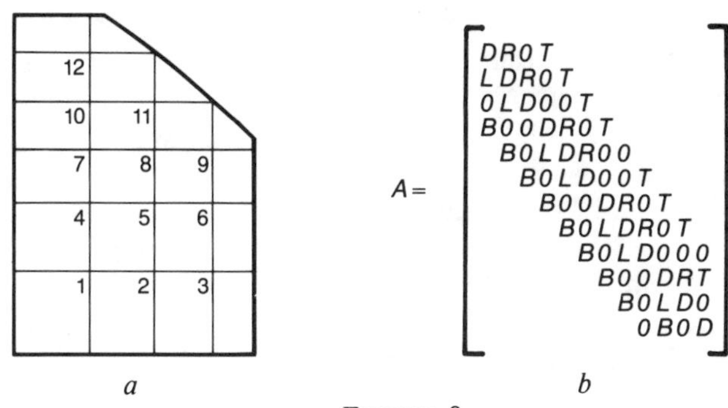

FIGURE 8

For (4.2)–(4.2c) in the domain of Fig. 8a with 12 interior mesh-points, this 'natural order' gives rise to the 12×12 matrix A displayed in Fig. 8b in which we have suppressed the subscripts and signs of the coefficients of (4.2). The matrix A is *symmetric* and *banded* with *band half-width three*, in the sense that all nonzero entries occur on diagonals within three of the main diagonal; i.e., $a_{r,s} \neq 0$ implies $|r-s| \leqslant 3$. Similarly, the 'natural' ordering (by mesh rows) of (4.2) with Dirichlet boundary conditions gives a symmetric banded matrix with band half-width I, the maximum number of interior mesh-points in any mesh row.

Property A. A second order-independent property of A in Fig. 8b is that its nonzero off-diagonal entries $a_{r,s}$ have subscripts r and s which are in disjoint sets:

$$S: \{1,3,5,7,9,11,12\} \quad \text{and} \quad T: \{2,4,6,8,10\}.$$

Similarly, (4.2) with Dirichlet boundary conditions leads to a matrix with Young's 'Property A'.

DEFINITION. An $N \times N$ matrix $A = \|a_{r,s}\|$ has *Property A* when the integers $1, 2, \cdots, N$ can be divided into two complementary (i.e., nonvoid disjoint) subsets S and T such that if $r \neq s$, then $a_{r,s} \neq 0$ implies $r \in S$ and $s \in T$ or vice versa.

Stieltjes matrices. A third such property is that A is a Stieltjes matrix.

DEFINITION. A *Stieltjes matrix* is a positive definite symmetric matrix whose off-diagonal elements are negative or zero [V, p. 85].

It can be shown that for a connected grid of mesh-points, the Stieltjes matrix A is *irreducible* (cf. Chapter 4, §7), from which it follows that all entries of A^{-1} are positive [V, p. 85], for any ordering of mesh-points.

This result is the discrete analogue of the positivity of the Green's function of the linear source problem (Chapter 2, §7, and Chapter 8, §8).

Again we emphasize that the symmetry of A requires that (4.2) be given on a regular polygonal domain and that an ordering (5.2) be used. Other domains and other orderings lead to nonsymmetric matrices. However, A does have Property A in such cases.

Tridiagonal block structure. One can visualize very simply the pattern of 0's in the matrix arising from (4.2) when the unknowns and associated equations are indexed in their 'natural' order. We partition the unknowns by the mesh rows: $\mathbf{U} = (\mathbf{U}_1, \cdots, \mathbf{U}_J)^T$ where the entries in $\mathbf{U}_j$ are the unknowns $U_{i,j}$ at interior mesh-points in the j-th row. Partitioning the equations in the same way, we obtain a subdivision of A into an $J \times J$ array of blocks. Display (5.3) shows this partitioning of the matrix in Fig. 8b. The blocks T_3, D_4, T_4, and D_5, are 3×2, 2×2, 2×1, and 1×1, respectively.

$$(5.3)\quad A = \begin{bmatrix} \mathrm{D}_1 & \mathrm{T}_1 & 0 & 0 & 0 \\ \mathrm{B}_2 & \mathrm{D}_2 & \mathrm{T}_2 & 0 & 0 \\ 0 & \mathrm{B}_3 & \mathrm{D}_3 & \mathrm{T}_3 & 0 \\ 0 & 0 & \mathrm{B}_4 & \mathrm{D}_4 & \mathrm{T}_4 \\ 0 & 0 & 0 & \mathrm{B}_5 & \mathrm{D}_5 \end{bmatrix}, \qquad \begin{aligned} \mathrm{D}_j = \mathrm{D}_j^T &= \begin{bmatrix} D_{1,j} & -R_{1,j} & 0 \\ -L_{2,j} & D_{3,j} & -R_{2,j} \\ 0 & -L_{3,j} & D_{3,j} \end{bmatrix}, \quad j = 1,2,3, \\ \mathrm{T}_j = \mathrm{B}_j^T &= \begin{bmatrix} -T_{1,j} & 0 & 0 \\ 0 & -T_{2,j} & 0 \\ 0 & 0 & -T_{3,j} \end{bmatrix}, \quad j = 1,2. \end{aligned}$$

In general, for (4.2) and Dirichlet boundary conditions, all *non*zero entries occur in blocks on or adjacent to the diagonal: the matrix is *block tridiagonal*. This is obvious because all nonzero coefficients refer to equations (and unknowns) in the same or adjacent rows of mesh-points. Moreover, each diagonal block is itself a tridiagonal square matrix. In the simplest cases, all nonzero entries in blocks adjacent to the diagonal blocks occur along one diagonal (not necessarily the main diagonal); in general, there is at most one nonzero entry in each column and row of these blocks.

For the special case of the five-point approximation to the Poisson equation on a rectangle with a square mesh, the diagonal and off diagonal block have the form

$$\mathrm{D}_j = \begin{bmatrix} 4 & -1 & & & & \\ -1 & 4 & -1 & & & \\ & -1 & \cdot & \cdot & & \\ & & \cdot & \cdot & -1 & \\ & & & -1 & 4 & -1 \\ & & & & -1 & 4 \end{bmatrix}, \qquad \mathrm{T}_j = \mathrm{B}_j = \begin{bmatrix} -1 & & & & & \\ & -1 & & & & \\ & & \cdot & \cdot & & \\ & & \cdot & \cdot & & \\ & & & & -1 & \\ & & & & & -1 \end{bmatrix}$$

Gauss elimination. If ordinary Gauss elimination[16] were used to solve $AU = \mathbf{b}$ with a general ordering of the N equations and unknowns in (5.1), storage for all $N^2 = n^4$ entries of A and about $N^3/3 = n^6/3$ multiplications (and about as many additions) would be required. However, if the *natural* ordering is used, a great saving in both storage and computation is possible with so-called band elimination.

Band elimination. Namely, to 'triangularize A' by 'forward elimination', it suffices to multiply row s (beginning with $s = 1$) by $m_{r,s} = a_{r,s}/a_{s,s}$ and subtract the result from row r for $r = s+1, s+2, \cdots, s+I$; one then has zeros in every entry in column s below $a_{s,s}$. A straightforward operation count shows that band elimination for a matrix with band half-width I requires about NI^2 multiplications when $I \ll N$, and about as many additions (or subtractions) plus the computation of about NI divisions. (Obviously, one should label axes so that $I \leqslant J$.) Hence if $I = J = n$, making $N = n^2$, then $n^4 + O(n^3)$ multiplications (and about as many additions) are required. Furthermore, only the entries of A within the band have to stored, i.e., about $2NI = 2n^3$ entries. For symmetric A, the storage and the computation are halved.

LU-decomposition.[17] In general, Gauss elimination without rearrangement of equations (row 'pivoting') or unknowns (column 'pivoting') has a very helpful interpretation in terms of matrix algebra, which applies to 'dense' matrices with few or no zeros as well as to 'band' matrices. The 'forward' *elimination* phase reduces a given linear system $A\mathbf{x} = \mathbf{b}$ to the form $U\mathbf{x} = \mathbf{c}$, where $\mathbf{c} = L^{-1}\mathbf{b}$, U is upper triangular, and L^{-1} is lower triangular with unit main diagonal. Moreover, the entries of L can be computed at the same time; the entry $l_{r,s} = m_{r,s}$ is the multiplier of row s used above to make $a_{r,s}$ zero. The entries of L can be stored in place of the strictly lower diagonal entries of A, as these are eliminated. At the end, one has U and the entries of L below its unit diagonal stored and $A = LU$ (neglecting permutations); this gives the *LU-decomposition* (or 'factorization') of A. It requires about $N^3/3$ multiplications for a full $N \times N$ matrix, but only about NI^2 multiplications for a band matrix. If A has the band half-width I, so do L and U; however, the inverse $A^{-1} = U^{-1}L^{-1}$ of A is usually a *full* matrix.

By storing L and U while solving $A\mathbf{x}_0 = \mathbf{b}_0$ for one right side, we can solve $A\mathbf{x}_j = LU\mathbf{x}_j = \mathbf{b}_j$ for other right sides, $\mathbf{b}_j$, $j = 1, 2, 3, \cdots$, by first computing $\mathbf{y}_j = U\mathbf{x}_j$ as the solution of $L\mathbf{y}_j = \mathbf{b}_j$ and then solving $U\mathbf{x}_j = \mathbf{y}_j$ for $\mathbf{x}_j$. Even if A is dense, since L and U are (lower resp.

---

[16] For an elementary discussion of Gauss elimination, see Birkhoff-Mac Lane [77, Chap. 2, §3].

[17] For fuller accounts of the ideas presented here, including Cholesky decomposition, see Wilkinson [65], Forsythe-Moler [67], Stewart [73], and George-Liu [81].

upper) *triangular*, this takes only about N^2 multiplications for each new choice of $\mathbf{b}_j$, as contrasted with the $N^3/3$ operations required to solve $A\mathbf{x}_0 = \mathbf{b}_0$.

For band matrices A arising from the standard 5-point discretization of a linear source problem on a rectangle with $N = IJ$ interior mesh-points and equations $(I \leqslant J)$, the required number of multiplications is about $I^3J = NI^2$ for the LU-factorization and about NI for each right side.

Cholesky decomposition.[18] When A is *symmetric* and *positive definite*, as it is for good discretizations of most self-adjoint positive definite boundary value problems, one can improve substantially on band elimination and LU-decomposition, as efficient algorithms for solving $A\mathbf{x} = \mathbf{b}$. Such a matrix can be written as $A = LU = R^TR$, where $R = \sqrt{D}L^T$ is upper triangular and $D = \operatorname{diag} U$ is diagonal with positive diagonal entries. For example:

$$A = \begin{bmatrix} 2 & -1 & & \\ -1 & 2 & -1 & \\ & -1 & 2 & \cdot \\ & & \cdot & \cdot \end{bmatrix} = R^TR,$$

where $R_{j,j} = \sqrt{1 + 1/j}$, $R_{j,j+1} = 1/R_{j,j}$, and all the other entries of the *bidiagonal* matrix R are zero.

One computes successively the rows of R (i.e., the columns of R^T):

$$R_{1,1} = \sqrt{A_{1,1}}; \quad R_{1,j} = A_{1,j}/R_{1,1}, \; j = 2, 3, \cdots; \quad R_{2,2} = \sqrt{A_{2,2} - R_{1,2}^2};$$

and so on. Moreover, in carrying out this 'Cholesky decomposition,' it suffices to store the entries on and below the diagonal of A, and to perform about $N^3/6$ multiplications; hence the the cost of computation is halved.[19]

Similarly, if the symmetric positive definite matrix has band half-width I, then the number of multiplications can be reduced to about $NI^2/2$.

Symmetry. When a linear source problem has an axis of symmetry, the work needed to solve it can sometimes be reduced substantially. For example, consider

$$\text{(5.4)} \qquad \begin{aligned} -\nabla[p(\mathbf{x})\nabla u] + q(\mathbf{x})u &= f(\mathbf{x}) \quad \text{in} \quad \Omega, \\ u &= g(\mathbf{y}) \quad \text{on} \quad \Gamma, \end{aligned}$$

where Ω is the L-shaped region in Fig. 1a. Suppose that p, q, f, and g

---

[18] See Forsythe-Moler [67, Chap. 23] or Stewart [73, pp. 134–158].

[19] Since the cost of computation is (in principle) roughly proportional to the *product* of the storage needed and the time involved, one can argue that the cost is asymptotically divided by *four*.

are all symmetric with respect to the line $y=x$, so that $p(x,y)=p(y,x)$, $q(x,y)=q(y,x)$, $f(x,y)=f(y,x)$, and also $g(x,y)=g(y,x)$. Then $u(x,y)=u(y,x)$, and thus the solution u must satisfy the Neumann conditions $\partial u/\partial n=0$ on $y=x$.

Now consider the standard 5-point approximation (4.2) on a square mesh with spacing $h=1/I$; by symmetry $U_{i,j}=U_{j,i}$. The left side of the discrete approximation (4.2) simplifies when $j=i$ because, for example, $U_{i-1,i}=U_{i,i-1}$ and $L_{i,i}=B_{i,i}$. In order to preserve the symmetry of the resulting matrix A, we divide by 2 to get

$$D_{i,i}U_{i,i}/2-R_{i,i}U_{i+1,i}-B_{i,i}U_{i,i-1}=F_{i,i}/2.$$

This gives a boundary value problem for U on *half* of the L-shaped region. This reduced problem, with roughly half the number of unknowns, requires about half the storage of the original problem and one-*eighth* the arithmetic. This is because with the natural ordering, the band half-width is $I-1$ instead of $2I-1$.

A similar procedure can be used in the case $f(x,y)=-f(y,x)$, $g(x,y)=-g(y,x)$, of anti-symmetry; for this, $U_{i,j}=0$ on the line of symmetry. — The general case can be treated by resolving f and g into their 'even' and 'odd' parts by setting, for example

$$f(x,y)=\frac{1}{2}[f(x,y)+f(y,x)]+\frac{1}{2}[f(x,y)-f(y,x)],$$

and solving for each part separately. There are still savings of roughly half in storage and of 75% in arithmetic.

LINPACK. Much as the human labor of writing and 'debugging' discretization subroutines can be greatly reduced by using the 'canned' programs in ELLPACK, so the labor of solving general linear systems can be greatly reduced by using the carefully designed and debugged Fortran programs in various standard program libraries such as HARWELL, IMSL, NAG, SL-MATH, and so on. At least one such library is available in compiled form at most computing centers.

If A is a positive definite (symmetric) band matrix of the kind described above, the LINPACK subroutines SPBFA and SPBSL factor and solve $A\mathbf{x}=\mathbf{b}$ very efficiently.[20] However, very few of the other algorithms specifically designed for solving discretized elliptic problems efficiently, and described in the later portions of this book, have been implemented in LINPACK.

[20] A description of these may be found in J. J. Dongarra, C. B. Moler, J. R. Bunch, and G. W. Stewart, *LINPACK Users' Guide*, SIAM 1979, Chap. 4; listings are given there in Appendix C.

Today, programs for solving specified classes of linear systems in such packages rarely differ by large factors in running time, or by orders of magnitude in accuracy. Thus, for positive definite symmetric banded matrices derived from the standard 5-point discretizations of the Poisson equation on rectangular domains, preliminary tests run at Harvard in 1978 by Wadi' Juraydini indicate that the LINPACK package developed at the Argonne National Laboratory is about 20% faster and has about half the error of the earlier IMSL code he tried.

Virtual memory. For large problems, disk storage (or 'virtual memory') is used to store matrix entries; moreover, several 'pages' of virtual memory may be needed to store an entire matrix. In Fortran, two-dimensional arrays are stored sequentially by *columns*: A(2,1) follows A(1,1), A(1,2) follows A(N,1), and so on. To reduce the time of transfer of these pages into and out of core, matrix processing with Fortran should be done by columns instead of by rows. This is done in LINPACK.

6. Error estimation. In this and the next two sections, we will consider error estimation more carefully. On a (uniform) *square* mesh, expanding in Taylor series, we have, as in (4.3),

$$\mathbf{L}_h[u]-\mathbf{L}[u]=h^2[(pu_{xxxx}+4p_x u_{xxx}+3p_{xx}u_{xx}+p_{xxx}u_x) + \text{similar terms in } y]/24+O(h^4), \tag{6.1}$$

assuming that all functions are 'smooth'. On a *non*uniform mesh, the difference approximation has formally only $O(h)$ accuracy. Thus, if $u''(x)$ is approximated by the *divided difference* quotient

$$\delta=2[u(x-h)-(1+\theta)u(x)+\theta u(x+\theta h)]/h^2\theta(1+\theta), \tag{6.2}$$

as in (3.1), then expansion in Taylor series with remainder gives, for any $u \in C^4$,

$$\delta=u''(x)+h(\theta-1)u'''(x)/3+O(h^2). \tag{6.2'}$$

Hence, on a *non*uniform rectangular mesh, (6.1) becomes

$$\mathbf{L}_h[u]-\mathbf{L}[u]=O(h). \tag{6.3}$$

Truncation error estimates like those above are obviously related to *discretization* error estimates of the form $U_h(x_i,y_j)-u(x_i,y_j)=O(h^2)$, but the relation is complicated. Just because the *exact* solution of a given DE *nearly* satisfies an approximating ΔE, it does not follow that the exact solution of this ΔE nearly approximates the solution of the given DE.

However, on any fixed mesh, one *can* establish a basic connection between the truncation error vector $\mathbf{d}=\mathbf{L}_h[u]-\mathbf{L}[u]$, evaluated at mesh-points, and the discretization error[21] vector $\mathbf{e} = \mathbf{U} - \mathbf{u}$. If the DE $\mathbf{L}[u] = f$ is linear and roundoff is negligible, then evidently

$$(6.4)\qquad \mathbf{L}_h[\mathbf{e}] = \mathbf{L}_h[\mathbf{U}_h] - \mathbf{L}_h[\mathbf{u}] = \mathbf{f} - \mathbf{L}_h[\mathbf{u}] = \mathbf{L}[\mathbf{u}] - \mathbf{L}_h[\mathbf{u}] = -\mathbf{d}.$$

This formula, unfortunately, does not establish a simple connection between the *magnitude* of $\mathbf{d}$ and the *magnitude* of $\mathbf{e}$.

Norms. The 'magnitude' of the error $\mathbf{e}$ is usually measured by some specified 'norm', $\|\cdot\|$, having the properties listed in Chapter 2, (5.5). As was stated in Chapter 2, §5, if $\langle \mathbf{v},\mathbf{w}\rangle$ is any inner product, then $\langle \mathbf{v},\mathbf{v}\rangle^{1/2}$ is a norm. For any such norm (e.g., for the Euclidean norm $\|\mathbf{v}\|_2 = (\sum v_k^2)^{1/2}$), the distance function $d(\mathbf{v},\mathbf{w}) = \|\mathbf{v}-\mathbf{w}\|$ is a *metric* as usual. When $\mathbf{L}_h$ is symmetric and positive definite as it is in (4.4), we can bound $\|\mathbf{e}\|_2$ in terms of $\|\mathbf{d}\|_2$ and the *extreme eigenvalues* $\lambda_{\max}(\mathbf{L}_h)$ and $\lambda_{\min}(\mathbf{L}_h)$:

$$(6.5)\qquad \frac{\|\mathbf{d}\|_2}{\lambda_{\max}(\mathbf{L}_h)} \leqslant \|\mathbf{e}\|_2 \leqslant \frac{\|\mathbf{d}\|_2}{\lambda_{\min}(\mathbf{L}_h)}.$$

It follows from (6.5) that if $\lambda_{\min}(\mathbf{L}_h)$ is bounded away from zero and the truncation error tends to zero as $h \downarrow 0$, then the discretization error tends to zero at least as fast as the truncation error.

For the Poisson ΔE, $-\nabla_h^2 U_{i,j} = f_{i,j}$, in a domain Ω of area A, with a square mesh of side h and $N \approx A/h^2$ mesh-points, the rescaled norm $(A \sum U_{i,j}{}^2/N)^{1/2}$ is approximately the L_2-norm

$$\|u\|_2{}^2 = \iint u(x,y)^2\,dx\,dy,$$

as $h \downarrow 0$. Moreover, it is usually of the same order of magnitude as $\|U\|_\infty = \max|U_{i,j}|$ (and $\|u\|_\infty = \max_{\bar{\Omega}}|u(\mathbf{x})|$). For Dirichlet-type boundary conditions, $\mathbf{d} \equiv \mathbf{e} \equiv 0$ on Γ_h, and the inequality (6.5) applies. Moreover, as we will show in §8, $\lambda_{\min} = O(1)$ and $\lambda_{\max} = O(1/h^2)$. Hence by (6.5) $\|\mathbf{e}\| = O(\|\mathbf{d}\|)$ or less.

Moreover, similar results hold for most linear source problems. For these, it is most appropriate to use a Sobolev-type inner product $\mathbf{P}_h\langle \mathbf{u},\mathbf{v}\rangle$, like that of (4.8), approximating

$$(6.6)\qquad \mathbf{P}\langle u,v\rangle = \iint [p(x,y)(u_x v_x + u_y v_y) + q(x,y)uv]\,dx\,dy;$$

see Chapter 6 for more details.

Likewise, if the max-norm of the truncation error of the linear operator

$$(6.7)\qquad \begin{aligned} \mathbf{L}_h[U_{i,j}] = [&D_{i,j}U_{i,j} - L_{i,j}U_{i-1,j} - R_{i,j}U_{i+1,j} \\ &- T_{i,j}U_{i,j+1} - B_{i,j}U_{i,j-1}]/\alpha_{i,j} \end{aligned}$$

---

[21] For brevity, we will often refer to the *discretization* error below as "*the* error".

is $O(h^2)$, then so is the max-norm of the discretization error. Here the coefficients and $\alpha_{i,j} = (x_{i+1} - x_i)(y_{j+1} - y_j)/4$, are as in (4.2)–(4.2c). The proof of this result depends on a special property of the operator $\mathbf{L}_h$.

Monotonicity. It can be shown (Collatz [60, p. 45, Theorem 2]) that for a connected grid of mesh-points, $\mathbf{L}_h$ is of *monotone type* (Chapter 2, §7). In other words, for Dirichlet boundary conditions, all entries in the inverse A^{-1} of the corresponding matrix are positive for any ordering of mesh-points. As in the case that A is a Stieljes matrix, this is the discrete analogue of the positivity of the Green's function of the linear source problem (Chapter 2, §7, Chapter 8, §8). The next result is a corollary.

THEOREM 3. *Let* $\mathbf{L}_h$ *be the difference operator* (6.7) *associated with the standard* 5*-point discretization of the linear source problem, and let* $U_{i,j}$ *be nonnegative at all boundary mesh-points,* Γ_h*, of a connected grid* Ω_h*. Then* $\mathbf{L}_h[U_{i,j}] \geqslant 0$ *in* Ω_h *implies* $U_{i,j} \geqslant 0$ *in* Ω_h.

We now use the method at the end of Chapter 2, §7, to bound the discretization error.

Suppose Ω is in $[-1,1] \times [-a,a]$, and set $V_i = \cosh\alpha - \cosh\alpha x_i$. For simplicity, we assume a uniform mesh, as in (4.4′). We then obtain

$$\mathbf{L}_h[V_i] = \left[\frac{p_{i-1/2,j} + p_{i+1/2,j}}{2}\,\frac{4\sinh^2\alpha h/2}{h^2} + q_{i,j}\right]\cosh\alpha x_i$$
$$+\frac{p_{i+1/2,j} - p_{i-1/2,j}}{2h}\,\frac{4\sinh\alpha h/2\cosh\alpha h/2}{h}\sinh\alpha x_i .$$

Because $2[\sinh\alpha h/2]/h = \alpha + O(h)$, we can choose α large enough to make $\mathbf{L}_h[V_i] \geqslant 1$ for all $h < 1$ provided p has a continuous second derivative. From (6.4), we then get $\mathbf{L}_h[\|\mathbf{d}\|_\infty V_i - e_{i,j}] \geqslant 0$, from which $e_{i,j} \leqslant \|\mathbf{d}\|_\infty V_i$ is obtained and, similarly, $-\|\mathbf{d}\|_\infty V_i \leqslant e_{i,j}$. These give the next result.

THEOREM 4. *If* $p \in C^2$*, the discretization error* $\mathbf{e}_h$ *of the standard* 5*-point approximation to the linear source problem with Dirichlet boundary conditions satisfies* $\|\mathbf{e}_h\|_\infty \leqslant K\|\mathbf{L}_h[u] - f\|_\infty$*, where* K *is independent of* h.

Convection-diffusion equation. The simplest elliptic DE that is *not* self-adjoint is the time-independent convection-diffusion equation

$$u_x = u_{xx} + u_{yy} + f(x,y). \tag{6.8}$$

Therefore, a good model problem consists in solving (6.8) in a rectangle $[0,a] \times [0,b]$, for Dirichlet boundary conditions. Any elliptic DE of the form

$$Uu_\xi = \nu u_{\xi\xi} + \mu u_{\eta\eta} + r(\xi,\eta), \qquad \nu > 0, \quad \mu = k^2\nu, \tag{6.9}$$

in a rectangle with sides parallel to the coordinate axes, although apparently more general, can be reduced to (6.8) by a rescaling of coordinates: set $\xi = \nu x/U$, $\eta = k\nu y/U$.

The obvious 5-point central divided difference approximation to (6.8), on a square mesh of side h, is

$$-\nabla_h^2 U_{i,j} + (U_{i+1,j} - U_{i-1,j})/2h = f_{i,j}. \tag{6.10}$$

This is easily verified to have $O(h^2)$ truncation error, and it is satisfactory if $h \ll 1$.

One-dimensional case. But in most cases, it is better *not* to approximate (6.8) or (6.9) by a central difference approximation. One can see why easily if one sets $u = \rho$, and thinks of the one-dimensional convection-diffusion equation

$$U\rho_x = \nu\rho_{xx} + f(x), \qquad x \in [0,a], \tag{6.11}$$

as determining the variation in the mean *concentration* $\rho(x)$ of a chemical, dissolved in a fluid with longitudinal diffusivity ν, flowing with mean velocity U through a pipe, and of $f(x)$ as the rate at which the chemical is being added to the solution at a distance x downstream.

Rewriting (6.11) as

$$\rho_x = \epsilon\rho_{xx} + g(x), \qquad \epsilon = \nu/U, \tag{6.11'}$$

suppose that $\epsilon \ll 1$. The one-dimensional analogue of (6.10) is

$$-(1+\theta)\rho_{j-1} + 2\rho_j - (1-\theta)\rho_{j+1} = h^2 g_j/\epsilon, \qquad \theta = h/2\epsilon, \tag{6.12}$$

When $\epsilon \ll 1$, to set $h \leqslant 2\epsilon$ makes the operation count unnecessarily high; moreover, (6.12) becomes 'lower triangular' when $h = 2\epsilon$, and unstable when $h > 2\epsilon$.

Instead of (6.12), one should use either Allen's approximation[22]

$$[-e^{\theta}\rho_{j-1} + 2\cosh\theta\,\rho_j - e^{-\theta}\rho_{j+1}]/2h\sinh\theta = g_j, \tag{6.13}$$

where $\theta = h/2\epsilon$, or 'upwind differencing':[23]

$$-(1+2\theta)\rho_{j-1} + (2+2\theta)\rho_j - \rho_{j+1} = 2h\theta g_j.$$

The ΔE (6.13) is *exact* when $g(x) \equiv 0$, having 1 and $e^{x/\epsilon}$ as a basis of solutions, which is the same as for (6.11') with $g(x) \equiv 0$.

By discretizing an appropriate mean value theorem, E. C. Gartland has obtained a different improvement of (6.10);[24] his improvement is, however, not exact when $f(x,y) = 0$.

---

[22] D. N. de G. Allen and R. V. Southwell, Quart. J. Mech. Appl. Math. 8 (1955), 129–145.

[23] See P. J. Roache, *Computational Fluid Dynamics*, Hermosa.

[24] E. C. Gartland, SIAM J. Sci. Stat. Comp. 3 (1982), 460–472. We are indebted to E. C. Gartland for improvements in the above discussion.

Roundoff errors. The preceding discussion assumes 'exact arithmetic'. But really, except for a very few DE's and meshes leading to ΔE's expressible exactly in terms of 'machine representable' (i.e., binary) numbers, machine computations refer to 'rounded' equations[25] of the form $\mathbf{L}_h[\mathbf{U}] = \mathbf{f} + \boldsymbol{\epsilon}$, whose (local) relative 'roundoff error' $|\epsilon_{i,j}|/|f_{i,j}|$ is usually some moderate multiple of the 'machine precision' or 'macheps' $\epsilon = 2^{-t}$. Thus, unlike (local) truncation and (global) discretization errors, roundoff errors are *machine dependent*.

During the numerical solution of systems $A\mathbf{x} = \mathbf{b}$ of linear algebraic equations by elimination methods, roundoff errors accumulate because computers use finite precision arithmetic. Depending on the method of solution used,[26] these errors can be amplified through division by numbers which are computed as the difference of two nearly equal values. As a result, the computed solution $V_{i,j}$ of a difference equation at a mesh-point (x_i, y_j) may differ appreciably from the exact solution $U_{i,j}$. The vector $\mathbf{E}$ of the differences $U_{i,j} - V_{i,j}$ at (x_i, y_j) is called the (global) vector *roundoff error*.

Forsythe and Wasow emphasize that "roundoff is probably negligible in comparison with the discretization error in ordinary computation with finite-difference methods for elliptic problems" [FW, p. 324]. Their statement applies with especial force to second-order, self-adjoint problems like the linear source problem, because they lead to positive definite symmetric matrices. For these, no pivoting is needed.[27] Moreover, for second-order problems like (4.2), the condition number (see §8) is typically moderate, except in the case of Neumann problems without absorption.

Generally speaking, whereas the discretization error tends to zero with h, the roundoff error *increases* as h decreases. Hence there is a *crossover* point in mesh refinement, below which the roundoff error is dominant. For example, this crossover point would occur when $h \approx 2^{-t/4} \approx 10^{-t/13}$ if the roundoff error were $2^{-t}/h^2$ and the discretization error were h^2. Computers using floating point numbers with about 7 or 8 (decimal) digits of precision (such as IBM and DEC single precision) would normally give 3 or 4 significant decimal digits of accuracy for the solution of the *discretized* linear source problem for $h < 1/100$; computers using numbers with twice as many digits (such as IBM and DEC double or CDC single precision) should yield 10 to 12 significant digits of accuracy. To make the

[25] See Wilkinson [65], Forsythe-Moler [67], and Stewart [73]; see also G. Birkhoff and S. Gulati, ZaMP 30 (1979), 148-158.

[26] In Chapter 4 we will discuss various alternatives to the method of band elimination.

[27] Wilkinson [65, p. 230]; see also Forsythe-Moler [67,p. 114].

discretization error equally small with the standard 5-point approximation, one must have I and J of the order of 10^5.

7. Orders of accuracy. Theorem 1 of §4 displays the order of magnitude of the *truncation error* $\mathbf{d}$ of the discretization (4.4′) of the linear source problem, as being $O(h^2)$ on a *uniform* mesh, while (6.3) indicates it to be $O(h)$ for a *non*uniform mesh. Much more generally, for smooth $u(\mathbf{x})$, the truncation error of an elliptic operator with smooth coefficients is typically of the form

$$d(\mathbf{x};h) = h^{\nu} d_{\nu}(\mathbf{x}) + h^{\nu+1} d_{\nu+1}(\mathbf{x}) + O(h^{\nu+2}). \tag{7.1}$$

Furthermore, Theorem 4 of §6 shows that, in linear source problems, $\mathbf{d}(\mathbf{x})$ and the discretization error $\mathbf{e}(\mathbf{x})$ usually have the same order of magnitude. In this section and the next, we will examine the orders of magnitude of the discretization errors $\mathbf{e}$, and explain the technique of Richardson extrapolation. In §9, we will summarize without proof some basic theorems about discretization errors.

As in (6.4), the discretization errors are related to the truncation errors by the equation $\mathbf{L}_h[\mathbf{e}] = -\mathbf{d}$. To make our discussion of discretization errors more concrete, and to help clarify the facts, we will next present the results of some specific numerical experiments.

The disk. The effect on the error of 'irregular stars' near curvilinear boundaries is illustrated in Table 2 of §3. The ratio $e_{\max}/h^2$ is .196 for $h = 1/2$; it increases slowly as h decreases, and tends to about .25 until roundoff effects become significant at about $h = 1/55$ (see §6).[28]

Example 6. Consider the standard 5-point approximation to the Poisson DE $-\nabla^2 u = 1$, for the boundary condition $u \equiv 0$ on Γ, in the following domains:

(a) the unit square S: $[0,1] \times [0,1]$;

(b) the L-shaped region R: $([0,1] \times [0,2]) \cup ([1,2] \times [0,1])$;

(c) the slit square H with boundary the boundary of $[-1,1] \times [-1,1]$ and the line segment $x = 0$, $-1 \leqslant y \leqslant 0$.

As was stated in Chapter 2, §8, the 'model problem' of (a) has a mild $(r^2 \log r)$ singularity at each corner of S, while the problem of (b) has an $O(r^{2/3})$ singularity at the reentrant corner. Problem (c) illustrates an even worse $r^{1/2}$ singularity associated with a 360° interior angle. Note that the *local* truncation errors $\nabla_h^2 u - \nabla^2 u$ are $O(1)$, $O(h^{-4/3})$, and $O(h^{-3/2})$, respectively, as $h \downarrow 0$ for the solutions of these problems near their singularities; they are *unbounded* for (b) and (c).

---

[28] We have computed this ratio for many more values of h than are listed in Table 2. We used 'short words' with about 7–8 decimal digits of precision and band elimination without iterative refinement (cf. §8).

We have made numerical experiments, with $2^{-56} \simeq 10^{-17}$ precision arithmetic, for these problems using the standard 5-point discretization, for which

$$e(\mathbf{x},h) = h^2 e_1(\mathbf{x},h) + O(h^4) \tag{7.2}$$

might be expected (in exact arithmetic). The facts are as follows.

Square. By symmetry, the solution satisfies $u_x = 0$ on $(0.5, y)$ and $u_y = 0$ on $(x, 0.5)$. Hence, as in §2, we can reduce the number of unknowns by a factor of nearly four by taking $h = 1/2I$ and setting $U_{I+1,j} = U_{I-1,j}$ and $U_{i,I+1} = U_{i,I-1}$. As Table 3a shows,[29] the expectation of an $O(h^2)$ error, $e(\mathbf{x},h) = u(\mathbf{x}) - U_h(\mathbf{x})$, so that

$$e(\mathbf{x},h)/e(\mathbf{x},h/2) \approx 4, \qquad e(\mathbf{x},h/2)/e(\mathbf{x},h/4) \approx 4 \tag{7.3}$$

was quite well fulfilled by $E_h = e(\mathbf{x},h)$ at $\mathbf{x} = (1/8, 7/8)$.

TABLE 3a. *Error and ratio for Example* 6a.

h	1/8	1/16	1/32	1/64	1/128
$E_h \times 10^7$	4260.	1150.	296.	74.4	18.6
$R_h = E_{2h}/E_h$		3.69	3.90	3.97	3.99

We obtained similar results for the max norm of the error and concluded that the global discretization error in Problem (a) is not affected seriously by the mild singularites at the corners. The results below show that the stronger singularities of (b) and (c) do affect the error $e_h(\mathbf{x})$ globally, but the effect is concentrated near them.

Error bound. The problem of Example 6a, whose solution has mild singularities at the four corners, also illustrates the 'direct product' concept defined in Chapter 2, §5. We will now use this method to bound the error in the L_2-norm of the trigonometric interpolant to U_h, by expanding $1, u$, and U_h in double sine series. We will prove rigorously that the error is $O(h^2)$.

For $0 < x, y < 1$, the expansion of 1 is

$$1 = \sum_{r=1}^{\infty} \sum_{s=1}^{\infty} b_{r,s} \sin(r\pi x) \sin(s\pi y), \tag{7.4}$$

where $b_{r,s} = 0$ if r or s is even, and $b_{r,s} = 16/\pi^2 rs$, if r and s are odd. Thus we have:

$$u(x,y) = \sum_{k \text{ odd}}^{\infty} \sum_{l \text{ odd}}^{\infty} c_{k,l} \sin(k\pi x) \sin(l\pi y), \quad c_{k,l} = 16/[\pi^4 kl(k^2 + l^2)]. \tag{7.5}$$

[29] To compute the error for (a), we used an approximation to u accurate to about 13 digits supplied by S. C. Eisenstat of Yale University; see Chapter 7, §10. We solved (a) by FFT (see Chapter 4, §6); band elimination was used for (b) and (c).

On the other hand, for $h = 1/N$ and $(x_i, y_j) = (ih, jh)$, $1 \leqslant i, j \leqslant N-1$, an interior mesh-point, we have

$$1 = \sum_{r=1}^{N-1} \sum_{s=1}^{N-1} B_{r,s} \sin(r\pi x_i) \sin(s\pi y_j),$$

where $B_{r,s} = 0$ if r or s is even, and otherwise

$$B_{r,s} = \frac{4h^2 \cos(rh\pi/2) \cos(sh\pi/2)}{\sin(rh\pi/2) \sin(sh\pi/2)}.$$

Thus we have at all mesh-points $(x_i, y_j) = (ih, jh)$,

$$U_h(x_i, y_j) = \sum_{k \text{ odd}}^{N-1} \sum_{l \text{ odd}}^{N-1} C_{k,l} \sin(k\pi x_i) \sin(l\pi y_j) \tag{7.6}$$

where

$$C_{k,l} = \frac{h^2 B_{k,l}}{4[\sin^2(k\pi h/2) + \sin^2(l\pi h/2)]}, \qquad k, l \text{ odd}.$$

In (7.6), replace x_i, y_j, with x, y, to obtain the trigonometric interpolant to the nodal values of U_h in the square $0 \leqslant x, y \leqslant 1$. The square of the L_2 norm of the error is then

$$\|e_h\|^2 = \sum_{k=1}^{N-1} \sum_{l=1}^{N-1} (C_{k,l} - c_{k,l})^2 + \sum_{k \geqslant N}^{\infty} \sum_{l \geqslant N}^{\infty} c_{k,l}{}^2 \qquad (k, l \text{ odd}). \tag{7.7}$$

The second sum is bounded by $256/\pi^8$ times

$$\int\int_{N-1}^{\infty} \frac{dx\, dy}{x^2 y^2 (x^2+y^2)^2} + 2\int_{N-1}^{\infty} dx \int_1^N \frac{dy}{x^2 y^2 (x^2+y^2)^2} + 2\int_{N-1}^{\infty} \frac{dx}{x^2(x^2+1)^2},$$

and thus it is $O(h^5)$. To obtain a bound on the first sum in (7.7), set $p = k\pi h/2 = \rho \cos\phi$, $q = l\pi h/2 = \rho \sin\phi$, and note that

$$\frac{C_{k,l}}{c_{k,l}} = \frac{pq(p^2+q^2) \cos p \cos q}{\sin p \sin q\, [\sin^2(p) + \sin^2(q)]}$$

is analytic for $|\rho| < \pi$. Expanding in powers of ρ, we have

$$C_{k,l}/c_{k,l} = 1 + \rho^2[-2\cos^2(\phi)\sin^2(\phi)/3 + \cdots].$$

On the square $0 \leqslant p, q \leqslant \pi/2$, the quantity in square brackets has maximum absolute value $M \simeq .4$, independent of k, l, and h. Consequently,

$$|C_{k,l} - c_{k,l}| \leqslant |c_{k,l}| \rho^2 M = 4h^2 M/\pi^2 kl,$$

and the sum of the squares of these terms is bounded by

$$\frac{16M^2h^4}{\pi^4}\left[1 + 2\int_1^N \frac{dx}{x^2} + \int\int_1^N \frac{dx\, dy}{x^2 y^2}\right] = O(h^4);$$

therefore, $\|e_h\| = O(h^2)$.

Richardson extrapolation. The fact that the (local) truncation error of the 5-point difference approximation to the Poisson DE can be expanded in a power series of the form

$$d_h(\mathbf{x}) \equiv \nabla_h^2 u(\mathbf{x}) - \nabla^2 u(\mathbf{x}) = h^2 d_2(\mathbf{x}) + h^4 d_4(\mathbf{x}) + \cdots \tag{7.8}$$

suggested to L. F. Richardson [27] the idea that the same ought to be true of the discretization error $e_h(\mathbf{x})$. Indeed, he assembled in 1927 impressive numerical evidence to support his conjecture,[30] which is obviously much sharper than the simple assertion that $e_h(\mathbf{x}) = O(h^2)$.

As Richardson pointed out, if in fact

$$e_h(\mathbf{x}) = h^2 e_2(\mathbf{x}) + h^4 e_4(\mathbf{x}) + \cdots, \tag{7.9}$$

then by comparing $U_h(\mathbf{x})$ with $U_{2h}(\mathbf{x})$ at any $\mathbf{x} \in \Omega_{2h}$, one can eliminate the $O(h^2)$ error term, getting

$$u(\mathbf{x}) = U_h(\mathbf{x}) + [U_h(\mathbf{x}) - U_{2h}(\mathbf{x})]/3 + O(h^4). \tag{7.10}$$

The approximation $u(\mathbf{x}) \approx V_h(\mathbf{x}) = U_h(\mathbf{x}) + [U_h(\mathbf{x}) - U_{2h}(\mathbf{x})]/3$, which Richardson called "deferred approach to the limit", is today usually called "Richardson extrapolation". It will be compared with other higher-order methods in §10. The $O(h^4)$ term in (7.10) cannot ordinarily be replaced by $M(\mathbf{x})h^4 + O(h^6)$.

Unfortunately, although valid for most ordinary DE's,[31] (7.9) only holds globally in special domains (whose boundaries happen to fall on mesh lines) *and* for special partial DE's not having corner singularities. For this reason, little use has been made of Richardson extrapolation for obtaining more accurate approximate solutions to elliptic problems.[32]

We now apply these ideas to results for (b) and (c) in Example 6.

Ell. For the L-shaped region of Example 6b, symmetry in the line $x = y$ implies $u(jh, jh+h) = u(jh+h, jh)$, allowing us to halve the number of unknowns. Here the deviation from (7.3) is substantial. Asymptotically, $e(\mathbf{x}, h) = h^\nu e_1(\mathbf{x}) + o(h^\nu)$ implies[33]

$$D_h = U_h - U_{2h} = e_h - e_{2h} = Ch^\nu + o(h^\nu),$$

and so $R_h = D_h / D_{2h} \approx 2^\nu$. As Table 3b shows, the observed values of

[30] Richardson [27]; in 1938, Shortley and Weller [38] accepted (7.9) as an established fact.

[31] See [FW, p. 307]; Joyce [71]; Birkhoff-Rota [78, p. 214]; H. J. Stetter, Apl. Mat. 13 (1968), 187–190; Pereyra [68]; and others. For the Laplacian on rectangular domains, see F. Hofmann [67].

[32] However, it has been used successfully with the generalized marching algorithm (see Chapter 4, §5, and Bunch-Rose [76]), and with the capacitance matrix method (see Chapter 8, §9, and V. Pereyra, W. Proskurowski, and O. Widlund, Math. Comp. 31 (1977), 1–16).

[33] A function $E(h)$ is $o(h^\nu)$ as $h \downarrow 0$ if $E(h)/h^\nu$ tends to zero.

(7.11) $\quad \nu_h = \log(R_h)/\log(2) = \log_2[(U_{2h} - U_{4h})/(U_h - U_{2h})]$

at (1/2,1/2) are nearer to 1.5 than to 2.

TABLE 3b. *Approximation, difference, ratio, and exponent for Example* 6b.

h	$U_h(1/2,1/2)$	D_h	R_h	ν_h
1/2	.120192			
1/4	.127055	.006863		
1/8	.129590	.002535	2.707	1.437
1/16	.130509	.000919	2.758	1.464

Slit rectangle. By symmetry again, for $h = 1/I$ and a $2I \times I$ mesh, we have $U_{I+1,j} = U_{I-1,j}$ for $j > I/2$, thus reducing the number of unknowns to about I^2. The results in Table 3c indicate that $\nu \approx 1.2$.

TABLE 3c. *Approximation and exponent for Example* 6c; $h = 1/32$.

(x,y)	U_{16h}	U_{8h}	U_{4h}	U_{2h}	U_h	ν_{4h}	ν_{2h}	ν_h
(0,.75)		.101855	.106291	.108047	.108825		1.337	1.174
(0,.50)	.123702	.145426	.155215	.159156	.160915	1.150	1.313	1.164
(0,.25)		.124839	.142212	.150478	.154088		2.290	1.195

Example 6′. As a test on the reliability of the data analysis, we also solved $\nabla^2 v = 0$ on the domains of Examples 6b and 6c with boundary conditions chosen so that the solutions were

$$v(r,\theta) = r^{2/3}\sin(-2\theta/3), \qquad r = 0 \text{ at } (1,1), \quad \text{for (b)},$$
$$v(r,\theta) = r^{1/2}\sin([\theta - \pi/2]/2), \qquad \text{for (c)};$$

then v and u behave the same asymptotically as $r \downarrow 0$. Knowing the exact solution, we determined ν_h from (7.11).

For (b) and the point (1/2,1/2) we obtained $\nu_h \approx 1.5$, close to the values in Table 3b. But, when we used $\|\mathbf{e}_h\|_\infty$, we obtained $\nu_h \approx 0.64$ and for each h, the maximum error was at the mesh-point closest to the reentrant corner.

Similarly, for (c) and the point (0,.05), $\nu_h \approx 1.1$ and for $\|\mathbf{e}_h\|_\infty$, $\nu_h \approx 0.38$.

8. Effects of smoothness. Example 6c in the last section shows that a straightforward expansion into Taylor series of a difference approximation $\mathbf{L}_h$ to a differential operator $\mathbf{L}$ can lead one to *over*estimate its order of accuracy (i.e., to *under*estimate the order of magnitude of the *solution error* $\|U_h - u\|$). We will now present an example in which such an expansion might lead one to *under*estimate the order of accuracy (i.e., to *over*estimate the error).

Example 7. Consider the DE

$$\mathbf{L}[u] = (xu')' = 4x \tag{8.1}$$

on [0.05,1], with boundary conditions $u(0.05) = u(1) = 0$, whose solution is $u(x) = x^2 - 1 + .9975 \log x / \log 0.05$. A 3-point discretization can be obtained by specializing (4.2) to one variable.

A formal Taylor series expansion of the truncation error suggests that the discretization error $e_h(x)$ will be $O(h^2)$ on a *uniform* mesh, but $O(h)$ on a *nonuniform* mesh — and hence that a uniform mesh should be used. Actually, a nonuniform mesh with mesh-points $x_j = 20^{(j-I)/I}$ for $j = 0, 1, \cdots, I$ gives twice the accuracy of a uniform mesh; both have $O(h^2)$ accuracy. Specifically, for this mesh, the truncation error $\mathbf{L}[v] - \mathbf{L}_h[v]$ is zero for v a constant and $v(x) = \log x$, and it is $O(h^2)$ for $v(x) = x^p$, $p = 1, 2, \cdots$, for the *one*-dimensional case. We have solved (8.1) with both $2^{-24} \approx 10^{-7}$ and $2^{-56} \approx 10^{-17}$ precision arithmetic. For the first of these, roundoff becomes significant when $I > 64$; if iterative refinement had been used, the roundoff would have been at the level of the machine precision.

This example indicates that the estimate of $O(h)$ accuracy for the standard 5-point approximation to ∇^2 on a *non*uniform mesh is unduly conservative. Though true locally at each *individual* point with $\Delta x_i = \theta \Delta x_{i+1}$, if θ is constant and $h_{i+1/2} = (x_{i+1} - x_i) \downarrow 0$, the *global* effect on $\|U_h - u\|$ is $O(h^2)$. This is always true for one-dimensional linear source problems if the total variation in the $\log h_i$ is uniformly bounded.

The DE $-u_{xx} = f(x)$ on [0,1], with $u(0) = u(1) = 0$, and the ΔE $-h^{-2}\delta^2 U_j = f_j$, also illustrate the effect of *smoothness* (compared with the mesh spacing $h = 1/I$) on the ratio $\|\mathbf{e}\|/\|\mathbf{d}\|$ of the discretization error to the truncation error.

For $f(x) = \sin \pi x$, the eigenfunction having the lowest eigenvalue, we have

$$\frac{u}{U} = \frac{\mu_1}{\lambda_1} = \frac{4 \sin^2(\pi h/2)}{h^2 \pi^2} = 1 - \frac{\pi^2 h^2}{12} + \cdots.$$

Hence the *relative error* (R.E.) in the solution is

$$\text{R.E.} = \frac{\|\mathbf{e}\|}{\|\mathbf{U}\|} = \frac{\|\mathbf{U} - \mathbf{u}\|}{\|\mathbf{U}\|} \approx \frac{\pi^2 h^2}{12}. \tag{8.2}$$

Since $\mathbf{d} = (\lambda_1 - \mu_1)\mathbf{u}$ while $\mathbf{u} = \lambda_1 \mathbf{f}$, the *relative residual* (R.R.) is

$$\text{R.R.} = \frac{\|\mathbf{d}\|}{\|\mathbf{f}\|} = \frac{\|u'' - h^{-2}\delta^2 u\|}{\|\mathbf{f}\|} = 1 - \frac{\mu_1}{\lambda_1} \approx \frac{\pi^2 h^2}{12}. \tag{8.2$'$}$$

Consequently, $||\mathbf{e}||/||\mathbf{d}|| \approx 1$; the discretization and truncation errors are nearly equal.

For $f = \sin[\pi(I-1)x]$, the eigenfunction of $-h^{-2}\delta^2$ having the largest eigenvalue, in contrast,

$$\frac{u}{U} = \frac{\mu_{I-1}}{\lambda_{I-1}} = \frac{4\cos^2(\pi h/2)}{[h\pi(I-1)]^2} \approx \frac{4}{\pi^2}.$$

Although the error, $||\mathbf{e}|| \approx h^2(1/4 - 1/\pi^2)$, is $O(h^2)$, so is u, and, in contrast to (8.2) and (8.2′), both the relative error and relative residual are $O(1)$:

$$\text{(8.3)} \qquad \text{R.E.} \approx \frac{1}{4} - \frac{1}{\pi^2}, \qquad \text{R.R.} \approx 1 - \frac{4}{\pi^2},$$

and $||\mathbf{e}||/||\mathbf{d}|| \approx h^2/4$.

Similar formulas hold for intermediate eigenfunctions, and show that if f is *smooth* (i.e., if the coefficients of the Fourier sine series tend to zero sufficiently rapidly), then the R.E. is $O(h^2)$. However, for highly irregular f, the R.E. can tend to zero very slowly.

Since the range of the eigenvalues of the Model Problem ($-\nabla^2 u = f$ in the unit square) is the same, analogous conclusions hold for it.

Example 8. Let R be the rectangle $[0,a]\times[0,b]$, and set $h = a/I$, $k = b/J$, for positive integers I and J. Let $\mathbf{L}_{h,k}$ be the standard 5-point central difference approximation (4.4) (with $p \equiv 1, q \equiv 0$) to the Poisson DE $\mathbf{L}[u] = -\nabla^2 u = f(x,y)$ on a *uniform* mesh. Let R be subdivided into $I \times J$ equal rectangles, with the boundary condition $u \equiv 0$ on ∂R.

In this model problem, the eigenfunctions of $\mathbf{L}_{h,k}$ are just the restrictions of those of $-\nabla^2 = \mathbf{L}$ to the grid R_h of mesh-points:

$$\text{(8.4)} \qquad \phi_{m,n} = \sin(\pi m x/a)\sin(\pi n y/b),$$

with $m = 1, \cdots, I-1$ and $n = 1, \cdots, J-1$. Moreover,

$$\mathbf{L} = -\frac{\partial^2}{\partial x^2} - \frac{\partial^2}{\partial y^2} \approx \mathbf{L}_{h,k} = -\frac{1}{h^2}\delta_x^2 - \frac{1}{k^2}\delta_y^2.$$

As a result, the eigenvalues of each $\phi_{m,n}(x,y) = \phi_m(x)\psi_n(y)$ are sums,

$$\lambda_{m,n} = \lambda'_m + \lambda''_m \quad \text{for } \mathbf{L} \quad \text{and} \quad \mu_{m,n} = \mu'_m + \mu''_n \quad \text{for } \mathbf{L}_{h,k},$$

where

$$\text{(8.5)} \qquad \lambda'_m = \pi^2 m^2/a^2 \quad \text{and} \quad \mu'_m = (4/h^2)\sin^2(\pi m h/2a).$$

Since $4\sin^2(\alpha/2) = \alpha^2(1 - \alpha^2/12 + \cdots)$, the ratios μ'_m/λ'_m are given by

$$\text{(8.6)} \qquad \mu'_m/\lambda'_m = 1 - \pi^2 m^2 h^2/12a^2 + \cdots;$$

and similarly for λ''_m and μ''_n.

Tensor products. Like Example 6a (see *Error bound* in §7), Example 8 utilizes the direct product concept introduced in Chapter 2, §5. If Φ and Ψ are two vector spaces of univariate functions on intervals $[a,b]$ and $[c,d]$ with bases $\phi_i(x)$ and $\psi_j(y)$, then the products $\phi_i(x)\psi_j(y)$ form a basis for the 'tensor product' space $S=\Phi\otimes\Psi$ of linear combinations of these products on the rectangle $[a,b]\times[c,d]$. Moreover, from any operators $\mathbf{L}$ and $\mathbf{M}$ on Φ and Ψ, respectively, the formulas

$$(\mathbf{L}\otimes\mathbf{I})[\phi_i\psi_j]=\mathbf{L}[\phi_i]\psi_j, \qquad (\mathbf{I}\otimes\mathbf{M})[\phi_i\psi_j]=\phi_i\mathbf{M}[\psi_j]$$

construct an analogous 'tensor product' operator on $\Phi\otimes\Psi$.

Thus, in Example 8, the difference quotient approximations $\mathbf{L}_h = h^{-2}\delta_x^2$ and $\mathbf{L}_k = k^{-2}\delta_y^2$ to the ordinary differential operators $\mathbf{L}_x = d^2/dx^2$ and $\mathbf{L}_y = d^2/dy^2$, respectively, when applied to $\Phi\otimes\Psi$ by the formula

$$\begin{aligned}(\mathbf{L}_h\otimes\mathbf{L}_k)[u]_{i,j} = [&u_{i+1,j+1}+u_{i-1,j+1}+u_{i-1,j-1}+u_{i+1,j-1}\\ &-2(u_{i+1,j}+u_{i,j+1}+u_{i-1,j}+u_{i,j-1})+4u_{i,j}]/h^2k^2,\end{aligned}$$

give an approximation $\mathbf{L}_h\otimes\mathbf{L}_k\approx\partial^4/\partial x^2\partial y^2$ to the differential operator $\mathbf{L}_x\otimes\mathbf{L}_y$. Likewise, the linear combination

$$\mathbf{L}_{h,k}=\mathbf{L}_h\otimes\mathbf{I}+\mathbf{I}\otimes\mathbf{L}_k \tag{8.7}$$

approximates $\partial^2/\partial x^2+\partial^2/\partial y^2=\nabla^2$.

More generally, the linear differential operator $\mathbf{L}$ defined by

$$\mathbf{L}[u]=-[p(x)u_x]_x-[q(y)u_y]_y+[r(x)+s(y)]u,$$

is the formal sum $\mathbf{L}=\mathbf{L}'+\mathbf{L}''$ of the operators

$$\mathbf{L}'[u]=-[p(x)u_x]_x+r(x)u, \qquad \mathbf{L}''[u]=-[q(y)u_y]_y+s(y)u,$$

However, to express the action of $\mathbf{L}$ on the space of functions $u(x,y)$ of *two* variables in terms of actions of $\mathbf{L}'$ and $\mathbf{L}''$ on differentiable functions $v(x)$ and $w(y)$ of *one* variable, we must write $\mathbf{L}=\mathbf{L}'\otimes\mathbf{I}''+\mathbf{I}'\otimes\mathbf{L}''$, where $\mathbf{I}'$ and $\mathbf{I}''$ are the identity operators on the spaces of $v(x)$ and $w(y)$. An analogous difference approximation to the operator $\mathbf{L}$ is obtained as $\mathbf{L}_{h,k}=\mathbf{L}'_h\otimes\mathbf{I}''+\mathbf{I}'\otimes\mathbf{L}''_k$, where

$$\mathbf{L}'_h[v]=-\delta_x[p\delta_x[v]]/h^2+rv, \qquad \mathbf{L}''_k[w]=-\delta_y[q\delta_y[w]]/k^2+sw.$$

If ϕ,ψ,ϕ_h,ψ_k are eigenfunctions of $\mathbf{L}',\mathbf{L}'',\mathbf{L}'_h,\mathbf{L}''_k$ with eigenvalues $\lambda',\lambda'',\mu',\mu''$, respectively, then

$$\mathbf{L}[\phi(x)\psi(y)]=(\lambda'+\lambda'')\phi(x)\psi(y), \tag{8.8}$$

$$\mathbf{L}_{h,k}[\phi_{h,i}\psi_{k,j}]=(\mu'+\mu'')\phi_{h,i}\psi_{k,j}. \tag{8.9}$$

Condition number. Formula (6.5) gives a standard (but very loose) *a priori* bound on the ratio $e(\mathbf{x};h)/d(\mathbf{x};h)$. We shall now reinterpret this bound in terms of the *condition number* of the linear operator $\mathbf{L}_h$ (or matrix), which we shall continue to assume to be symmetric and positive definite. The Euclidean (or spectral) condition number $\kappa(\mathbf{L}_h)$ is the ratio of its largest to its smallest eigenvalue:

$$\kappa(\mathbf{L}_h)=\lambda_{\max}/\lambda_{\min}=||\mathbf{L}_h||\cdot||\mathbf{L}_h^{-1}||. \tag{8.10}$$

The *relative error* (R.E.) $||\mathbf{e}||/||\mathbf{U}||$ and the *relative residual* (R.R.) $||\mathbf{d}||/||\mathbf{f}||$ of the solution are therefore related to the condition number by

$$||\mathbf{f}||/||\mathbf{U}||\leqslant\lambda_{\max}\quad\text{and}\quad||\mathbf{e}||/||\mathbf{d}||=1/\lambda_{\min},$$

all in the Euclidean norm $||\cdot||$. Multiplying these inequalities, we get

$$||\mathbf{e}||/||\mathbf{U}||\leqslant\kappa(\mathbf{L}_h)\,||\mathbf{d}||/||\mathbf{f}||. \tag{8.11}$$

This can be generalized to any norm $||\cdot||_\nu$, since[34]

$$\frac{||\mathbf{e}||_\nu/||\mathbf{U}||_\nu}{||\mathbf{d}||_\nu/||\mathbf{f}||_\nu}=\frac{||\mathbf{f}||_\nu/||\mathbf{U}||_\nu}{||\mathbf{d}||_\nu/||\mathbf{e}||_\nu}=\frac{||\mathbf{L}_h[\mathbf{U}]||_\nu/||\mathbf{U}||_\nu}{||\mathbf{L}_h[\mathbf{e}]||_\nu/||\mathbf{f}||_\nu}\leqslant\kappa_\nu(\mathbf{L}_h). \tag{8.12}$$

This gives the following result.

THEOREM 5. *In any norm*, $||\cdot||_\nu$, *the relative error* $||\mathbf{e}||_\nu/||\mathbf{U}||_\nu$ *is at most* $\kappa_\nu(\mathbf{L}_h)$ *times the relative residual* $||\mathbf{d}||_\nu/||\mathbf{f}||_\nu$, *where* $\mathbf{d}$ *is the residual* $\mathbf{L}_h[\mathbf{u}]-\mathbf{L}[\mathbf{u}]$.

It is classic (see Chapter 4, §11) that the first ratio in (8.12) can vary from $1/\kappa_\nu(\mathbf{L}_h)$ to $\kappa_\nu(\mathbf{L}_h)$, *depending on* $\mathbf{f}$ *and* $\mathbf{d}$. Indeed,

$$\kappa_\nu(\mathbf{L}_h)=\kappa_\nu(\mathbf{L}_h^{-1})=\kappa_\nu(\mathbf{G}_h)$$

is *defined* as the maximum value of this ratio. Hence, if $\mathbf{f}$ and $\mathbf{d}$ are uncorrelated, the inequality (8.11) (in any norm) is 'best possible'.

We now return to the case of symmetric, positive definite $\mathbf{L}_h$ and Euclidean norms. For linear source problems, the Model Problem of $-\nabla^2u=1$ in the unit square is fairly typical. In that case,

$$\lambda_{\min}=2\pi^2+O(h^2)\quad\text{and}\quad\lambda_{\max}=8/h^2+O(1); \tag{8.13}$$

the estimates $\lambda_{\min}=O(10)$ and $\lambda_{\max}=O(10/h^2)$ are thus typical of linear source problems. Combining them, we have the estimate

---

[34] In any norm, $\kappa_\nu(\mathbf{L}_h)$ is a measure of the maximum *distortion* of $\mathbf{L}_h$:

$$(||\mathbf{L}_h[u]||_\nu/||u||_\nu)_{\max}/(||\mathbf{L}_h[v]||_\nu/||v||_\nu)_{\min}.$$

for the Euclidean condition number. Substituting back into (6.5), moreover, we have the estimate

$$\frac{h^2}{10}\,||\mathbf{d}|| \leqslant ||\mathbf{e}|| \leqslant \frac{1}{10}\,||\mathbf{d}||. \tag{8.15}$$

Note that the situation discussed above is *opposite* to that commonly found in the literature on computation with ill-conditioned matrices, because there it is typical to normalize (equilibrate) the matrix so its rows have length $O(1)$ in which case $\lambda_{\min} << 1$ and $\lambda_{\max} = O(1)$.[35] In contrast, the matrices associated with *divided* difference approximations, such as $\mathbf{L}_h$ in (4.4), have $O(h^{-2})$ entries and have their minimum eigenvalue bounded away from zero and so $||\mathbf{e}||$ in (8.15) is $O(h^2)$ when u is smooth. The situation closer to that of numerical linear algebra occurs when the standard 5-point approximation of (4.2) is used; it is 'scaled' by the areas $\alpha_{i,j}$ and entries in corresponding matrices are $O(1)$. For this we have

$$\frac{1}{10}\,||\mathbf{d}^*|| \leqslant ||\mathbf{e}|| \leqslant \frac{1}{10h^2}\,||\mathbf{d}^*||, \tag{8.16}$$

where $||\mathbf{d}^*|| = ||h^2\mathbf{L}_h[u] - h^2\mathbf{L}[u]||$ is $O(h^4)$. In both cases, empirical values of $||\mathbf{e}||$ tend to be nearer the upper bound when u is smooth.

Example 9. Consider next the deflection of a simply supported rectangular flat plate with domain $\mathrm{R} = [0,a] \times [0,b]$, as in Chapter 1, Example 6. This problem gives rise to the elliptic DE $\nabla^4 u = F(x,y)$, with the boundary conditions $u \equiv \partial^2 u/\partial n^2 \equiv 0$ on $\partial\mathrm{R}$. The DE can be approximated by the difference equation $\mathbf{L}_{h,k}^2[U] = \mathbf{L}_{h,k}[\mathbf{L}_{h,k}[U]] = F$, with $\mathbf{L}_{h,k}$ as in (8.7).

This ΔE has a 13-point diamond-shaped stencil; the boundary conditions permit periodic continuation by *odd* reflection across the sides $x = 0$ and $y = 0$: $u(-x,y) = -u(x,y)$ and $u(x,-y) = -u(x,y)$. This allows us to apply the stencil to *all* interior points. The eigenfunctions are the same as in Example 8. Indeed, all operators are the *squares* of the corresponding operators in Example 8. Hence the eigenvalues are also squared.

Thus, in Example 9, if $I = J = 100$, we have

$$\kappa(\mathbf{L}_{h,h}^2) = \kappa(\nabla_h^4) = [\kappa(\nabla_h^2)]^2 \approx [2 \times 10^4]^2 = 4 \times 10^8. \tag{8.17}$$

This makes roundoff troublesome in IBM or VAX single precision.

For the simply supported square plate of Example 9, with $a = b = 1$, $\mathbf{L}_h = \nabla_h^4$ and similar calculations give

[35] See, for example, Wilkinson [66, pp. 192–194]; Forsythe-Moler [67, Chap. 11]. The Hilbert matrices are typical examples of ill-conditioned matrices (Forsythe-Moler [67, Chap. 19]).

(8.18) $$\frac{h^4}{100}||\mathbf{d}|| \leqslant ||\mathbf{e}|| \leqslant \frac{1}{100}||\mathbf{d}||,$$

since for this problem $\nabla_h^4 = \nabla_h^2 \nabla_h^2$.

Factorization. Note that the simply supported plate problem of Example , and more generally for any boundary conditions of the form $U = g(\mathbf{y})$ and $\nabla^2 U = h(\mathbf{y})$, can be factored into two second-order problems:

$$\nabla_h^2 V_{i,j} = f_{i,j} \quad \text{on } R_h, \qquad V = h(\mathbf{y}) \quad \text{on } \partial R_h,$$
$$\nabla_h^2 U_{i,j} = V_{i,j} \quad \text{on } R_h, \qquad U = g(\mathbf{y}) \quad \text{on } \partial R_h.$$

To solve these problems in 'tandem' with two 5-point stencils requires much less effort than solving $\nabla_h^4 U = f$ with a 13-point stencil having the same order of accuracy.

Unlike the simply supported plate problem, the *clamped* plate problem, for which u and $\partial u/\partial n$ are given on ∂R, cannot be factored in this way.

Roundoff error, II. Although roundoff errors are seldom troublesome in linear source problems, they can be more serious with convection-diffusion equations and fourth-order biharmonic problems. This is because the resulting matrices can have very large condition numbers.

To see when roundoff errors might become troublesome, we solved one-dimensional problems, $D_h^{2k} U_i = 0$, $k = 1, \cdots, 5$, with band elimination for boundary conditions which make $U_i \equiv 1$, where $D_h^2 = (1/h^2)\delta^2$ is the second divided central difference operator. The condition number[36] is $O(1/h^{2k})$ and, as expected, the roundoff error grows like the condition number.

Iterative refinement. If the relative roundoff error and $\kappa(A)/2^t$ are substantially less than 100%, then the relative error in the computed solution of $A\mathbf{x} = \mathbf{b}$ can be reduced significantly by the following procedure of *iterative refinement* due to J. H. Wilkinson; see Wilkinson [65, pp. 255–263].

Having computed the *residual* $\mathbf{r}^{(j)} = \mathbf{b} - A\mathbf{x}^{(j)}$ of any approximate solution $\mathbf{x}^{(j)}$ in *double precision*, one then computes an improved solution $\mathbf{x}^{(j+1)} = \mathbf{x}^{(j)} + \mathbf{e}^{(j)}$ by adding the approximate error $\mathbf{e}^{(j)}$ obtained by solving $A\mathbf{e} = \mathbf{r}^{(j)}$ in single precision. Once a (single precision) LU decomposition of A has been made (e.g., by band elimination), this is easy (see p. 65). This process can be repeated to obtain the exact solution $\mathbf{x}$ with relative error the order of the machine precision 2^{-t}, unless A is extremely ill-conditioned. Since both A and its factors must be saved, twice as much storage is required when iterative refinement is used.

Although the errors $||\mathbf{x} - \mathbf{x}^{(j)}||$ decrease, the residuals $||\mathbf{r}^{(j)}||$ usually all have about the same size, of the order of $2^{-t}||\mathbf{b}||$.

Table 4 lists those $N_1 = 1/h$ for which the roundoff error of band elimination is h^2 and $N_2 = 1/h$ when the roundoff error is 100%. Also listed are the number of iterative refinements N_3 to achieve full machine

[36] The condition number of D_h^{2k} is the same as that of ∇_h^{2k}; see Chapter 4, §11.

precision, 10^{-7}. Experiments with ∇_h^2 and ∇_h^4 for n up to 80 agree well with the one-dimensional results.

TABLE 4. *Roundoff error for* D_h^{2k} *with precision* 10^{-7}.

$2k$	N_1	N_2	$N_3 = 10$	20	30	50	100	1000	5000
2	150	≈10000	1	1	1	1	1	3	8
4	40	200	1	1	2	2	4		
6	30	80	1	2	2	6			
8	20	55	1	3	7	25			
10	11	30	2	8					

9. Error bounds. To obtain rigorous general *a priori* pointwise error estimates for numerical solutions of elliptic problems is very difficult. In practice, it is hard enough to make reliable error estimates even *a posteriori* (see §7), after inspecting the computer printout, without quadrupling the execution time and storage by such devices as mesh-halving. Indeed, if one could estimate the error to within $\pm 10\%$ (say), one could obtain a more accurate approximation by an order of magnitude, by subtracting this estimate! This was the principle that motivated Richardson extrapolation, and also Fox's method of difference correction, to be discussed next.

Fox's difference correction. Since

$$\nabla_h^2 e = h^2(u_{xxxx} + u_{yyyy})/12 + O(h^4),$$

one can reduce the error by adding to the solution of $\nabla_h^2 U = f$ the solution of

$$\nabla_h^2 E = (\delta_x^4 U + \delta_y^4 U)/12h^2. \tag{9.1}$$

Because deriving good *a priori* estimates of discretization errors is so difficult, we will devote little space to deriving them in this book. Instead, we will simply mention a few important results, and list some supplementary references in which such discretization errors are analyzed more thoroughly. In this chapter, we will discuss only discretization errors for *difference* approximations on rectangular meshes. However, other error bounds derived in Chapters 6 and 7 may also be applicable. As we will show there, piecewise bilinear finite element approximations on general subdivisions into quadrilaterals are equivalent to 9-point difference formulas.

Discrete Green's function. One can combine remainder formulas with *a priori* knowledge of the derivatives of the exact solution, obtained by analytic considerations (see Chapter 2), to bound the truncation error, $\mathbf{d} = \mathbf{L}_h[u] - \mathbf{L}[u]$. Since the actual error vector $\mathbf{e} = \mathbf{U}_h - \mathbf{u}$ satisfies

$\mathbf{e} = \mathbf{G}_h[\mathbf{d}]$ where $\mathbf{G}_h = A^{-1}$, this leads to an *a priori error bound* in terms of norms of $\mathbf{G}_h$ and $\mathbf{d}$. Here $\mathbf{G}_h$ may be called the *Green's matrix* because it acts like a *discrete Green's function* [FW, pp. 315–318] for the source problem being solved. It is a positive matrix for (4.4). Finally, again using analytical considerations to be discussed in Chapter 8, §2, one can often bound a norm of $\mathbf{G}_h$.

Using other arguments, global convergence as $h \downarrow 0$ was first proved for the Laplace ΔE on a square mesh by R. G. D. Richardson in 1917 and by Phillips and Wiener in 1922; the aim of these authors was to establish *existence theorems* for solutions of the Dirichlet problem for $\nabla^2 u = 0$ from algebraic existence theorems for $\nabla_h^2 u = 0$. In 1928, Courant, Friedrichs, and Lewy showed that, in addition, all difference quotients of given order converge to the appropriate derivatives, as $h \downarrow 0$. Actually, this follows from the global convergence of u and Harnack-type theorems; see [K, Chap. X].

The maximum principle of Chapter 2, §7, was applied to the Poisson equation by Gerschgorin [30] to prove $O(h)$ global accuracy. Using linear interpolation on the boundary, Collatz[37] sharpened this result in 1933, under appropriate differentiability assumptions, to prove $O(h^2)$ accuracy. Further work was also done by Wasow in 1952–7, and by P. Laasonen, who discussed carefully the loss of accuracy introduced by corners.[38] Walsh and Young studied the effect on the error of the smoothness of the [illegible] on [illegible] perimeter of a square. For the Dirichlet problem, [illegible] Mh [illegible] us and their first results in a series of papers written in 1964–5; see especially Bramble-Hubbard [64], [64a], and the references given there.[41] For linear source problems, A is 'monotone' (i.e., the inverse G of A is nonnegative; see Chapter 2, §7). On this point, see Bramble-Hubbard [64b] and the work by Harvey Price.[42] The preceding authors have also shown that higher order differences can give higher order accuracy (for $\nabla^2 u = f$ and $\nabla^4 u = f$ on a square mesh).

The accuracy of the 5-point difference approximation with variable coefficients has been studied by Bramble, Hubbard, Kellogg, and Thomée, under weakened assumptions of smoothness on the boundary.[43] For $u \in C^4(\Omega) \cap C^2(\bar{\Omega})$, for example, they proved $O(h^2)$ accuracy. Finally, the $O(h^2)$ convergence of *all* difference quotients to the appropriate derivatives was proved for the Laplace DE on a square mesh by V. Thomée in Birkhoff-Varga [70, pp. 186–200], and by Achi Brandt.[44] Making stronger smoothness assumptions, Thomée [69] also showed that difference quotients converge at the same rate as the solution in the interior (giving discrete Schauder-type inequalities).

The effect of singularities was also studied by Bramble, Hubbard, and Zlamal, who obtained error bounds for the Poisson equation.[45] They showed that for singularities like r^ν, the global error is $O(h^{\nu-\epsilon})$, ϵ any positive number. The numerical results of Example 6′ agree with this: for singularities $r^{2/3}$ and $r^{1/2}$, our results indicate $O(h^{0.64})$ and $O(h^{0.38})$ error, respectively.

Also, V. Thomée has proved convergence of order $O(h^{1/2})$ for simple difference approximations to the Dirichlet problem for any linear, constant-coefficient equation of elliptic type, and McAllister has obtained global error bounds for difference approximations to certain mildly nonlinear elliptic problems.[46] Later, Kuttler treated the eigenvalue problem.[47] Finally, in Birkhoff-Varga [70, pp. 201–209], Bramble has shown that one can reduce the error of difference approximations to $\mathbf{L}[u] = f$ for uniformly elliptic $\mathbf{L}$,[48] by appropriately smoothing f.

---

[41] Including J. H. Bramble and B. Hubbard, Contributions to Differential Equations 2 (1963), 229–252; ibid. 3 (1963), 319–340; SIAM J. Numer. Anal. 2 (1965), 1–14; J. ACM 12 (1965), 114–123; Numer. Math. 4 (1962), 313–332; ibid. 9 (1966), 236–249.

[42] H. Price, Math. Comp. 22 (1968), 489–516. See also P. G. Ciarlet, Aequat. Math. 4 (1970), 338–352.

[43] J. H. Bramble, R. B. Kellogg, and V. Thomée, BIT 8 (1968), 154–173; Math. Comp. 23 (1969), 695–710. Further extensions are given by J. H. Bramble in Birkhoff-Varga [70, pp. 201–209]; V. Thomée, Contributions to Differential Equations 3 (1964), 301–324.

[44] A. Brandt, Math. Comp. 20 (1966), 473–499.

[45] J. H. Bramble, B. E. Hubbard, and M. Zlamal, SIAM J. Numer. Anal. 5 (1968), 1–25.

[46] V. Thomée, Contributions to Differential Equations 3 (1964), 301–324. G. T. McAllister, J. Math. Anal. Appl. 27 (1969), 338–366.

[47] J. R. Kuttler, SIAM J. Numer. Anal. 7 (1970), 206–232.

[48] The DE (1.1) in Chapter 1 is said to be *uniformly* elliptic when there is a positive lower

10. Higher-order accuracy. In most cases, standard difference approximations give the highest ('optimal') order of accuracy that is attainable without going to a larger or more complicated stencil. Thus it is impossible to achieve more than $O(h^2)$ accuracy by a 5-point approximation to ∇^2: one cannot match more than the five coefficients corresponding to u, u_x, u_y, u_{xx}, and u_{yy} in the Taylor series expansion of u. Moreover, for $u(x,y) = c(x-x_i)(y-y_j)$ with c an arbitrary constant, $u_{xy} = c$ and u vanishes at all five stencil points (x_i, y_j), $(x_i \pm h, y_j)$, and $(x_i, y_j \pm h)$. Hence one cannot express u_{xy} even approximately in terms of the 5 values of u in (1.1). For this reason, difference approximations to elliptic problems in which u_{xy} enters (such as (1.1) of Chapter 1) must normally use a 9-point formula. With such a stencil, one has

$$u_{xy} = (u_{i+1,j+1} - u_{i-1,j+1} + u_{i-1,j-1} - u_{i+1,j-1})/(4h^2) + O(h^2).$$

On the other hand, one can always approximate any derivative of a very smooth function with arbitrarily high-order accuracy by using stencils with enough mesh-points;[49] this follows from Taylor's formula with remainder. We now illustrate this general principle by a few useful examples.

Laplace operator. Nine-point formulas having $O(h^6)$ accuracy on a square stencil are especially simple and easy to construct for the Laplace equation. Because the Laplacian is invariant under rotations, the difference quotient operator X_h, constructed by rotating the standard 5-point formula through 45°, satisfies

$$\text{(10.1)} \quad \begin{aligned} X_h[u_{i,j}] &= (u_{i+1,j+1} + u_{i-1,j+1} + u_{i-1,j+1} + u_{i+1,j-1} - 4u_{i,j})/2h^2 \\ &= \nabla^2 u(x_i, y_j) + h^2(u_{xxxx} + 6u_{xxyy} + u_{yyyy})_{i,j}/12 + O(h^4). \end{aligned}$$

The operator X_h also approximates ∇^2 with $O(h^2)$ accuracy (truncation error), hence so does any linear combination $\alpha\nabla_h^2 + \beta X_h$, $\alpha + \beta = 1$; but none has $O(h^4)$ accuracy. Nevertheless, one can achieve $O(h^6)$ accuracy for the *Laplace equation* with the 9-point difference approximation N_h defined by

$$\text{(10.2)} \quad \begin{aligned} N_h[u_{i,j}] &= [2\nabla_h^2 u_{i,j} + X_h u_{i,j}]/3 \\ &= [4(u_{i+1,j} + u_{i,j+1} + u_{i-1,j} + u_{i,j-1}) + u_{i+1,j+1} \\ &\quad + u_{i-1,j+1} + u_{i-1,j-1} + u_{i+1,j-1} - 20u_{i,j}]/6h^2, \end{aligned}$$

which has the stencil

bound for the difference $AC - B^2$ for all points of the region Ω. If $A, B, C \in C^2(\bar{\Omega})$, then elliptic implies uniformly elliptic.

[49] A useful compendium of such approximations is contained in the tables at the end of Collatz [60]; see also W. G. Bickley et al., Proc. Roy. Soc. London A262 (1961), 219–236; D. J. Panov, *Formulas for the Numerical Solution of Partial Differential Equations*, Ungar, 1963.

$$\mathbf{N}_h = \frac{1}{6h^2}\begin{bmatrix} 1 & 4 & 1 \\ 4 & -20 & 4 \\ 1 & 4 & 1 \end{bmatrix}.$$

One achieves $O(h^6)$ accuracy because, as in [KK, p. 184],

$$\text{(10.3)}\quad \mathbf{N}_h[u_{i,j}] = \left[\nabla^2 u + \frac{h^2}{12}\nabla^4 u + \frac{h^4}{360}\left(\nabla^6 u + 2\frac{\partial^4 \nabla^2 u}{\partial x^2 \partial y^2}\right)\right]_{i,j} + O(h^6),$$

which simplifies to $N_h[u] = O(h^6)$ if $\nabla^2 u = 0$.

Birkhoff-Gulati [74] observe that $u = \mathrm{Re}\{z^8 - 3h^4 z^4 - 4h^8\}$ satisfies the Laplace equation with $u(0,0) = -4h^8$ and $u = 0$ at the eight mesh-points nearest the origin. Consequently, no 9-point approximation to the Laplace equation on a square mesh can have $O(h^8)$ accuracy.

Poisson equation. The following 9-point formula [KK, p. 186, (34)] approximates the Poisson DE $-\nabla^2 u = f(x,y)$ with $O(h^4)$ accuracy:

$$-\mathbf{N}_h[U] = f + h^2(\nabla_h^2 f)/12;$$

see also Birkhoff-Gulati [74, §11]. By including values of f at four more mesh points, Rosser [75] obtained an $O(h^6)$ approximation.

Mehrstellenverfahren. Similarly, (10.3) shows that if $u \in C^8$ is a solution of the *Poisson equation* $\nabla^2 u = f$, then the solution of

$$\text{(10.4)}\qquad \mathbf{N}_h U_{i,j} = f_{i,j} + \frac{h^2}{12}\nabla^2 f_{i,j} + \frac{h^4}{360}\left[\nabla^4 f_{i,j} + 2\frac{\partial^4 f_{i,j}}{\partial^2 x\, \partial^2 y}\right]$$

is an $O(h^6)$ approximation to u. Collatz [60, p. 384] calls an approximation such as (10.4), which includes values of derivatives of the source term, a 'Mehrstellenverfahren'.[50]

Using analogous power series expansions, Young-Dauwalder [65][51] derived difference approximations for general second-order linear DE's

$$\text{(10.5)}\qquad \mathbf{L}[u] = Au_{xx} + 2Bu_{xy} + Cu_{yy} + Du_x + Eu_y + Fu = G,$$

with a 9-point stencil of points (ih, jh), $i,j = -1,0,1$. To get the coefficients of

$$\text{(10.6)}\quad \frac{1}{h^2}\sum_{i,j}\alpha_{i,j}U_{i,j} = \sum_{k,l}\beta_{k,l}\,\partial^{k+l}G(0,0)/\partial x^k \partial y^l, \qquad \beta_{0,0} = 1,$$

having highest-order accuracy, they replaced G with $\mathbf{L}[u]$, and expanded

[50] 'Mehrstellenverfahren' is translated as 'Hermitian method' in Collatz [60].

[51] See also D. M. Young and R. T. Gregory, *A Survey of Numerical Mathematics*, vol. 2, Addison-Wesley, 1973, pp. 988–991.

u and the coefficient functions $A, \cdots, F$ in Taylor series. After rearrangment and collecting terms, this gives a series of the form

$$\gamma_{0,0}u(0,0)+\gamma_{1,0}u_x(0,0)+\gamma_{0,1}u_y(0,0)+\gamma_{2,0}u_{xx}(0,0)+\cdots=0.$$

For smooth u, one can write

$$\gamma_{mn}=\sum_{i,j}\delta_{m,n,i,j}\alpha_{i,j}+\sum_{k,l}\epsilon_{m,n,k,l}\beta_{k,l}=0, \tag{10.7}$$

with coefficients δ, ϵ depending linearly on derivatives of $A, \cdots, F$. Young and Dauwalder obtained $O(h^M)$ difference approximations by solving these equations for $m+n=0,1,\cdots,M$ to obtain $O(h^M)$ estimates of the coefficients α, β. They also give tables expressing the coefficients α, β in terms of derivatives of $A, \cdots, F$.

They showed that in general, if $A \neq C$ and A, B, C are nonzero, then only $O(h^2)$ accuracy can be achieved with such a 9-point stencil. However, if $A \equiv C$ or if $B \equiv 0$, then $O(h^4)$ accuracy can be obtained. Finally, if $A \equiv C \neq 0$, while $B \equiv 0$ *and* $(D/A)_y=(E/A)_x$, then $O(h^6)$ accuracy can be secured. In the last case, the DE is equivalent to the self-adjoint equation

$$\nabla\cdot[\mu A\nabla u]+\mu Fu=\mu G,$$

where the integrating factor μ satisfies $(\mu A)_x=\mu D$ and $(\mu A)_y=\mu E$.

One can avoid evaluating derivatives of f in (10.4) by using divided difference approximations [KK, p. 185]; $O(h^4)$ approximation of the Laplacian of f and $O(h^2)$ approximations to the fourth derivatives of f are needed to maintain the $O(h^6)$ accuracy. Rosser [75] has constructed such approximations in terms of values of f at mesh-points in the interior and on the boundary of a rectangular domain. A more systematic approach is discussed in the next section.

Other formulas having $O(h^4)$ accuracy are known. For example [KK, p. 179], the formula

$$\frac{1}{12h^2}[16(u_{i+1}+u_{i-1})-(u_{i+2}+u_{i-2})-30u_i]=u''(x_i)+O(h^4)$$

leads to a 9-point cross-shaped stencil (see Bramble-Hubbard [64a]) giving $O(h^4)$ approximation to $\nabla^2 u(x_i, y_j)$. Unfortunately, although such higher-order approximations are mathematically attractive, the associated large stencils usually lead to serious practical complications near the boundary. This is especially true of formulas involving 9-point cross-shaped stencils, which only begin to be worthwhile on extremely fine meshes.

Richardson extrapolation. Instead of using larger stencils, one can often achieve improved accuracy by *a posteriori* analysis, using Richardson extrapolation (§7) or Fox's method of difference correction (§9). For example, one can sometimes compare nodal values from uniform meshes having different values of h. By comparing the computed values $U_{2h}(\mathbf{x}), U_h(\mathbf{x}), \cdots$ at the *same mesh-point* $\mathbf{x}$ in Ω, one can often reduce the error there.

Thus, in many cases, the approximate value at $\mathbf{x}$ satisfies

$$U_h(\mathbf{x}) = u(\mathbf{x}) + M(\mathbf{x})h^\nu + O(h^\mu), \qquad \nu < \mu, \tag{10.8}$$

for some *known* ν, with $M(\mathbf{x})$ independent of h. In such cases, having computed $U_{2h}(\mathbf{x})$ and $U_h(\mathbf{x})$, one can eliminate the $O(h^\nu)$ error term because

$$u = U_h + (U_h - U_{2h})/(2^\nu - 1) + O(h^\mu), \tag{10.9}$$

where we have suppressed the dependence on $\mathbf{x}$. Hence, *if* we know ν, we can obtain a more accurate value at the grid points on a coarser mesh. $(x_{2,i}, y_{2,j})$ as with Richardson extrapolation.

Even if we do not know ν, we can estimate it experimentally from the formula

$$\nu \approx \log_2 [(U_{2h} - U_h)/(U_h - U_{h/2})], \tag{10.10}$$

i.e., by comparing the changes in the computed approximate values U_{2h}, U_h, and $U_{h/2}$ at a few points.

In practice, it is often preferable to weaken (10.8) to

$$e_h(\mathbf{x}) \approx M(\mathbf{x})h^\nu, \tag{10.11}$$

regarding both $M(\mathbf{x})$ and ν as *empirical* approximations. For any fixed ν and $h' = \theta h$ with $\theta > 1$, clearly (10.11) implies

$$u(\mathbf{x}) \approx U_h(\mathbf{x}) + [U_h(\mathbf{x}) - U_{h'}(\mathbf{x})]/(\theta^\nu - 1). \tag{10.12}$$

This formula, with $\theta > 1.25$ and a conservative guess for ν, almost always gives improved results. With three mesh sizes h, $h' = \theta h$, and $h'' = \theta' h$ $(\theta' > \theta)$, one can estimate ν empirically since

$$R_h = \frac{[U_{h'}(\mathbf{x}) - U_{h''}(\mathbf{x})]}{[U_h(\mathbf{x}) - U_{h'}(\mathbf{x})]} \approx \frac{[(\theta'/\theta)^\nu - 1]}{(1 - 1/\theta^\nu)} \approx \left[\frac{\theta'}{\theta}\right]^\nu,$$

from (10.11), whence

$$\nu \approx [\log(R_h)]/[\log(\theta'/\theta)]. \tag{10.13}$$

Variations with $\mathbf{x}$ in the value of $\nu = \nu(\mathbf{x})$ computed using (10.13) are obvious signs of unreliability.

11. The HODIE method. Finally, we describe a method for constructing higher-order difference approximation to linear elliptic DE's, named the HODIE method by its authors.[52] It yields 9-point difference approximations to linear DE's like (10.5):

$$\mathbf{L}[u] = Au_{xx} + 2Bu_{xy} + Cu_{yy} + Du_x + Eu_y + Fu = G. \tag{11.1}$$

We first describe how HODIE approximates (11.1) at an interior mesh-point of a square mesh, surrounded by four mesh squares of side h as in Fig. 9a.

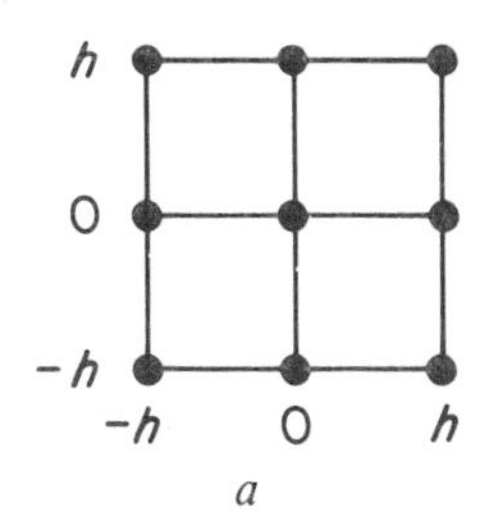

a

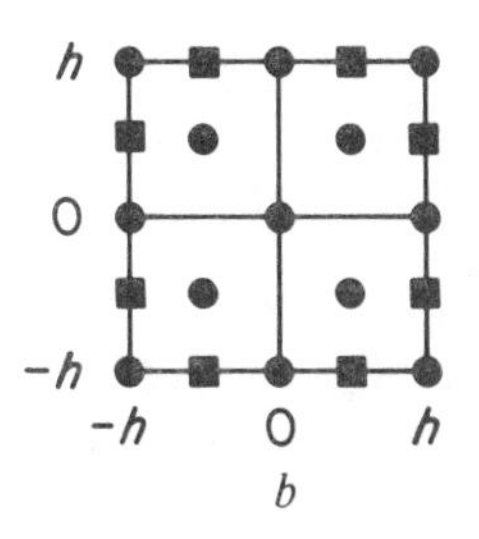

b

FIGURE 9

For unknowns, the HODIE method uses the approximate values $U_{i,j}$ of u at the nine mesh-points dotted in Fig. 9a; these are called 'stencil' points. In addition, the average value of the source term is approximated by a suitable linear combination of values of G at K 'evaluation' (or 'auxiliary') points (x_k, y_k), $k = 1, \cdots, K$, located in the four squares of Fig. 9a: $x_{i,k} = (i + \xi_k)h$ and $y_{j,k} = (j + \eta_k)h$, where $|\xi_k| \leqslant 1$ and $|\eta_k| \leqslant 1$.

The equation associated with the mesh-point $(x_i, y_j) = (ih, jh)$ is

$$\mathbf{H}_h[U_{i,j}] = \mathbf{I}_h[G_{i,j}], \tag{11.2}$$

where $\mathbf{H}_h$ and $\mathbf{I}_h$ have the forms

$$\mathbf{H}_h[U_{i,j}] = (1/h^2) \sum_{m=-1}^{1} \sum_{n=-1}^{1} \alpha_{m,n} U_{i+m,j+n}, \tag{11.2'}$$

and

$$\mathbf{I}_h[G_{i,j}] = \sum_{k=0}^{K} \beta_k G(x_{i,k}, y_{j,k}). \tag{11.2''}$$

The coefficients $\alpha_{m,n}$ and β_k are chosen to make $\mathbf{H}_h[s] = \mathbf{I}_h[\mathbf{L}[s]]$ for all $s \in S$, some $(K+8)$-dimensional linear space of functions.

---

[52] Lynch-Rice [78]; also, R. E. Lynch and J. R. Rice in de Boor-Golub [78, pp. 143–175]. See also R. F. Boisvert, SIAM Sci. Stat. Comp. 2 (1981), 268–284; P. Henrici in *Studies in Numerical Analysis*, G. H. Golub (ed.), Math. Assoc. Amer., 1984.

As a special case, the HODIE method gives some difference approximations to the Poisson DE $-\nabla^2 u = f(x,y)$ (on a square mesh) derived earlier by Robin Esch[53] and S. Gulati. Esch matched terms in Fourier series. Gulati's HODIE-type scheme used 9 evaluation points in the four mesh squares of Fig. 9a, located at the centroid (ih, jh) and at the corners and midpoints of a square of side $\sqrt{2/5}\,h$. His approximation used the 9-point stencil $\mathbf{N}_h$ of (10.2), and is exact on the space of 7th degree polynomials. This gives

$$-\mathbf{N}_h[U_{i,j}] = \frac{1}{72}\left[32f(ih,jh) + 5\sum_{k=2}^{9} f(x_k, y_k)\right].$$

The preceding formula has $O(h^6)$ accuracy if $u \in C^8$, but requires about four times as many evaluations of f as the following scheme.

Example 10. For the Poisson equation, $-\nabla^2 u = f(x,y)$, we take $\mathbf{H}_h = -\mathbf{N}_h$ to be the 9-point approximation of (10.2) and $\mathbf{I}_h$ as

$$\mathbf{I}_h[f] = \frac{1}{360}\left[148f + 4\sum_{P(1)} f + 48\sum_{P(1/2)} f + \sum_{P(2)} f\right].$$

Here $P(m)$ denotes the set of all grid points, on a mesh with spacing $h/2$ at a distance $\sqrt{m}\,h$ from (ih, jh), i.e., the 13 evaluation points denoted by circles in Fig. 9b. Since $\mathbf{N}_h[u] = \mathbf{I}_h[\nabla^2 u]$ for all polynomials of degree seven, and $\mathbf{N}_h[u] = \mathbf{I}_h[\nabla^2 u] + O(h^6)$ in general, the discretization error

$$(11.4)\qquad \sum_{k=2}^{K} \beta_k \mathrm{L}[s_l(\xi_k,\eta_k)] = -\mathrm{L}[s_l(\xi_1,\eta_1)], \quad l=1,\cdots,K-1.$$

Having the β's, one then calculates the α's by solving

$$(11.4')\quad \sum_{m=-1}^{1}\ \sum_{n=-1}^{1} \alpha_{m,n}(s_l)_{i+m,j+n} = h^2(\mathrm{I}_h[\mathrm{L}[s_l]])_{i,j}, \quad l=K,\cdots,K+8.$$

Example 11. Consider the one-dimensional convection-diffusion equation

$$-\epsilon u_{xx} + u_x = f(x);$$

discussed in §6. The general solution of the homogeneous DE is $a+b\,e^{x/\epsilon}$. On a uniform mesh, the $O(h^2)$ divided central difference approximation is given by Eq. (6.12) of §6, and $U_j = A + Br^{jh/\epsilon}$ is the general solution of the homogeneous ΔE. For $h < 2\epsilon$, the coefficients of U_{j-1} and U_{j+1} are negative and $r > 1$ is approximately equal to e, with the accuracy increasing as h decreases. However, for $h > 2\epsilon$, the coefficients have different signs and $r < 1$, so that, in general, U_j does not give a good approximation of $u(jh)$.

E. C. Gartland[55] observed that with the HODIE method, an $O(h^2)$ (unconditionally stable) ΔE of *monotone type* can be obtained. For a single evaluation point, the equations which give the coefficients are

$$\alpha_{-1}s_k(-h) + \alpha_0 s_k(0) + \alpha_1 s_k(h) = s_k''(\tau h) + s_k'(\tau h), \qquad k=0,1,2,3.$$

With the basis 1, x, x^2, $x^3 - xh^2$, the equation for the third basis element is

$$0 = 3\tau^2h^2 - 6\epsilon\tau h - h^2$$

The choice $\tau = \epsilon - \epsilon\sqrt{1+h^2/3\epsilon^2}$ makes the right side zero (the choice $\tau = 0$ makes the right side $O(h^2)$ and leads to the divided central difference approximation). The difference equation is

$$(11.5)\qquad \begin{gathered}(-\sigma-\eta)\,U_{i-1} + 2\sigma\,U_i + (-\sigma+\eta)\,U_{i+1} = f(jh+\tau h),\\ \sigma = (\epsilon/h^2)\sqrt{1+h^2/3\epsilon^2}, \qquad \eta = 1/2h,\end{gathered}$$

with $-|\sigma|+|\eta| < 0$ for all h and $\epsilon \neq 0$.

Either equation (6.13) or (11.5) leads immediately to a monotone approximation to the two-dimensional equation

$$-\epsilon\nabla^2 u + u_x = f(x,y)$$

on a rectangular mesh; thus (11.5) gives

[55] See E. C. Gartland, SIAM J. Sci. Stat. Comput. 3 (1982), 460–472, for related work for the two-dimensional convection-diffusion equation.

$$(11.6)\qquad \begin{aligned}(-\sigma-\eta)\,U_{i-1,j}+(-\sigma+\eta)\,U_{i+1,j}-(\epsilon/k^2)(U_{i,j-1}+U_{i,j+1})\\ +(2\sigma+2\epsilon/k^2)\,U_{i,j}=f(jh+\tau h,jk).\end{aligned}$$

We note that this cannot, in general, have discretization error less than $O(h^2)$. Because, with

$$\begin{aligned}U(x,y)&=u(x,y)\\ &=x^4-6x^2y^2+y^4+4\tau^3h(h^2-x^3+3xy^2)/(1-3\tau^2)-h^4,\end{aligned}$$

and $k=h$, the left side of (11.6) is $2\sigma h^4+2\epsilon h^2=O(h^2)$, whereas the right side is zero.

Example 12. Similar methods give an approximation for the three-dimensional Poisson equation $-\nabla^2 u=f(x,y,z)$. One achieves $O(h^6)$ accuracy on a cubic mesh with mesh-points (ih,jh,kh) by setting

$$\mathbf{H}_h[U]=\frac{1}{30h^2}\left[14\sum_{P(1)}U+3\sum_{P(2)}U+\sum_{P(3)}U-128\,U\right],$$

$$\mathbf{I}_h[f]=\frac{1}{720}\left[8\sum_{P(1)}f+48\sum_{P(3/4)}f+\sum_{P(3)}f+280f\right],$$

where $P(m)$ denotes the set of all grid points, on a mesh with spacing $h/2$, at a distance $\sqrt{m}\,h$ from (ih,jh,kh).

Variable coefficients. In the variable coefficient case, different difference equations (11.2) are required for each of the $N=O(I^2)$ interior mesh-points. That is, systems of the form (11.4)-(11.4′) must be solved $O(I^2)$ times to obtain numerical values for the coefficients $\alpha_{m,n}$ and β_k. One then gets a system of $O(I^2)$ difference equations (11.2) to be solved for the nodal values U_{ij}; the work (number of arithmetic operations) in solving this system in $O(I^2)$ unknowns asymptotically dominates the work required to compute its coefficients. Because the resulting system is not symmetric, the total work to compute the unknowns $U_{i,j}$ is about twice that of solving the (symmetric) standard 5-point discretization.

Example 13. Consider the DE

$$-(e^{xy}u_x)_x-(e^{xy}u_y)_y+2(x^2+y^2)e^{xy}u=G(x,y)$$

on the unit square $[0,1]\times[0,1]$ with Dirichlet boundary conditions. After the indicated differentiations are carried out, one has an equation of the form (11.1). The $K=21$ evaluation points indicated by circles and squares in Fig. 9b were used to construct HODIE schemes with S the space of polynomials of degree at most 6. To compute the error $e_h=\max_{i,j}|u(ih,jh)-U_{i,j}|$, Lynch-Rice [78] set $u=e^{xy}$ in numerical experiments, thus making $G\equiv 0$ in the DE displayed above. Table 5 shows some of their results for the HODIE scheme and for the standard 5-point approximation; these results indicate $O(h^7)$ discretization error.

Many of their other experiments, with different operators and exact solutions, showed similar results, generally with at most $O(h^6)$ accuracy.[56]

TABLE 5. *Error* $e_h = \|u - U_h\|_\infty$ *for Example* 13.

	Standard 5-point		HODIE	
h	e_h	e_h/h^2	e_h	e_h/h^7
1/4	4.2×10^{-4}	0.0068	4.9×10^{-6}	0.085
1/6	2.2×10^{-4}	0.0078	3.3×10^{-7}	0.092
1/8	1.3×10^{-4}	0.0082	4.6×10^{-8}	0.097
1/10	8.5×10^{-5}	0.0085	9.8×10^{-9}	0.098

Example 14. Consider the convection-diffusion equation

$$\mathbf{L}[u] = -\frac{1}{50}(u_{xx} + u_{yy}) + u_x = f(x,y) \tag{11.7}$$

on the unit square, with

$$f(x,y) = -\pi^2(1 - e^{50x}/e^{50}) \sin \pi y.$$

For appropriate Dirichlet boundary conditions, the solution is

$$u(x,y) = (1 - e^{50x}/e^{50}) \sin \pi y. \tag{11.8}$$

Along $y = 1/2$, u decreases from unity to about 0.9 as x increases from 0 to 0.9. Then in a 'boundary layer', it decreases sharply to zero at $x = 1$. Consequently, one does not expect much accuracy on a square mesh for h larger than 1/10.

In self-adjoint form, (11.7) is

$$\mathbf{M}[u] = -(p(x)u_x)_x - (p(x)u_y)_y = 50p(x)f(x,y), \tag{11.9}$$

where $p(x) = e^{-50x}$. On a square mesh, the standard 5-point approximation is, after multiplying by h,

$$\begin{aligned}\mathbf{M}_h[U] = {}& e^{-50jh+25h}(-U_{j,k-1} + U_{j,k}) + e^{-50jh-25h}(U_{j,k} - U_{j+1,k}) \\ & + e^{-50jh}(-U_{j,k-1} + 2U_{j,k} - U_{j,k+1}) = 50h^2e^{-50jh}f(jh,kh).\end{aligned} \tag{11.10}$$

Direct substitution shows that 1, e^{50x}, y, and $y\,e^{50x}$ are in the null spaces of $\mathbf{L}$, $\mathbf{M}$, and $\mathbf{M}_h$.

We also note that a two-dimensional analogue of Southwell's scheme (6.13) can be obtained by replacing the right side of (11.10) with

$$2h \sinh(25h)\, e^{-50jh} f(jh,kh), \tag{11.11}$$

[56] R. E. Lynch and J. R. Rice in de Boor-Golub [78, pp. 143–175].

and then dividing the equation by e^{-50jh}. Since

$$50h^2 e^{-50x} = -(e^{-50x})' = 2h \sinh(25h)\, e^{-50x}(1+O(h^2)),$$

(11.11) is an $O(h^2)$ approximation to the right side of (11.10).

Table 6 lists errors for the divided central difference approximation of (11.7), Allen's scheme, the standard 5-point approximation to (11.9), and the HODIE approximation to (11.7).

TABLE 6. *Max* $|u_{j,k} - U_{j,k}|$ *for Example* 14.

$n = 1/h$	divided difference	Allen's scheme	Standard 5-point	HODIE
3	.521	.113	.000376	.623
6	.661	.142	.000475	1.53
12	.362	.0803	.000507	.0292
24	.144	.0265	.000197	.000905
48	.0376	.00718	.0000555	.0000167

As predicted theoretically, the asymptotic rates of convergence ($O(h^2)$ for the first three schemes, $O(h^6)$ for HODIE) are not observed unless h is significantly less than 1/10.

An examination of more data shows that as h decreases, the error first oscillates, and then decays monotonically. For example, for the divided difference approximation, $||e_h|| = 2.9, 0.52, 1.1, 0.49, 0.66$, for $1/h = 2, 3, 4, 5, 6$, respectively; $||e_h||/h^2 = 24, 52, 72, 89, 87$, for $1/h = 6, 12, 18, 36, 48$, respectively.

Chapter 4

Direct and Iterative Methods

1. Direct methods. In Chapter 3, §5, we noted that difference approximations to linear elliptic problems can often be solved in a straightforward way on high-speed computers by *band elimination*. This chapter will describe some alternative, usually more efficient methods for solving them. Broadly speaking, they can be classified into *direct* methods, which (neglecting roundoff) give the solution in a finite number of steps, and *iterative* methods, which compute an infinite sequence of better and better approximations.

On the first generation of large-scale computers, 'direct' methods such as band elimination were seldom used for solving large linear systems $A\mathbf{u}=\mathbf{b}$ arising from elliptic difference equations. (At the time, 'large' meant that the number N of unknowns was over 1000.) This was partly because of the limited storage of contemporary computers: on an $I\times J$ mesh, to store even the symmetric matrix associated with a 5-point difference approximation requires, after fill-in by band elimination, storing about $I^2J=NI$ numbers. Thus, if $I=30$, and $J=50$, then $I^2J=45000$. Computers having that much core storage were rare until 1965 or so.

In contrast, *iterative* methods to be described in §§7–15, such as Gauss-Seidel and SOR, require storing only about $7N$ numbers (or $4N$ if the matrix is symmetric and has diagonal entries unity). Therefore, large elliptic problems were almost always solved by *iterative* (or 'semi-iterative') methods.

However, especially for the 'finite element' approximations widely used since 1965 in structural mechanics (see Chapter 7), *direct* methods have become adopted increasingly for solving medium-sized elliptic problems in the last 15 years. For larger problems, 'hybrid' combinations of direct *and* iterative methods (such as 2-line overrelaxation, conjugate gradient acceleration, strongly implicit, etc.; see Chapter 5) are often more efficient than either exclusively 'direct' or exclusively 'iterative' methods.

Because of this, whereas older books on the numerical solution of elliptic problems paid little attention to direct elimination methods, nearly half of this chapter will be devoted to them. In Chapter 3, §5, we discussed the (direct) methods of band and band-Cholesky elimination for solving $A\mathbf{u} = \mathbf{b}$; in this chapter we will describe other direct methods. The methods described are all faster than band elimination for typical 'sparse' matrices arising from linear source problems. Actually, in the range $300 < N < 3000$, the costs of good direct and good iterative methods usually differ by a factor less than two (in our experience).[1] When higher-order (difference *or* finite element) approximations are used, the gain in efficiency tends to be greater for direct than for iterative methods, because their operation count is less sensitive to matrix sparsity.

Cost estimation. Much of the cost of scientific computing consists in performing arithmetic operations, so-called 'number crunching'. When *band-Cholesky elimination* is used to solve $A\mathbf{u} = \mathbf{b}$ for a large elliptic problem on an $I \times J$ mesh, for example, the number of multiplications is nearly I^3J, or I^2 per unknown. The number of additions is about the same, while the number of divisions is about the number of unknowns, $IJ = N$. In a computer renting for \$900/hour (25¢/second), and capable of performing 10^6 multiplications or additions per second, this suggests that the time required to solve by band-Cholesky the difference equations associated with a 25×40 rectangle might therefore be about 1.5 seconds, and the cost about 40¢.

However, such estimates are very rough. There is unfortunately no simple formula for estimating reliably the cost of solving such equations. In particular, the cost of 'number-crunching' may be much less than that of storage (memory allocation), transferring data (fetching and storing), input/output, 'connect time' for a terminal, etc. Therefore, the actual cost of solving the above problem might well be nearer \$4 than 40¢.

Comparative operation counts. Just as band elimination is a very simple, moderately efficient way to solve systems of simultaneous linear equations arising from elliptic problems *directly*, so the Young-Frankel SOR method to be discussed in §13 is a well-tested and quite efficient way to solve them *iteratively*. We will next make some (necessarily very rough; cf. also §4) comparisons of the efficiencies of band elimination and SOR.

The comparative efficiency of direct and iterative methods is highly problem-dependent. For example, it depends greatly on the dimension n of the domain $\Omega \subset \mathbb{R}^n$ in which a given elliptic problem is to be solved. Thus, if band elimination is used to solve $A\mathbf{u} = \mathbf{b}$, the operation count per unknown is roughly β^2, where β is the band half-width. Hence this count is asymptotically $O(1)$ for two-endpoint problems ($n = 1$); it is $O(N)$ in

[1] Somewhat lower 'crossover' values of N were reported by G. J. Fix and K. Larsen in SIAM J. Numer. Anal. 8 (1971), 345–363.

the plane (where $n = 2$ and β is proportional to $1/h$, or $N^{1/2}$), and $O(N^{2-2/n})$ for linear elliptic problems in $\mathbb{R}^n$. Since $N = O(h^{-n})$, the 'operation count' per unknown is $O(h^{2-2n})$.

In two dimensions, for solving larger linear source problems (with $N > 5000$, say), iterative methods become increasingly advantageous (see also §7). Thus band elimination requires $O(N^2) = O(I^4)$ multiplications to solve $A\mathbf{U} = \mathbf{b}$, for A the $I^2 \times I^2$ matrix of band half-width I associated with a second-order elliptic problem on an $I \times I$ mesh (see Chapter 3, §4). In contrast, some iterative schemes, such as the SOR method of §13, require only $O(I^3) = O(N^{3/2})$ multiplications (and far less storage) to solve the same problem. Hence the ratio of costs should change by a factor of about $10^{1/2} \approx 3$ over the range $300 < N < 3000$: from 40% less to 80% more, say, depending very much on the problem.

In three dimensions, the band-width is $O(N^{2/3})$, and so iterative methods become preferable to band elimination for much smaller N. This is because, for an $I \times aI \times bI$ mesh, with $1 \leqslant a \leqslant b$, the band-width is aI^2. Hence, the multiplication count is a^3bI^7, or a^2I^4 per mesh-point. This contrasts with 7 multiplications per mesh-point per iteration if SOR is used. If m iterations are required to give sufficient accuracy, the resulting asymptotic multiplication count is therefore $O(mI)$; the dependence of m on I (and the iteration scheme) will be discussed in the second half of this chapter; it is independent of dimension.

Moreover, for solving even the simplest *nonlinear* problems, it is almost always most effective (and even necessary!) to use iterative methods.[2]

On the other hand, a simple comparison of operation counts for band elimination and SOR by no means tells the whole story. For most problems, there are more efficient direct methods than band elimination and more efficient iterative methods than SOR. This chapter will describe some of these more efficient methods: §§2–6 will be devoted to direct, and the rest of the chapter to some simple iterative methods.

We will begin, in the next section, with 'profile' elimination (a pedestrian but useful improvement on band elimination). In §3, we will discuss some general ideas such as the graph of a matrix, reordering equations and unknowns, minimizing fill-in, and dissection methods. In particular, we will describe Alan George's method of nested dissection, which has the same order of complexity as the SOR method to be discussed later in this chapter.

Asymptotic formulas, useful for estimating the order of magnitude of the cost of achieving p-digit accuracy for large p, are the subject of §4. In §5, an unconventional 'marching' method will be discussed, which has a

[2] See, for example, Ortega-Rheinboldt [70]; Rheinboldt [74].

very low operation count but is also unstable (roundoff errors are amplified exponentially). A modified form of it is very efficient and less unstable for convection-diffusion problems.

In §6, we will discuss tensor product methods, the fast Fourier transform, and cyclic reduction. From §7 on, we will take up iterative methods.

2. Profile elimination. In the 1970's, much ingenuity was devoted to devising 'faster' direct algorithms for solving linear systems $A\mathbf{u} = \mathbf{b}$ having 'sparse' coefficient matrices.[3] (The 'sparsity' of a matrix A is, by definition, the fraction of its entries that are zero.) Generally speaking, the aim of such algorithms is to minimize the number of arithmetic operations ('operation count') and storage required to solve a linear system $A\mathbf{u} = \mathbf{b}$. We have already shown in Chapter 3, §5, how the operation count for Gauss elimination is reduced by band elimination; we will next show how to reduce it further.

The elimination process for a symmetric matrix requires about half as many multiplications as for a nonsymmetric one, and about half as much storage.

Consider the band elimination process for the standard 5-point discretization of the linear source problem on an $I \times J$ rectangular mesh, with the nodes in their 'natural' order. Although there are only 5 nonzero entries per unknown in the original system, after the elimination phase there are about $2I$ nonzero entries per unknown necessary to record the L and U factors of $A = LU$. (For three-dimensional problems on an $I \times J \times K$ mesh, the initial 7 entries per unknown grows to JK entries per unknown.) The coefficient-matrix of the problem originally had only $5N$ nonzero entries (or 0.5% of all entries for a 25×40 rectangle). 'Fill-in' during band elimination increases this fraction by a large factor (to about $50N$, or 5%, in the example cited); the operation count and storage requirements increase correspondingly. In this section, we will describe some algorithms that have been proposed for reducing them.

Profile elimination. One such algorithm is called *profile elimination* (or the 'envelope' method). In it, the nonzero entries of $A = \|a_{r,s}\|$ are bounded on either side ('enveloped') by a *profile*, as indicated in Fig. 1b. Namely, for each r, let $s'(r)$ be the least s and $s''(r)$ the greatest s such that $a_{r,s} \neq 0$. Then $a_{r,s} \neq 0$ implies $s'(r) \leqslant s \leqslant s''(r)$.

In rectangular domains, band elimination is reasonably efficient: relatively few zero multiples of rows are 'subtracted' from later rows during elimination. But this can happen quite often for a 5-point approximation on a $(2n-1) \times (2n-1)$ diamond-shaped domain with $2n^2 - 2n + 1$ interior mesh-points (see Fig. 1a for $n = 3$; the stars in Fig. 1b indicate the

[3] See Rose-Willoughby [72], Bunch-Rose [76]; Duff [77]; George-Liu [81].

nonzero entries in the matrix). For $i < n$, nodes $k = (i,j)$ in the i-th mesh row with $(i-1)^2+2 \leqslant s \leqslant i^2-1$, the first nonzero matrix entry $a_{r,s'}$ has $s' = r-2i-2$; the last nonzero matrix entry $a_{r,s''}$ has $s'' = r+2i$.

Therefore, a substantial saving is achieved by subtracting multiples of only those m-th rows of the matrix with $s'(r) \leqslant m < r$ from the r-th row, when $i \leqslant n$ (the subtraction has to be done only for columns k with $m < k \leqslant \min\{s''(m), s''(r)\}$). The case $i > n$ is similar, and the nonzero entries (obtained by fill-in and recording L of the LU factorization) occupy a thin region bounded by two *envelopes* (a 'left profile' and a 'right profile'), visible in Fig. 1b.

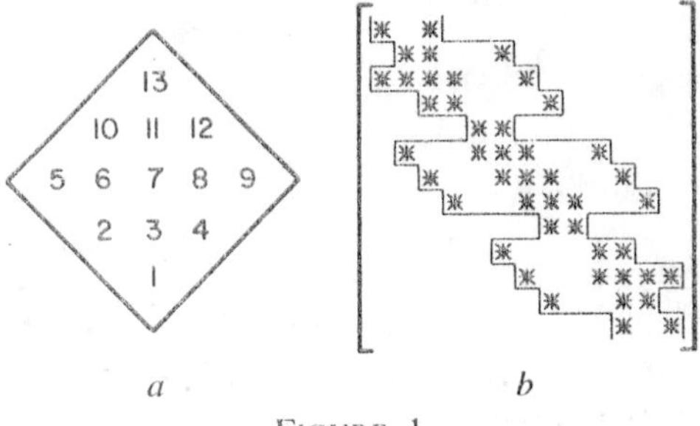

FIGURE 1

The preceding method of profile elimination, which is treated more thoroughly in George-Liu [81, Chap. 4], can be easily applied to any matrix; the case of linear source problems whose nodes are listed (or 'indexed') in their natural order is very special.

Node reordering. For many source problems, much greater savings (as compared with simple band elimination) can be achieved by properly reordering ('indexing') the nodes of the domain.

Consider for example the I-shaped region of Fig. 2, with the standard 5-point approximation to the linear source problem. Suppose that a rectangular mesh subdivides each of the regions A, B, C, D, into 10×6 mesh squares, while F is a 10×6 crossbar. Under the natural ordering by

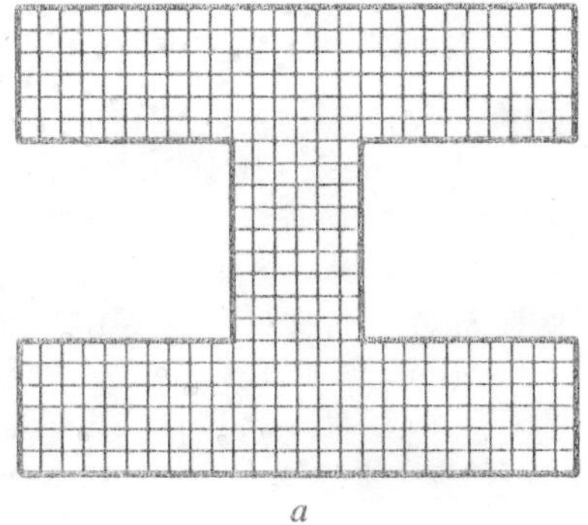

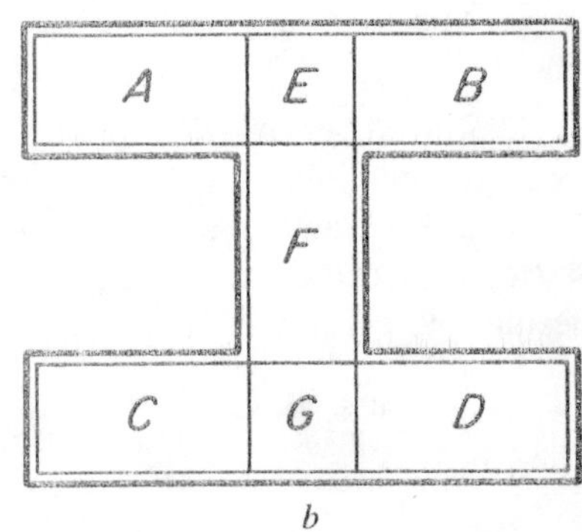

FIGURE 2

columns, the band half-width would be 21 for Dirichlet boundary conditions. Hence straightforward band elimination would require about

441 multiplications per mesh-point, or 134405 in all. Profile elimination would require about 40% of this.

Now let the nodes be *reordered* so that the $(10+5+10)\times 5\times 2$ interior nodes of A, E, C and B, G, D are eliminated first after about $250\times 5^2 = 6250$ multiplications (and as many additions). To eliminate the 45 interior nodes of F next, since the band half-width is also 5, will only require about $45\times 5^2 = 1125$ multiplications. This leaves 10 unknowns on the two interior segments separating E and G from F. These can then be eliminated after about $10^3/3 \approx 333$ multiplications (and as many additions). Summing, we get about 7708 multiplications, or an average of 25 per unknown, 6% of 441.

Stated another way, by using profile elimination after *reordering* the unknowns and corresponding equations of a linear system $A\mathbf{x}=\mathbf{b}$, one can sometimes achieve very large reductions in the operation count for Gauss elimination.

Many reordering schemes for reducing fill-in with profile elimination have been proposed. For reordering the nodes of subdivided regions, an algorithm of Cuthill-McKee has been quite successful; a 'reverse Cuthill-McKee' more so. Moreover, a more recent algorithm of Gibbs, Poole, and Stockmeyer seems to give even smaller profiles.[4] For a carefully worked out reordering algorithm (and related computer programs), see Chap. 4 of George-Liu [81].

Example 1. The choice of node order may also affect the *accuracy* of the process of elimination. Consider for example the DE $-u''+10u = x^3-x$ on $[0,1]$. Its exact solution is $0.1x^3-0.94x$ for the boundary conditions $u(0)=0$, $u(1)=-0.84$. One can reduce the roundoff error substantially on a mesh with 2^k intervals ($h=2^{-k}$) by the dyadic ordering[5]

$$1/2;\ 1/4, 3/4;\ 1/8, 3/8, 5/8, 7/8;\ 1/16, 3/16, \cdots,$$

if $h = 2^{-10}$ or so.

3. Graph of a matrix; nested dissection. In the last section, we showed by example that the 'natural ordering' (by mesh rows) of Chapter 3, §5, could be far from optimal. Even for an $I\times J$ mesh with J rows and $I \geqslant J$ columns, it is more efficient to index mesh-points using the 'natural ordering by columns,' because the 'fill-in' is less. This suggests the

---

[4] E. Cuthill and J. McKee, Proc. ACM 24th Nat. Conf. (1969), 157–172. See also E. Cuthill in Rose-Willoughby [72, pp. 157–166]; N. E. Gibbs, W. G. Poole, Jr., and P. K. Stockmeyer, SIAM J. Numer. Anal. 13 (1976), 236–250; and I. S. Duff and J. K. Reid, TOMS 5 (1979), 18–35.

[5] I. Babuska, SIAM J. Numer. Anal. 9 (1972), 53-77. For the corresponding *eigenvalue* problem, see Max D. Gunzburger and R. A. Nicolaides, ICASE Rep. 81-39 (Dec., 1981).

question: for a general difference approximation on a general domain, which ordering of the nodes (mesh-points) will *minimize* the fill-in? For the systematic study of this question, the concept of the (symmetric) 'graph' of a matrix is very helpful.

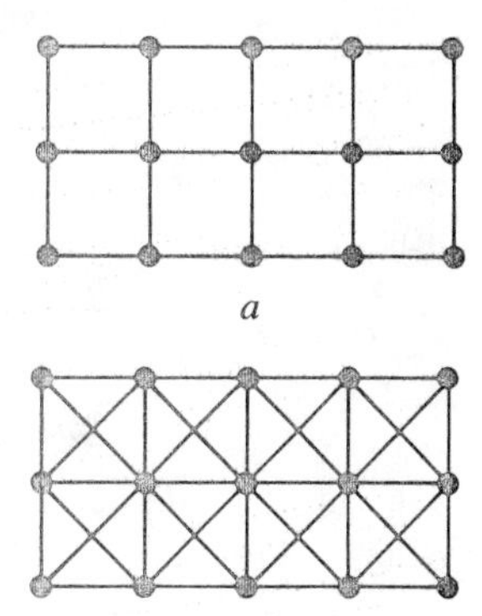
FIGURE 3

Graph of a matrix. In the elliptic ΔE's we have considered so far, each *equation* was clearly associated with a particular *unknown*, because the unknown was at the center of the 'stencil' giving the equation. Typically, this unknown U_k was a 'nodal value' $U_{i,j}$ at some mesh-point or 'node', and the ΔE was obtained from it by expanding in Taylor series about that point. In such cases, one can associate with the discretization a *graph* whose vertices (or 'nodes') are by definition the *unknown*, two vertices being joined by edges (or 'links') if and only if they occur in the same equation. Graphs of matrices associated with centered 5-point and 9-point stencils on a 3×5 rectangular network are depicted in Figs. 3a and 3b, respectively.

For the usual 5-point approximations to DE's of the form

$$Au_{xx} + Cu_{yy} + Du_x + Eu_y + Fu = G$$

on a rectangular mesh, the graph of the matrix is the same as the network of mesh lines and nodes. The same is true of the triangular subdivision of the hexagon of Chapter 3, §2, with the 7-point approximation used there.

In general, let A be any $N \times N$ square *matrix* in which $a_{i,j} \neq 0$ implies $a_{j,i} \neq 0$. Diagonally symmetrizable matrices, in which $S = DAD^{-1}$ is symmetric for some diagonal matrix D, are of this type. The *graph* of A is then the (symmetric) graph in which node i and node j are joined by an edge if and only if $a_{i,j} \neq 0$. In such cases, 'fill-in' by band elimination can often be reduced by *reordering* the nodes of this graph. This replaces $A\mathbf{u} = \mathbf{b}$ by a new vector equation $B\mathbf{v} = \mathbf{c}$, with $B = PAP^{-1}$, $\mathbf{v} = P\mathbf{u}$, and $\mathbf{c} = P\mathbf{b}$, P being a permutation matrix. Making such *reorderings* of the unknowns and corresponding equations of a linear system $A\mathbf{x} = \mathbf{b}$, one can sometimes achieve very large reductions in the operation count for Gauss elimination.

Red-black orderings. Five-point approximations to second-order elliptic problems are amenable to 'red-black' orderings that are useful for iterative (e.g., the CCSI method to be discussed in Chapter 5) as well as for direct methods. By a red-black ordering, we mean one in which all the nodes (x_i, y_j) with $i+j$ even precede all the nodes with $i+j$ odd. Since nodes with $i+j$ even can be thought of as the red squares of a checkerboard of red and black squares, this is called a *red-black ordering*. It is perhaps most 'natural' to order each of the two preceding sets of nodes by rows, but clearly *any* 'red-black' ordering will give rise to a matrix of the form

$$(3.1)\qquad A = D - G = \begin{bmatrix} D_1 & -G_2 \\ -G_1 & D_2 \end{bmatrix}, \quad D = \begin{bmatrix} D_1 & 0 \\ 0 & D_2 \end{bmatrix}, \quad G = \begin{bmatrix} 0 & G_2 \\ G_1 & 0 \end{bmatrix},$$

with D_1 and D_2 square diagonal matrices.

Cyclic reduction. Evidently, when Gauss elimination is applied to a matrix of the form (3.1), only the block D_2 is filled in. Hence solving $A\mathbf{x} = \mathbf{b}$ can be very economical, especially if D_1 is made larger than D_2. This is the simplest example of a class of processes referred to loosely as 'cyclic reduction'; another example will be discussed near the end of §6.

Bipartite graph. In (3.1), the diagonal blocks D_1 and D_2 evidently correspond to *complementary* sets R and B of nodes, with the property that every edge of the *graph* of A links a node of R with a node of B. Such graphs are called *bipartite*, and, conversely, every matrix whose graph is bipartite can be put in the form (3.1) by permuting rows and columns.

Dissection methods. We next describe some methods of 'dissection' developed by Alan George,[6] and explain how, for very fine rectangular meshes on plane domains, they can reduce by orders of magnitude the operation count of Gauss elimination over that for 'natural' ordering. They generalize the technique applied in §2 to an I-shaped region.

Dissection methods use a 'divide and conquer' strategy which first cuts a domain into two or more nearly equal subdomains (often along parallel mesh lines). They then eliminate variables corresponding to the interior of each subdomain separately, storing the equations which express them in terms of boundary values of that subdomain. Each subdomain boundary consists of this 'cut', together with part of the original boundary. Finally, nested dissection 'mops up', by solving for the unknown values on this dividing line, and obtaining from them the interior values in each subdomain by back-substitution.

Nested dissection. The preceding *one-way dissection* process can be iterated. Each of the above subdomains can again be cut in two or more pieces (usually by lines orthogonal to the initial dividing line or lines) and the process repeated as many times as desired, before elimination begins. This procedure is called 'nested dissection'. Figure 4 shows a

---

[6] See George [73] and George's paper in Rose-Willoughby [72, pp. 101–114]. We shall follow here the exposition of G. Birkhoff and A. George in Traub [73, pp. 221-270]; for a fuller exposition see George-Liu [81, Chaps. 7–8].

possible resulting order for eliminating the 21 unknowns corresponding to interior nodes in a rectangle divided into 4×8 squares, and the nonzero entries (marked by stars) in the resulting matrix, for a 9-point formula.

1	11	2	19	5	16	6
12	9	10	20	17	14	15
3	13	4	21	7	18	8

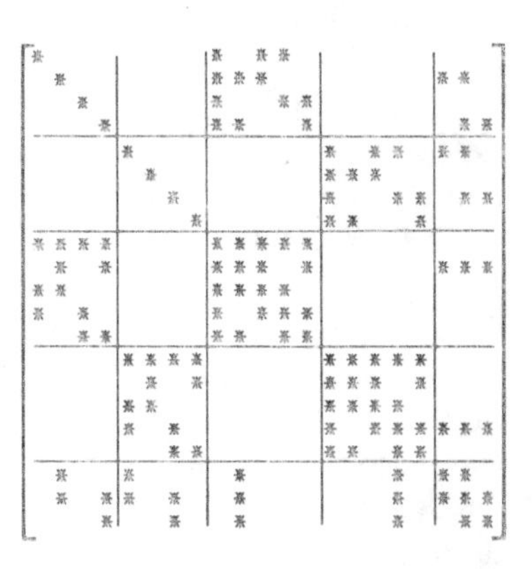

FIGURE 4

Like fast Poisson solvers (§6), nested dissection works most simply for partitions of rectangles into $2^m \times 2^{m+n}$ subrectangles. In that case, successive *bisections* lead in $m+n$ steps to a subdivision into 2×2 squares, each having just one interior node, whose unknown value $U_{i,j}$ with i and j odd is expressed in terms of the surrounding values. Next, in each 4×4 square, five more neighboring nodal values are 'eliminated' (i.e., expressed by an explicit stored equation in terms of boundary values). This process is iterated: for $k = 3, \cdots, m$, each $2^k \times 2^k$ square S contains $(2^k - 1)^2$ interior nodal values, of which all but $2^{k+1} - 3$ have been 'eliminated' (i.e., expressed as linear combinations of the boundary values). At the k-th step, the 'nested dissection' method expresses these $2^{k+1} - 3$ values (on vertical and horizontal bisecting mesh lines) in terms of the 2^{k+2} nodal values on the boundary of S. After the m-th step has been completed, the coefficient-matrix of the remaining unknowns is *block-tridiagonal*, with $2^n - 1$ diagonal $(2^m - 1) \times (2^m - 1)$ square blocks. These can be eliminated with a dyadic ordering, as in Example 1.

A typical example of the reduction in operation count that can be achieved by nested dissection is provided by the standard 5-point approximation to a linear source problem on an $I \times J$ mesh with J roughly equal to $\sqrt{2}I$. It is advantageous to tilt the rectangle at a 45° angle, subdividing into tilted squares. Each tilted square (mesh diamond) S having vertical and horizontal 'diagonals' of $(2^k + 1)$ points will then have only 2^{k+1} boundary mesh-points. In the final step, we must solve I simultaneous linear equations by Gauss elimination. Just before that, we must solve *two* sets of $I/\sqrt{2}$ equations; before that, *four* sets of $I/2$ equations, and so on. The total operation count is therefore approximately

$$(I^3/3)[1 + 2/2^{3/2} + 4/4^{3/2} + \cdots],$$

which is about $\sqrt{2}I^3/3(\sqrt{2} - 1)$, or about $2.41I$ per unknown. This compares with an operation count of about $I^3J = \sqrt{2}I^4$ for band elimination, or about I^2 per unknown.

Note that for centered 5-point general stencils on an $I \times I$ mesh, slicing along 45° and 135° diagonals minimizes the multiplication count, which is at most by $12I$ per mesh-point; the storage is $7.75 \log_2 I + O(1)$ per mesh-point as $I \uparrow \infty$ (for symmetric A, it is half as much).[7] For centered 9-point stencils, however, horizontal and vertical slicing is best.

Nested dissection has a remarkable theoretical property: its impressive asymptotic operation count is proportional to $\sqrt{N}$ per unknown, where N is the number of unknowns. This differs by a bounded factor from the asymptotic operation count for SOR, and makes nested dissection theoretically attractive for very fine subdivisions.

One-way dissection. In spite of its impressive asymptotic operation count, nested dissection is not used much in practice. This is primarily because to program it in a way that is satisfactory for general regions is extremely difficult. For this reason, 'one-way dissection' with multiple slices is generally recommended for complicated regions, especially if these regions are long and thin.[8]

Optimal orderings. Many papers have been written about finding an 'optimal ordering' of the nodes of the *graph* of a general matrix A which would *minimize* the fill-in that would occur if $B\mathbf{v} = \mathbf{c}$ was solved by Gauss elimination. We will not attempt to review here the large literature concerned with this fascinating, but very difficult general problem. Noteworthy papers were written around 1970 by F. G. Gustavson, Bruce Irons, and Donald Rose; moreover, the literature has been carefully reviewed by Rose and Bunch in Rose-Willoughby [72, pp. 177–187] and by Tarjan in Bunch-Rose [76, pp. 3–32].[9]

Neither will we discuss in detail the packages of subroutines (such as SPARSPACK (Univ. of Waterloo) and the Yale University Sparse Matrix Package) that have been developed for solving linear systems $A\mathbf{u} = \mathbf{b}$ with sparse coefficient-matrices. These often enable users to solve problems at a cost which is an order of magnitude less than that of band elimination.[10]

---

[7] See Birkhoff-George [73, Appendix C]. Also see D. J. Rose and G. F. Whitten in Bunch-Rose [76], and R. J. Lipton, D. J. Rose, and R. E. Tarjan, Siam J. Numer. Anal. 16 (1979), 346–358.

[8] See Alan George, SIAM J. Num. Anal. 17 (1980), 740–751, and the references given there.

[9] More generally, see the papers in Part 1 of Bunch-Rose [76]. Influential were the papers by B. Irons in Int. J. Num. Methods Engng. 2 (1970), 5–32, and by D. J. Rose in Read [72, pp. 183–217]. An error analysis of direct 'sparse matrix' methods has been given by J. R. Bunch in SIAM J. Num. Anal. 11 (1974), 847–873.

[10] See D. J. Rose, G. F. Whitten, A. H. Sherman, and R. E. Tarjan in Computers and Structures 11 (1980), 597–608; A. George and J. W-H. Liu TOMS 5 (1979), 139–162; S. C. Eisenstat, H. C. Elman, M. H. Schultz, and A. H. Sherman in Birkhoff-Schoenstadt [84], and I. S. Duff in Cowell [84].

General graphs. So far, we have concentrated our attention on *rectangular* meshes arising from difference approximations. However, many network flow problems, as well as problems in structural engineering solved by finite element methods to be considered in Chapter 7, lead to matrices having other kinds of graphs (triangular, etc.).

This suggests the following problem: given the (symmetric) graph of a matrix, find an ordering of its equations and unknowns which will minimize the operation count of Gauss (or Cholesky) elimination. This problem has been studied by Gustavson and others;[11] it is very difficult. *Minimum degree* algorithms, which first eliminate unknowns associated with nodes of least degree, are often advantageous. However, since all interior nodes of rectangular networks are of the same degree (four), this principle is not very helpful for them

Alternatively, it may be advantageous to first dissect or 'cut' the graph into disconnected components. For discussions of this old idea of 'domain decomposition' from different points of view, see Part 1 of Bunch-Rose [76], and George-Liu [81], and the papers by Glowinski and Widlund in Birkhoff-Schoenstadt [84].[12]

4. Asymptotic cost estimates. From a theoretical standpoint, the most basic question about any numerical scheme for 'arithmetizing' the solution of elliptic problems concerns its *asymptotic rate of convergence*. We will define this as the least number r such that one can reduce the error of an approximation to less than η by performing $O(\eta^{-r})$ arithmetic operations, in exact arithmetic. Restated, for any $\eta > 0$, let $\sigma(\eta)$ be the number of arithmetic operations required to reduce the discretization error to less than η.[13] Then the asymptotic rate of convergence is the least r such that $\sigma(\eta) = O(\eta^{-r})$. The exponent r is called the order of *computational complexity* of the method.[14]

To determine the order of complexity of a given computation scheme is evidently basic, if one accepts the concept of 'arithmetizing analysis' introduced in Chapter 3, §1. Namely, one imagines letting the mesh length h

[11] F. G. Gustavson et al., J. ACM 17 (1970), 87–109; R. K. Brayton, F. G. Gustavson, and R. Willoughby, Math. Comp. 24 (1970), 937–954. The *degree* of a node is the number of adjacent nodes (or equivalently, the number of edges at that node).

[12] See also J. R. Bunch and D. J. Rose, J. Math. Anal. Appl. 48 (1975), 574–593.

[13] The precise number of operations needed depends, of course, on the *norm* or other criterion used to measure the *magnitude* of the error, but the relevant *exponent* (the 'order of complexity') is often norm-independent.

[14] We are essentially adopting here the conventions suggested by S. C. Eisenstat and M. H. Schultz in Traub [73, pp. 271-283]. See also M. H. Schultz in SIAM-AMS Proceedings 7 (1974), p. 135.

tend to zero for any given problem, and considers the asymptotic operation count.

For an n-dimensional domain $\Omega \subset \mathbb{R}^n$, with an average mesh length $O(h)$, the number N of unknowns is typically $O(h^{-n})$. Therefore, for difference or finite element (see Chapter 7) approximations having *order of accuracy* q, so that $\eta = O(h^q)$, clearly $N = \eta^{-n/q}$.

If the *operation count per unknown* of a method for solving $A\mathbf{u} = \mathbf{b}$ is $O(N^s)$, we will therefore have

$$\sigma(\eta) = O(\eta^{-(1+s)n/q}). \tag{4.1}$$

Consequently, $r = (1+s)n/q$ for problems that fit this description.

For smooth functions, $q = 2$ for piecewise linear and bilinear approximations on a uniform mesh, while $q = 4$ for piecewise cubic and bicubic approximations. (For piecewise quadratic and biquadratic approximations, $q = 3$.) Hence in these cases, the required h is $O(\eta^{1/q})$ for a known positive integer q.

The asymptotic *cost* of achieving a given accuracy η, according to the above conventions, is therefore $O(N^{1+s})$, where $s = s(n)$ depends on the number of multiplications (and additions) per unknown required to solve the linear system $A\mathbf{u} = \mathbf{b}$ given by the discretization. This depends on the choice of linear system solver.

Thus, if *band elimination* is used, the operation count per unknown is roughly b^2, where b is the band-width. Hence this count is asymptotically $O(1)$ for two-endpoint problems ($n = 1$); it is $O(N)$ in the plane (where $n = 2$ and b is proportional to $1/h$, or $N^{1/2}$), and is $O(N^{2-2/n})$ in n-dimensions. Consequently, for linear elliptic problems in $\mathbb{R}^n$, the exponent s is $2-2/n$ if band-solvers are used. Since $N = O(h^{-n})$, the total 'operation count' is $O(N^{1+s}) = O(h^{2-3n})$.

In contrast, if nested dissection is used to solve $A\mathbf{u} = \mathbf{b}$, the number of operations is asymptotically proportional to $N^{1/2}$ for $n = 2$. Hence, if we use an approximation method having $O(h^2)$ accuracy, the (asymptotic) order of computational complexity is reduced from 2 to 1.5, by using nested dissection instead of band elimination when solving two-dimensional problems.

Cost estimates. Formula (4.1) suggests the asymptotic cost estimate

$$\$ = O(\eta^{-(1+s)n/q}). \tag{4.2}$$

One can interpret this in more practical terms as suggesting that the *product*

$$\$ \times \eta^{(1+s)n/q} \tag{4.3}$$

should be nearly constant experimentally (or at least slowly varying) as η is changed for a given s, dimension n, and discretization scheme with accuracy $O(h^q)$. In plotting computing time (or operation count) against η, one expects straight-line plots on log-log graph paper of predictable slope, to the extent that the assumptions made above are justified. From a practical standpoint, the reliability (worth?) of *a priori* theoretical estimates must be judged by the extent to which such predictions are confirmed by observation.

For example, a linear source problem might be approximated by the standard 5-point formula on a 20×20 mesh, with a relative error of 4%. To solve the resulting linear system $A\mathbf{u} = \mathbf{b}$ by band-Cholesky elimination would require about 3×10^5 arithmetic operations. To do this on a VAX takes about 3 seconds and costs about 3¢ *neglecting overhead*.[15] To reduce the relative error to 1% would normally require a 40×40 mesh. This would multiply N by 4 and the cost of solving the linear system by a factor of 16. To achieve the same error reduction would multiply the cost by 8 if nested dissection were used.

Storage costs.[16] We emphasize that the conventions adopted above neglect storage costs. These are usually roughly proportional to the product of the storage *time* and the number of quantities that have to be stored. As a result, if band elimination is used, the asymptotic cost of solving very large problems accurately is proportional not to N^{1+s} but rather to bN^{2+s}, where b is the band-width, of order $N^{1-1/n}$ in n-dimensions.

[illegible] *costs*. The preceding asymptotic analysis ignores many significant

5. Marching methods. We will conclude our review of direct methods for solving elliptic difference equations by discussing algorithms having especially small 'computational complexity'. We will first describe, in this section, a fast 'marching method' which essentially replaces boundary value problems by related initial value problems. In the next section, we will treat 'tensor product' methods applicable to DE's and boundary conditions having separable variables. We will also explain there how tensor product methods can be combined with the FFT to construct fast Poisson solvers.

A much simpler algorithm than nested dissection, for achieving an $O(I^3)$ operation count with 5-point and 9-point stencils in a rectangle, is the following 'marching method'.[17] For simplicity, we assume Dirichlet boundary conditions on an $I \times J$ rectangle, so that the $U_{0,j}$ are known.

The underlying idea is to replace the *boundary* value problem to be solved (for example, a Dirichlet problem for $-\nabla_h^2 U = f$) by an *initial value* problem for an equivalent ΔE (e.g., $\delta_{xx} U = -\delta_{yy} U - h^2 f$), even though only one line and not two lines of 'initial data' are available. Accordingly, the $U_{1,j}$ $(j = 1, \cdots, J-1)$ are left in symbolic form, and then simple recursion is applied to express $U_{i+1,j}$ in terms of these symbolic $U_{1,j}$ and known (computed) coefficients, for $i = 1, \cdots, I-2$:

$$U_{i+1,j} = a_{i,j} U_{i,j} + b_{i,j} U_{i,j-1} + c_{i,j} U_{i,j+1} + d_{i,j} U_{i-1,j} + e_{i,j} f_{i,j}. \tag{5.1}$$

Thus for the Poisson equation, all $a_{i,j} = 4$, $b_{i,j} = c_{i,j} = d_{i,j} = -1$, and $e_{i,j} = -h^2$.

Finally, for $i = I-1$, (5.1) gives a set of equations for determining the $U_{I-1,j}$. The whole process requires only $O(I^2 J)$ arithmetic operations.[18] Indeed, a more careful study shows that in an $I \times I$ square, for the Poisson equation, at most $14 \times I^2 + O(I)$ multiplications (and additions) are required. In the general case, $10 \times I^3/3 + O(I^2)$ multiplications are needed.

This 'marching' procedure can also be interpreted as Gaussian elimination applied to a reordering of the unknowns (i.e., of the columns of A), *without* reordering the equations: the variables $U_{1,1}, \cdots, U_{1,J-1}$ are simply listed last.

Unfortunately, this procedure is highly *unstable*. Thus, for the Poisson equation, roundoff errors grow exponentially, by a factor of

[17] The method goes back to von Mises; see G. Birkhoff and A. George in Traub [73, pp. 221–270 (esp. pp. 230–233)].

[18] This contrasts with $O(I^3 J)$ for band elimination. For $J = O(I)$, A. J. Hoffman, M. S. Martin, and D. J. Rose, SIAM J. Numer. Anal. 10 (1973), 364–369, have shown that $O(I^3)$ is a 'best possible' result.

$3+2^{3/2} \approx 5.828$ per row. Consequently, in an $m \times n$ rectangle, the condition number is about $O(5.828^m)$, so that if $m > 10$ or so, the amplification of roundoff errors is very serious.

Methods for controlling the growth of roundoff in the constant-coefficient case were studied in depth by Bank-Rose, [75] and [76], then Bank [77] analyzed the variable coefficient case. By utilizing these methods, one obtains the so-called 'generalized marching algorithm'. For the Poisson equation, this algorithm first uses the marching algorithm of (5.1) to solve several small problems, whose solutions are then combined to construct the solution of a larger problem. For further results, see Bank in Bunch-Rose [76, pp. 293–307], where the Poisson DE with mixed boundary conditions is treated.

Convection-diffusion equation. The preceding marching method is closely related to so-called 'upwind differencing' (see Chapter 3, §6). This is applicable to time independent convection-diffusion equations, such as

$$Uu_x = \nu(u_{xx} + u_{yy}) + f(x,y), \tag{5.2}$$

as well. For small ν/U, the solution of the Dirichlet problem for (5.2) in a rectangle is given very accurately, except in a thin downstream boundary layer, by 'marching' from the upstream end. Many applications of marching methods to such convection-diffusion problems are described in Roache [78],[19] together with detailed formulas for treating various boundary conditions.

For simplicity, we consider the case of a rectangle $[0,a] \times [0,b]$, and Dirichlet boundary conditions. The most relevant case arises when $U >> \nu$, so that $\epsilon = \nu/U >> 1$.

An effective procedure consists in making a 'first pass' by solving $v_x = \epsilon v_{yy} + f/U$ as an initial-boundary value problem for the initial conditions $v(0,y) = u(0,y)$ on the upstream edge, and $v \equiv u$ also on the two 'sides' $(x,0)$ and (x,b), $x \in [0,a]$. For this, the Crank-Nicolson scheme seems the most natural choice.

Having solved this problem, one can take $w = u - v$ as a new unknown function, satisfying $w \equiv 0$ on three sides of the rectangle, and having known values $w = W(x,y)$ on the downstream edge. Physically, $w(x,y)$ represents the 'boundary layer' correction. One can now compute w by taking as unknowns the values $w(a-h, y_j) = w_j$ and then 'marching' upstream.

[19] See also (for example) D. B. Spalding, Int. J. Num. Methods Eng. 4 (1972), 551–559 or J. C. Heinrich et al, ibid. 11 (1977), 131–143.

6. Tensor product solvers. We now describe a 'fast' method for solving elliptic problems associated with separable DE's and separable boundary conditions.[20] We have already considered several such problems in Chapter 2, §1, and their discretizations in Chapter 3, §8. We also recall from Chapter 2, §5, the concept of the *direct product* of two (real) vector spaces $\mathbb{R}^m$ and $\mathbb{R}^n$ with bases $\boldsymbol{\alpha}_1, \cdots, \boldsymbol{\alpha}_m$ and $\boldsymbol{\beta}_1, \cdots, \boldsymbol{\beta}_n$. The *direct product* or *tensor product* of $\mathbb{R}^m$ and $\mathbb{R}^n$ is the vector space $\mathbb{R}^m \otimes \mathbb{R}^n \cong \mathbb{R}^{mn}$ with basis $\boldsymbol{\gamma}_{i,j}$, $i=1, \cdots, m$, $j=1, \cdots, n$. It is suggestive to think of its basis elements as symbolic products $\boldsymbol{\alpha}_i \otimes \boldsymbol{\beta}_j = \boldsymbol{\gamma}_{i,j}$.

Example 2. Compute that solution u of the Poisson equation in an annulus with u vanishing identically on the boundary:

$$(6.1) \quad -\nabla^2 u = f(r,\theta), \; 1 < r < a, \quad u(1,\theta) \equiv u(a,\theta) \equiv 0, \; 0 \leqslant \theta < 2\pi.$$

We can multiply the DE by r^2, thus reducing it to

$$(6.1') \qquad -r(ru_r)_r - u_{\theta\theta} = F(r,\theta), \qquad F = r^2 f.$$

This is evidently equivalent to $-u_{\rho\rho} - u_{\theta\theta} = F$ in the log-polar coordinates $\rho = \log r$ and θ, which suggests discretizing it on a uniform rectangular grid in ρ and θ: $\rho_i = i(\log a)/I = i\Delta\rho$ $\theta_j = 2\pi j/J = j\Delta\theta$. The standard 5-point approximation then reduces to

$$(6.2) \qquad \mathbf{L}_\rho[U_{i,j}] + \mathbf{M}_\theta[U_{i,j}] = F_{i,j},$$

where $\mathbf{L}_\rho$ and $\mathbf{M}_\theta$ are the *ordinary* self-adjoint difference operators

$$\mathbf{L}_\rho[R_i] = (-R_{i-1} + 2R_i - R_{i+1})/\Delta\rho^2,$$

and

$$\mathbf{M}_\theta[\Theta_j] = (-\Theta_{j-1} + 2\Theta_j - \Theta_{j+1})/\Delta\theta^2,$$

acting on $\mathbb{R}^{I-1}$ and $\mathbb{R}^J$, respectively. The solution of (6.2) satisfies the Dirichlet boundary conditions $U_{0,j} = U_{I,j}$ and is periodic in j: $U_{i,0} = U_{i,J}$.

Kronecker products. Associated with the direct product $\mathbb{R}^m \otimes \mathbb{R}^n$ is the (direct or) 'Kronecker' product $A \otimes B$ of an $m \times m$ matrix A acting on $\mathbb{R}^m$ and an $n \times n$ matrix B acting on $\mathbb{R}^n$. If A and B correspond as usual to the linear transformations

$$\mathbf{x} \mapsto A\mathbf{x}, \quad (A\mathbf{x})_i = \sum_k a_{i,k} x_k, \quad \text{and} \quad \mathbf{y} \mapsto B\mathbf{y}, \quad (B\mathbf{y})_j = \sum_l a_{j,l} y_l,$$

[20] See Lynch-Rice-Thomas [64] and [64a]. For related earlier ideas, see G. A. Baker and T. A. Oliphant, Quart. Appl. Math. 17 (1960), 361–373; E. Egarváry, Acta Math. Acad. Sci. Hung. 11 (1960), 341–361. See also P. Rózsa in Miller [77, pp. 369–381].

of $\mathbb{R}^m$ and $\mathbb{R}^n$, respectively, then the corresponding linear transformation of $\mathbb{R}^m \otimes \mathbb{R}^n$ is evidently (interpreting $z_{k,l}$ as $x_k y_l$)

$$\mathbf{z} \mapsto C\mathbf{z}, \quad (C\mathbf{z})_{i,j} = \sum_k \sum_l c_{i,j,k,l} z_{k,l},$$

where $c_{i,j,k,l} = a_{i,k} b_{j,l}$.

The $mn \times mn$ matrix $C = A \otimes B$ defined by this formula is called the *Kronecker product* (or 'direct product') of A and B. Kronecker products satisfy the identities[21]

$$(A \otimes B)(C \otimes D) = AC \otimes BD, \quad (A \otimes B)^{-1} = A^{-1} \otimes B^{-1}, \quad (A \otimes B)^T = A^T \otimes B^T,$$
$$(A+B) \otimes C = A \otimes C + B \otimes C, \qquad A \otimes (B + C) = A \otimes B + A \otimes C.$$

Also, if $A\mathbf{x} = \lambda \mathbf{x}$ and $B\mathbf{y} = \mu \mathbf{y}$, then $(A \otimes B)(\mathbf{x} \otimes \mathbf{y}) = \lambda \mu \mathbf{x} \otimes \mathbf{y}$.

Clearly $A \otimes B \neq B \otimes A$ in general, although the two Kronecker products are 'similar' (i.e., $P(B \otimes A)P^{-1} = A \otimes B$ for a suitable permutation matrix P), and hence have the same eigenvalues.

In this Kronecker product notation, (6.2) assumes the form

$$(I \otimes L + M \otimes I)\mathrm{U} = \mathrm{F}, \tag{6.3}$$

where L and M are matrices corresponding to the difference operators $\mathbf{L}_\rho$ and $\mathbf{M}_\theta$ in (6.3). We now solve (6.3), where L and M are any square matrices having bases of eigenvectors $\boldsymbol{\phi}_k$ and $\boldsymbol{\psi}_l$ with eigenvalues λ_k and μ_l, respectively, by the following 'tensor product' method.

The method of solution essentially discretizes the classical 'separable variables' solution of Example 2. In Example 2, the eigenfunctions of the operator $\mathbf{L}_\rho + \mathbf{M}_\theta$ are products $\phi_k \psi_l$ of the univariate functions: $\phi_k(\rho_i) = \sin(ki\pi/I)$ and

$$\psi_{2,l}(\theta_j) = \cos(l\pi j/J), \qquad \psi_{2l-1}(\theta_j) = \sin(l\pi j/J).$$

Because the eigenfunctions ϕ_k and ψ_l are (orthogonal) *trigonometric* functions, $F(\rho_i, \theta_j)$ can be expanded into a discrete double Fourier series as

$$F_{i,j} = F(\rho_i, \theta_j) = \sum c_{k,l} \phi_k(\rho_i) \psi_l(\theta_j). \tag{6.4}$$

To solve the ΔE (6.2) formally, write $U_{i,j} = \Sigma\, a_{k,l} \phi_k(\rho_i) \psi_l(\theta_j)$. Evidently,

$$(\mathbf{L}_\rho + \mathbf{M}_\theta)[\phi_k \psi_l] = (\lambda_k + \mu_l) \phi_k \psi_l. \tag{6.5}$$

Comparing coefficients, we see that the $a_{k,l}$ are

$$a_{k,l} = c_{k,l} / (\lambda_k + \mu_l). \tag{6.5'}$$

---

[21] See P. R. Halmos, *Finite-dimensional Vector Spaces*, Van Nostrand, (1958), pp. 33–41 and 95–98; or P. Lancaster, *Theory of Matrices*, Acadmeic Press, 1969, §8.2.

In general, let $\Phi = \|\phi_{i,k}\|$ and $\Psi = \|\psi_{j,k}\|$ be the (square) matrices whose columns, $\mathbf{\Phi}_k$, $\mathbf{\Psi}_l$, have the components $\phi_{i,k}$ and $\psi_{j,l}$ of the ϕ_k and ψ_k, respectively. Then their inverses determine the coefficients $c_{k,l}$, making $\sum c_{k,l}\mathbf{\Psi}_k \otimes \mathbf{\Phi}_l = \mathbf{F}$ the array of nodal values $F_{i,j}$.

THEOREM 1. *Let Φ and Ψ diagonalize the matrices L and M, so that*

$$L = \Phi \Lambda_L \Phi^{-1} \quad \text{and} \quad M = \Psi \Lambda_M \Psi^{-1}, \tag{6.6}$$

where Λ_L and Λ_M are diagonal with entries λ_k and μ_l, respectively. Then

$$\mathrm{U} = \Psi \otimes \Phi [I \otimes \Lambda_L + \Lambda_M \otimes I]^{-1} \Psi^{-1} \otimes \Phi^{-1} \mathrm{F} \tag{6.7}$$

solves (6.3) *unless $I \otimes L + M \otimes I$ is singular.*

Remark. For $I \otimes L + M \otimes I$ to be singular, we must have $\lambda_k + \mu_l = 0$ for some k, l, which is impossible if one of L, M, is positive definite and the other positive semi-definite, as in Example 2.

Proof. Use (6.6) to replace L and M in the matrix in (6.3) and multiply it by the matrix in (6.7). This gives

$$\begin{aligned}
&(\Psi \otimes \Phi [I \otimes \Lambda_L + \Lambda_M \otimes I]^{-1} \Psi^{-1} \otimes \Phi^{-1})(I \otimes \Phi \Lambda_L \Phi^{-1} + \Psi \Lambda_M \Psi^{-1} \otimes I) \\
&\quad = \Psi \otimes \Phi [I \otimes \Lambda_L + \Lambda_M \otimes I]^{-1} (\Psi^{-1} \otimes \Lambda_L \Phi^{-1} + \Lambda_M \Psi^{-1} \otimes \Phi^{-1}) \\
&\quad = \Psi \otimes \Phi [I \otimes \Lambda_L + \Lambda_M \otimes I]^{-1} [I \otimes \Lambda_L + \Lambda_M \otimes I] \Psi^{-1} \otimes \Phi^{-1} = I \otimes I.
\end{aligned}$$

COROLLARY. *The nodal values of* U *are*

$$U_{i,j} = \sum_{l=1}^{J} \psi_{j,l} \sum_{i=1}^{I-1} \phi_{i,k} [\lambda_k + \mu_l]^{-1} \sum_{n=1}^{J} (\Psi^{-1})_{l,n} \sum_{m=1}^{I-1} (\Phi^{-1})_{k,m} F_{m,n}. \tag{6.8}$$

Operation count. If one knows the entries of $\Phi, \Psi, \Phi^{-1}, \Psi^{-1}$ and the eigenvalues λ_k, μ_l, all $(I-1)J$ nodal values $U_{i,j}$ can be evaluated by (6.8) in about $2IJ(I+J)$ multiplications and additions and IJ divisions. Namely, the last sum in (6.8), for all the IJ values of k and n taken, requires a total of about I^2J multiplications and I^2J additions, and similarly for each of the other sums.

In Example 2, the matrices Φ and Ψ are orthogonal; hence $(\Phi^{-1})_{k,m} = \phi_{m,k}$ and $(\Psi^{-1})_{l,n} = \psi_{n,l}$, so that all the matrices in (6.7) are easily inverted. Moreover, $\phi_{m,k}$, $\psi_{n,l}$, λ_k, and μ_l are known in terms of sine functions; consequently the $U_{i,j}$ are easily evaluated by (6.8).

In general, however, one must compute the eigenvectors and eigenvalues. For the 5-point approximation to separable linear source problems, these involve tridiagonal matrices (3-point difference operators).

Example 3. In Example 2, using ordinary polar coordinates with equal spacing in the radial direction, $r_i = 1 + i\Delta r$, $\Delta r = (a-1)/I$, the standard 5-point approximation gives the radial second difference operator

$$\mathrm{L}_r[R_i] = r_i[-r_{i-1/2}R_{i-1} + (r_{i-1/2} + r_{i+1/2})R_i - r_{i+1/2}R_{i+1}]/\Delta r^2.$$

The eigenvectors and eigenvalues must then be computed by solving

$$\frac{1}{r_i}\mathbf{L}_r[\phi_{i,k}] = \frac{\lambda_l}{r_i}\phi_{i,k}, \qquad i = 1, 2, \cdots, I-1,$$

with $\phi_{0,k} = \phi_{I,k} = 0$. This discrete Sturm-Liouville system has distinct eigenvalues and eigenfunctions orthogonal with respect to the inner product $\langle u,v\rangle = \Sigma\, u_i v_i / r_i$. That is, one solves the generalized eigenvalue problem $A\mathbf{\Phi}_k = \lambda_k D^{-1}\mathbf{\Phi}_k$. But since D is diagonal, $D_{i,i} = r_i$, equivalently one can solve the ordinary eigenvalue problem $B\mathbf{\Theta}_k = \lambda_k \mathbf{\Theta}_k$, where $B = D^{1/2}AD^{1/2}$ and $\mathbf{\Theta}_k = D^{-1/2}\mathbf{\Phi}_k$.

Each of these $I-1 \times I-1$ tridiagonal eigenproblems can be solved by EISPACK,[22] a systematized collection of subroutines available from IMSL.

FFT. The solution of the discrete approximation $-\nabla_h^2 U + k^2 u = f$ of the modified Helmholtz equation (the Poisson equation when $k^2 = 0$) with $u = 0$ on the boundary of a rectangle is given in Theorem 1, where the entries in Ψ and Φ are values of the sine function. In (6.7), $\Psi^{-1} \otimes \Phi^{-1}\mathbf{F}$ gives the discrete Fourier sine coefficients of $\mathbf{F}$. When they are divided by the appropriate eigenvalues, they become the Fourier coefficients of $\mathbf{U}$. For $I = J$, and $N = (I-1)^2$ unknowns, (6.8) gives a way of evaluating all the nodal values with about $4I^3$ multiplications, or $4I$ per unknown.

When I is a power of 2, the operation count can be reduced to $O(N \log_2 I)$, or $O(\log_2 I)$ per unknown, by evaluating the Fourier coefficients of $\mathbf{F}$ and the Fourier series for $\mathbf{U}$ using the *fast Fourier transform method* (FFT). By ingenious use of the addition formulas for the exponential function, Cooley and Tukey discovered this brilliantly successful method for computing Fourier coefficients.[23]

We outline FFT for functions of one variable and, for algebraic simplicity, we treat the case of $N = 2^n$ complex values $f_j = f(e^{2\pi ij/N})$, $j = 0, \cdots, N-1$, equally spaced on the unit circle. The Fourier coefficients c_k of f_j solve the linear system

$$\text{(6.9)} \qquad \sum_{k=0}^{N-1} c_k\, e^{2\pi ijk/N} = f_j, \qquad j = 0, \cdots, N-1, \quad i = \sqrt{-1}.$$

---

[22] B. T. Smith, J. M. Boyle, J. J. Dongarra, B. S. Garbow, Y. Ikebe, V. C. Klema, and C. B. Moler, *Matrix Eigensystems Routines* — EISPACK Guide, Springer Lecture Notes in Computer Science 6, 1976. Here §§1.15–1.22 concern tridiagonal matrices. For *generalized* eigenproblems, see B. S. Garbow, J. M. Boyle, J. J. Dongarra, and C. B. Moler, *Matrix Eigensystem Routines* — EISPACK Guide Extension, ibid, §51, 1977.

[23] J. W. Cooley and J. W. Tukey, Math. Comp. 19 (1965), 297–301; for a Fortran subroutine, see Conte-de Boor [80, pp. 277–301]. See also E. O. Brigham, *The Fast Fourier Transform*, Prentice-Hall, 1974; S. Winograd, Math. Comp. 32 (1978), 175–199; and Henrici [81, §§ 7.5, 7.6]. According to H. H. Goldstine, *A History of Numerical Analysis from the 15th through the 19th Century*, Springer, 1977, pp. *xiii* and 249, the idea was known to Gauss.

For any complex-valued periodic function u with integral period N, the formula

$$u_j = \sum_{k=0}^{N-1} a(k)\omega^{jk}, \qquad \omega = e^{2\pi i/N}, \tag{6.10}$$

defines its Fourier expansion. By the duality of the finite Fourier transform, we have reciprocally

$$a(k) = \frac{1}{N}\sum_{j=0}^{N-1} u_j\omega^{-jk}. \tag{6.10$'$}$$

When $N = 2^n$, then by proceeding as follows, we can evaluate the sums in (6.10) in $O(nN) = O(N\log_2 N)$ instead of $O(N^2)$ arithmetic operations.

First, express j and k in binary notation, as

$$j = j_{n-1}2^{n-1} + \cdots + j_1 2 + j_0, \quad k = k_{n-1}2^{n-1} + \cdots + k_1 2 + k_0, \tag{6.11}$$

and set

$$[j,k]_m = j_0 2^{n-1} + \cdots + j_{m-1}2^{n-m} + k_{n-m-1}2^{n-m-1} + \cdots + k_0. \tag{6.11$'$}$$

Then proceed from the array of $a(k) = a_0(k)$ to the $u_j = a_n([j,k]_n)$ through a sequence of $n-1$ intermediate arrays $a_m([j,k]_m)$, $m = 1, \cdots$ [illegible] fined recursively (as in Cooley-Tukey, op. cit., (19))

the result to the preceding and to the following equation. One gets

$$
\begin{array}{llll}
M^{(1)}\mathbf{U}_2 & -\mathbf{U}_4 & & = \mathbf{F}_2^{(1)}, \\
-\mathbf{U}_{2k-2} & +M^{(1)}\mathbf{U}_{2k} & -\mathbf{U}_{2k+2} & = \mathbf{F}_{2k}^{(1)}, \qquad 2k = 4, \cdots, I-4, \\
& -\mathbf{U}_{I-4} & +M^{(1)}\mathbf{U}_{I-2} & = \mathbf{F}_{I-2}^{(1)},
\end{array}
\tag{6.13}
$$

where

$$M^{(1)} = M^2 - 2I, \qquad \mathbf{F}_{2k}^{(1)} = \mathbf{F}_{2k-1} + M\mathbf{F}_{2k} + \mathbf{F}_{2k+1}. \tag{6.14}$$

After one obtains the $\mathbf{U}_{2k}$, one gets the $\mathbf{U}_{2k-1}$ by solving the tridiagonal systems

$$
\begin{aligned}
M\mathbf{U}_1 &= \mathbf{F}_1 + \mathbf{U}_2, \\
M\mathbf{U}_{2k+1} &= \mathbf{F}_{2k+1} + \mathbf{U}_{2k} + \mathbf{U}_{2k+2}, \qquad k = 0, 1, \cdots, I/2-2, \\
M\mathbf{U}_{I-1} &= \mathbf{F}_{I-1}.
\end{aligned}
$$

This is 'cyclic reduction' applied to *lines* of nodal values (the vectors $\mathbf{U}_{2k}$ being 'red' and $\mathbf{U}_{2k+1}$ being 'black'; see §3).

If I is divisible by 4, then half the unknowns can be eliminated similarly from the system (6.13). One gets a similar system with superscripts '(1)' replaced by '(2)' and appropriate changes in the subscripts; (6.14) is replaced with

$$M^{(2)} = (M^{(1)})^2 - 2I, \qquad \mathbf{F}_{4k}^{(2)} = \mathbf{F}_{4k-2}^{(1)} + M^{(1)}\mathbf{F}_{4k}^{(1)} + \mathbf{F}_{4k+2}^{(1)}. \tag{6.14'}$$

If $I = 2^m$, then one can continue this process $m-1$ times. One solves the single equation $M^{(m-1)}\mathbf{U}_{I/2} = \mathbf{F}_{I/2}^{(m-1)}$, and then the pair of equations

$$M^{(m-2)}\mathbf{U}_{I/4} = \mathbf{F}_{I/4}^{(m-2)} + \mathbf{U}_{I/2}, \qquad M^{(m-2)}\mathbf{U}_{3I/4} = \mathbf{F}_{3I/4}^{(m-2)} + \mathbf{U}_{I/2},$$

and so on.

Fast Poisson solvers. This scheme of *repeated cyclic reduction* is moderately unstable. However, Buneman has modified it so that it becomes stable. Hockney has combined repeated cyclic reduction with the FFT to obtain a method for solving the Poisson equation which is faster than either.[24]

Such 'fast Poisson solvers' play a very important role in obtaining approximate solutions of elliptic problems involving the Laplace operator (e.g., the Helmholtz, modified Helmholtz, and biharmonic equations). They can be easily extended to box-shaped regions in space. Moreover, the number of operations per unknown is asymptotically independent of

[24] See Dorr [70]; Hockney [65] and [70]; Buzbee-Golub-Nielson [70]; Swarztrauber [77]; Sweet [77]; C. Temperton, J. Comp. Phys. 34 (1980), 314–329; and E. N. Houstis, R. E. Lynch, T. S. Papatheodorou, Math. and Comput. in Simul. (1980), 91–97.

the number of dimensions for the tensor product, the FFT, and the cyclic reduction methods.

Finally, they can be modified so as to obtain greater efficiency in solving the Poisson and Helmholtz equations, etc., in *non*-rectangular regions — especially if they are nearly rectangular; see Chapter 8, §9.

7. Iterative methods. We now turn our attention to *iterative* methods. Many of these are 'one-step'[25] methods which compute from any initial $\mathbf{u}^{(0)}$ successive approximations $\mathbf{u}^{(1)}, \mathbf{u}^{(2)}, \cdots$, by *iterating* some process, so that $\mathbf{u}^{(m+1)} = \mathbf{S}[\mathbf{u}^{(m)}]$. Such a process is said to *converge* to $\mathbf{u}$ when, for any $\mathbf{u}^{(0)}$, $\lim_{m \to \infty} \mathbf{u}^{(m)} = \mathbf{u}$.

For sufficiently fine meshes (e.g., elliptic problems involving 10^4 or more unknowns), and for three-dimensional problems, iterative methods are usually more efficient than the direct elimination methods described in §§2–4 above. *Semi-iterative* methods will be treated in the next chapter, along with procedures for accelerating the convergence of iterative methods.

Jacobi method. There are many iterative algorithms for solving elliptic ΔE's such as the Poisson equation $-\nabla_h^2 u = f$, the biharmonic equation $\nabla_h^4 u = 0$, and the linear source problem of Chapter 3, (4.2); moreover, their efficiency varies over a wide range. Among these algorithms, the *Jacobi method* is one of the simplest. We will describe it first for solving systems of linear equations (vector equations) of the form

$$A\mathbf{u} = \mathbf{b}, \tag{7.1}$$

assuming that A has *positive diagonal entries*, as is necessarily the case if A is positive definite.

Premultiplying (7.1) by $D^{-1} = [\mathrm{diag}(A)]^{-1}$, one gets

$$(I - B)\mathbf{u} = \mathbf{k} \quad \text{with} \quad B = I - D^{-1}A \quad \text{and} \quad \mathbf{k} = D^{-1}\mathbf{b}. \tag{7.2}$$

This suggests the following iterative process: pick $\mathbf{u}^{(0)}$ and then compute $\mathbf{u}^{(m+1)}$ recursively from $\mathbf{u}^{(m)}$ by

$$\mathbf{u}^{(m+1)} = B\mathbf{u}^{(m)} + \mathbf{k}, \qquad b_{i,j} = -a_{i,j}/a_{i,i}, \quad k_i = b_i/a_{i,i}. \tag{7.3}$$

This is the *Jacobi* (or 'point-Jacobi') method.

The i-th equation of the system (7.3) is

$$u_i^{(m+1)} = \sum_{j \neq i} b_{i,j} u_j^{(m)} + k_i, \tag{7.4}$$

and it is by this formula that the i-th component of $\mathbf{u}^{(m+1)}$ is evaluated.

Storage savings. In general, iterative methods require much less storage to implement than direct methods. Thus, even for elliptic DE's with

---

[25] Such methods are also called "first-order" or of the "first-degree".

variable coefficients, it suffices to store $O(h^{-2})$ coefficients in two-dimensional problems, and $O(h^{-3})$ in space. Due to fill-in, the corresponding requirements for band elimination are $O(h^{-3})$ and $O(h^{-5})$, respectively.

Thus, when the system of N equations $A\mathbf{u} = \mathbf{b}$ arises from the 5-point discretization of a linear source problem, at most four of the coefficients in (7.4) are nonzero for each i. Hence, once the $-a_{i,j}/a_{i,i}$ and $b_i/a_{i,i}$ have been evaluated, $\mathbf{u}^{(m+1)}$ is computed from $\mathbf{u}^{(m)}$ in at most $4N$ multiplications and additions. Therefore, at most $5N$ nonzero coefficients $b_{i,j}$ and (temporarily during each m-th iteration cycle) the $2N$ numbers $U_{i,j}^{(m)}$ and $U_{i,j}^{(m+1)}$ need to be stored, giving $7N$ numbers in all. In the special case of regionwise constant coefficients, the components $U_{i,j}^{(m)}$ and the entries in the right side of the vector equation account for the bulk of the storage requirement, which is therefore about $2N$ and not $7N$ for the linear source problem. Whereas if band elimination is used with the natural ordering of the nodes (x_i, y_j) of an $I \times I$ square $(N = I^2)$, fill-in can force as many as NI nonzero $a_{i,j}$ to be stored, where I is the number of vertical mesh lines $x = x_i$.

If $A = D - C$ is *symmetric*, as is the case for any well-designed difference approximation to a *self-adjoint* elliptic problem, evidently

$$B = I - D^{-1}A = D^{-1}C = D^{-1/2}[D^{-1/2}CD^{-1/2}]D^{1/2}$$

is similar to the symmetric matrix $D^{-1/2}CD^{-1/2}$. In other words, B is *symmetrizable* with diagonal 'symmetrization matrix' $D^{1/2}$. This proves the following result.

LEMMA 1. *If $A = D - C$ is symmetric positive definite, then the Jacobi iteration matrix $B = I - D^{-1}A$ is symmetrizable with real eigenvalues.*

We will now prove that Jacobi iteration converges for A any 'irreducibly diagonally dominant' matrix. It thus converges for the standard 5-point discretization (Chapter 3, (4.2)) of the linear source problem for two independent variables,[26] and for the analogous 7-point discretization in $\mathbb{R}^3$. Specifically, it converges on any *connected mesh* (irreducible network, cf. §9), except when every node is a 'boundary' node where the external current is specified and there is no leakage (i.e., in the case of Neumann problems).

DEFINITION. An $N \times N$ matrix $A = \|a_{i,j}\|$ is *diagonally dominant* if

$$|a_{i,i}| \geqslant \sum_{j \neq i} |a_{i,j}|, \qquad i = 1, \cdots, N. \tag{7.5}$$

It is *strictly diagonally dominant* when the inequality is strict for all i.

[26] This applies to any ordering with the coefficient of $U_{i,j}$ in (4.2) of Chapter 3 on the diagonal of A.

LEMMA 2. *If A is nonsingular and diagonally dominant, then so is $D^{-1}A$. Furthermore, $B = I - D^{-1}A$ is a weak contraction in the l_1-norm and the l_∞-norm.*

Proof.[27] If A is nonsingular, it can have no row $\mathbf{0}$; hence $|d_{i,i}| > 0$ for all i and D is nonsingular. Since premultiplication by D^{-1} just multiplies each row of A by a nonzero constant, it does not affect the diagonal dominance (or graph) of A, which proves the first statement. It follows that

$$\sum_{j=1}^{N} |b_{i,j}| \leqslant 1, \qquad i = 1, \cdots, N,$$

whence, if $\mathbf{v} = B\mathbf{u}$,

$$\begin{aligned} ||\mathbf{v}||_1 = \sum_{i=1}^{N} |v_i| &\leqslant \sum_{i=1}^{N} \left[\sum_{j=1}^{N} |b_{i,j}| \cdot |u_j| \right] \\ &= \sum_{j=1}^{N} |u_j| \left[\sum_{i=1}^{N} |b_{i,j}| \right] \leqslant \sum_{j=1}^{N} |u_j| = ||\mathbf{u}||_1. \end{aligned} \tag{7.6}$$

The proof that $||B\mathbf{u}||_\infty \leqslant ||\mathbf{u}||_\infty$ is similar.

DEFINITION. An $N \times N$ matrix A is *reducible* if there is a permutation matrix P such that

$$PAP^{-1} = \begin{bmatrix} A_{11} & A_{12} \\ 0 & A_{22} \end{bmatrix},$$

where the (nonvoid) diagonal submatrices $A_{i,i}$ are square; otherwise, it is *irreducible*. A is *irreducibly diagonally dominant* when it is irreducible and diagonally dominant, with strict inequality holding in (7.5) for at least one i; see [V, p. 23].

It is easy to show that if A is irreducibly diagonally dominant, then A is nonsingular, and that $(D^{-1}A)^{N-1}$ is strictly diagonally dominant if $\operatorname{diag}(A)$ is positive (cf. [V], Thm. 1.8 and Lemma 1.7).

LEMMA 3. *If A is an $N \times N$ irreducibly diagonally dominant matrix, then B^{N-1} in (7.2) is a strong contraction in the l_1-norm.*

Proof. Since A is irreducibly diagonally dominant, some power of B (actually, B^{N-1}) must make strict inequality hold in (7.6). Hence $||B^{N-1}||_1 < 1$, by Lemma 2.

From Lemma 3, since $\mathbf{u}^{(m+1)} - \mathbf{u} = B(\mathbf{u}^{(m)} - \mathbf{u}) = B^{m+1}(\mathbf{u}^{(0)} - \mathbf{u})$, by (7.1) and (7.2), we obtain the following conclusion.

---

[27] See [V, p. 23]. For the original derivations, see S. Gerschgorin, Izv. Akad. Nauk. 7 (1931), 749–754, and G. Birkhoff and R. S. Varga, J. SIAM 6 (1958), 354–378.

THEOREM 2. *If A is an irreducibly diagonally dominant matrix, then the Jacobi iterative scheme with $B = I - D^{-1}A$ converges to the solution $\mathbf{u} = A^{-1}\mathbf{b}$ for any initial trial vector $\mathbf{u}^{(0)}$.*

It is easy to show that A is irreducibly diagonally dominant if it is the coefficient matrix associated by Eq. (4.2) of Chapter 3 with any linear source problem with leakage or absorption on a *connected* rectangular grid of mesh-points. More generally, this is true of 'network flow' problems with leakage or absorption on any connected net; cf. §9.

8. Other one-step iterative methods. In general, a *one-step stationary*[28] iterative method is a scheme of the form

$$\mathbf{u}^{(m+1)} = G\mathbf{u}^{(m)} + \mathbf{k}. \tag{8.1}$$

It is 'one-step' (or 'first-order') because it generates $\mathbf{u}^{(m+1)}$ from the single estimate $\mathbf{u}^{(m)}$, and 'stationary' because the *iteration matrix* G does not depend on m. The Jacobi method of §7 for solving $A\mathbf{u} = \mathbf{b}$ when all $a_{i,i} > 0$, with $G = B = I - D^{-1}A$ and $\mathbf{k} = D^{-1}\mathbf{b}$, is a very simple one-step iterative method. In this section, we shall introduce three others: the Gauss-Seidel, (first-order) Richardson, and SOR methods.

Gauss-Seidel. The Gauss-Seidel method is obtained from the Jacobi method of §8 by using *improved values* $u_i^{(m+1)}$ of $\mathbf{u}$ as soon as they are available. Thus, instead of (7.4), the i-th component of $\mathbf{u}^{(m+1)}$ is computed by the formula

$$u_i^{(m+1)} = \sum_{j<i} b_{i,j} u_j^{(m+1)} + \sum_{j>i} b_{i,j} u_j^{(m)} + k_i, \tag{8.2}$$

where, as in (7.3),

$$b_{i,j} = -a_{i,j}/a_{i,i}, \quad \text{and} \quad k_i = b_i/a_{i,i}. \tag{8.2$'$}$$

Evidently, the Gauss-Seidel method requires the same number of operations per iteration cycle (i.e., to compute $\mathbf{u}^{(m+1)}$ from $\mathbf{u}^{(m)}$) as the Jacobi method. By overwriting $u_i^{(m)}$ with $u_i^{(m+1)}$, the storage for unknowns is halved.

Thus, for $-\nabla_h^2 U_{i,j} = f_{i,j}$, sweeping through the unknowns *cyclically* in the natural ordering on a sequential machine, one iterates

$$U_{i,j}^{(m+1)} = \tfrac{1}{4}[U_{i-1,j}^{(m+1)} + U_{i,j-1}^{(m+1)} + U_{i+1,j}^{(m)} + U_{i,j+1}^{(m)} + h^2 f_{i,j}]. \tag{8.3}$$

This contrasts with Jacobi iteration, which computes the components of each new $\mathbf{U}^{(m+1)}$ by cyclically iterating

$$U_{i,j}^{(m+1)} = \tfrac{1}{4}[U_{i-1,j}^{(m)} + U_{i,j-1}^{(m)} + U_{i+1,j}^{(m)} + U_{i,j+1}^{(m)} + h^2 f_{i,j}]. \tag{8.4}$$

---

[28] In this book, all 'iterative' (as contrasted with 'semi-iterative') schemes will be assumed to be stationary, unless the contrary is stated.

More generally, let $A = D - E - F$, with $D = \text{diag}(A)$ a nonsingular diagonal matrix and E and F strictly lower and upper triangular, respectively. In this notation, the Gauss-Seidel method can be written as

$$(D-E)\mathbf{u}^{(m+1)} = F\mathbf{u}^{(m)} + \mathbf{b}. \tag{8.5}$$

This gives (8.1) with

$$G = (D-E)^{-1}F, \quad \text{and} \quad \mathbf{k} = (D-E)^{-1}\mathbf{b}. \tag{8.5'}$$

Remark 1. The Jacobi method can take full advantage of parallel processing, because with Jacobi iteration, $\mathbf{u}^{(m+1)}$ is obtained by (7.3) as $B\mathbf{u}^{(m)} + \mathbf{k}$. Each component of the new vector iterate is computed exclusively from those of the previous vector iterate, and so this method is a method of *simultaneous* 'displacements'. In general, Gauss-Seidel iteration is less well-suited to parallel processing, because it is a method of *successive* displacements in which different components of $\mathbf{u}^{(m+1)}$ are changed at each step of a single iteration cycle. Furthermore, the *ordering* of the components of $\mathbf{u}$ affects the rate of convergence of the Gauss-Seidel method but not that of the Jacobi method. On the other hand, Gauss-Seidel iteration is more efficient than Jacobi iteration on *sequential* computers.

Richardson method. Richardson's method[29] for solving $A\mathbf{u} = \mathbf{b}$ is

$$\mathbf{u}^{(m+1)} = \mathbf{u}^{(m)} + \omega(\mathbf{b} - A\mathbf{u}^{(m)}) = (I - \omega A)\mathbf{u}^{(m)} + \omega\mathbf{b}, \tag{8.6}$$

where the (positive) *parameter* ω must be specified. How to choose ω will be discussed in §§10–11.

Remark 2. For any matrix A whose diagonal entries $a_{i,i} = d$ are all equal, Richardson's method (8.6) with $\omega = 1/d$ is identical with the Jacobi method.

Example 4. Consider the ΔE (difference equation) of a loaded simply supported beam for a uniform mesh:

$$U_i = \frac{1}{6}[-U_{i-2} + 4U_{i-1} + 4U_{i+1} - U_{i+2} + h^4 f_i], \tag{8.7}$$

with endpoint conditions $U_0 = U_N = 0$ and $U_{-1} = U_1$, $U_{N-1} = U_{N+1}$. If N is even and the initial error is $\mathbf{e}^{(0)} = (1, 0, -1, 0, 1, \cdots, \pm 1)^T$, then the Jacobi method gives $\mathbf{e}^{(m)} = (4/3)^m \mathbf{e}^{(0)}$; hence it diverges. On the other hand, as will be shown in §10, the Richardson method converges for *any* positive definite matrix A and sufficiently small ω. Thus, for small enough ω, it converges for (8.7) and its two-dimensional analogue, the

[29] Richardson [10]. Following Frankel [50], this is usually called 'first-order' Richardson.

13-point discretization $\nabla_h^4 u = f$ of the loaded plate problem. As we shall see (§9), it also has a suggestive physical interpretation.

SOR. We now come to the *SOR* (*successive overrelaxation*) method. This differs from the Gauss-Seidel method of (8.2) in that it multiplies each Gauss-Seidel increment $u_i^{(m+1)} - u_i^{(m)}$ by a parameter ω called the *overrelaxation factor*. The components of $\mathbf{u}^{(m+1)}$ are computed at each step of the iterative cycle (loop) by

$$\begin{aligned} u_i^{(m+1)} &= u_i^{(m)} + \omega\left[\sum_{j=1}^{i-1} b_{i,j} u_j^{(m+1)} + \sum_{j=i+1}^{N} b_{i,j} u_j^{(m)} + k_i - u_i^{(m)}\right] \\ &= (1-\omega)\, u_i^{(m)} + \omega\left[\sum_{j=1}^{i-1} b_{i,j} u_j^{(m+1)} + \sum_{j=i+1}^{N} b_{i,j} u_j^{(m)} + k_i\right], \end{aligned} \tag{8.8}$$

with $b_{i,j}$ as in (8.2′). Like the Gauss-Seidel method, SOR uses improved values as soon as available. For $\omega = 1$, SOR reduces to the Gauss-Seidel method.

In general, for any $A = D - E - F$, as in (8.5) with $D = \operatorname{diag}(A)$ nonsingular and E and F strictly lower and upper triangular, respectively, the SOR method applied over a complete cycle of components is equivalent to

$$(D - \omega E)\mathbf{u}^{(m+1)} = [(1-\omega)D + \omega F]\mathbf{u}^{(m)} + \omega \mathbf{b}. \tag{8.9}$$

Setting $L = D^{-1}E$ and $U = D^{-1}F$, we get

$$\mathbf{u}^{(m+1)} = \mathbf{L}_\omega \mathbf{u}^{(m)} + \mathbf{k}, \tag{8.10}$$

with

$$\mathbf{L}_\omega = (I - \omega L)^{-1}[(1-\omega)I + \omega U], \qquad \mathbf{k} = \omega(I - \omega L)^{-1}D^{-1}\mathbf{b}. \tag{8.10′}$$

The SOR iteration matrix, $\mathbf{L}_\omega$, and $\mathbf{k}$ here depend on the relaxation factor ω; $\mathbf{L}_1$ (signifying $\mathbf{L}_\omega$ with $\omega = 1$) is the Gauss-Seidel iteration matrix.

Ostrowski-Reich theorem. The Ostrowski-Reich theorem[30] asserts that if A is a symmetric matrix with positive diagonal entries, then the SOR method converges if and only if A is positive definite and $0 < \omega < 2$.

It is a corollary that SOR converges (if $0 < \omega < 2$) for not only the standard 5-point approximation to the linear source problem in the plane and the 7-point approximation in space, but also for the highly accurate 9-point and HODIE approximations to the Poisson and modified Helmholtz equations on plane regions and 27-point approximations in space, as well as for the 13-point and 25-point approximations to the biharmonic equation. (For these approximations, see Chapter 3, §§10–11.)

[30] See [V, p. 78] or [Y, p. 113]. For a generalization, see R. S. Varga in Miller [73, pp. 329–343].

Matrix splittings.[31] The point-Jacobi, Gauss-Seidel, Richardson, and SOR methods can all be regarded as special cases of a single method. Namely, let $A = M - N$, where M is 'easily invertible' in the sense that $M\mathbf{u} = \mathbf{c}$ is easily solved for $\mathbf{u}$. Then $A\mathbf{u} = \mathbf{b}$ is equivalent to $M\mathbf{u} = N\mathbf{u} + \mathbf{b}$, and one can iterate

$$M\mathbf{u}^{(m+1)} = N\mathbf{u}^{(m)} + \mathbf{b}, \tag{8.11}$$

which is equivalent to

$$\mathbf{u}^{(m+1)} = G\mathbf{u}^{(m)} + \mathbf{k}, \quad \text{with} \quad G = M^{-1}N \quad \text{and} \quad \mathbf{k} = M^{-1}\mathbf{b}. \tag{8.11'}$$

As special cases of matrix splitting, we have

Jacobi:	$M = D$,	$N = D - A$,
Gauss-Seidel:	$M = D - E$,	$N = F$,
Richardson:	$M = \omega^{-1}I$,	$N = \omega^{-1}I - A$,
SOR:	$M = \omega^{-1}D - E$,	$N = (\omega^{-1} - 1)D + F$,

where, as usual, $D = \operatorname{diag}(A)$, and E and F are strictly lower and upper triangular matrices, respectively.

Nowadays, matrix splittings designed to speed up convergence are widely used with *preconditionings* (see Chapter 5, §6).

9. Physical analogies; relaxation methods. For any positive definite symmetric matrix A, Richardson's method (8.6) for solving

$$A\mathbf{u} = \mathbf{b} \tag{9.1}$$

is equivalent to the Cauchy polygon method with time step $\Delta t = \omega h^2$ for solving numerically the vector *ordinary* DE $\mathbf{u}_t = -A\mathbf{u} + \mathbf{b}$.

The solution of (9.1) is that (column) vector $\mathbf{u}$ which *minimizes* the quadratic functional

$$\mathbf{J}[\mathbf{v}] = \frac{1}{2}\mathbf{v}^T A\mathbf{v} - \mathbf{v}^T\mathbf{b}. \tag{9.2}$$

This is because, for the unique $\mathbf{u} = A^{-1}\mathbf{b}$ and for any $\mathbf{v}$, we have

$$2\mathbf{J}[\mathbf{v}] = \mathbf{v}^T A\mathbf{v} - 2\mathbf{v}^T A\mathbf{u} = (\mathbf{v} - \mathbf{u})^T A(\mathbf{v} - \mathbf{u}) - \mathbf{u}^T A\mathbf{u},$$

and so $2\mathbf{J}[\mathbf{v}] \geqslant -\mathbf{u}^T A\mathbf{u} = 2\mathbf{J}[\mathbf{u}]$, with equality holding if and only if $\mathbf{v} = \mathbf{u}$.

This interpretation often has suggestive physical analogies. For example, consider the *heat equation with source*

$$u_t = \alpha\nabla^2 u + f \tag{9.3}$$

[31] See [V, p. 88]; also R. S. Varga in Langer [60a, pp. 121–142], where 'partial factorization' is also discussed.

in a rectangular polygon Ω with the functions $u(\mathbf{y},t) = g(\mathbf{y})$, $\mathbf{y} \in \partial\Omega$, and $u(\mathbf{x},0) = u^{(0)}(\mathbf{x})$, $\mathbf{x} \in \Omega$, specified. For Ω subdivided by a square mesh Ω_h with mesh-length h, letting $u_{i,j} = U(x_i, y_j)$ as in Chapter 3, the *semi*-discretization of (9.3) is the (large) system of ordinary DE's

$$du_{i,j}/dt = \alpha \nabla_h^2 u_{i,j} + f_{i,j}, \qquad (ih, jh) \in \Omega_h.$$

The Cauchy polygon method with constant time step Δt for solving this is

$$U_{i,j}^{(m+1)} = U_{i,j}^{(m)} + \omega h^2 [f_{i,j} + \alpha \nabla_h^2 U_{i,j}^{(m)}], \qquad \omega = \Delta t / h^2,$$

where $U_{i,j}^{(m)} \approx u(ih, jh, m\Delta t)$; this is equivalent to the Richardson method for solving the time-*independent* problem

$$-\alpha h^2 \nabla_h^2 U_{i,j} = h^2 f_{i,j}, \; (ih, jh) \in \Omega, \quad U_{k,l} = g(kh, lh), \; (kh, lh) \in \partial\Omega_h.$$

Here the solution of the difference equation $-\alpha \nabla_h^2 U_{i,j} = f_{i,j}$ (in vector notation, $A\mathbf{U} = \mathbf{f}$) can be imagined as representing the temperature at thermal *equilibrium* in a network with heat sources.

Network analogy.[32] Stieltjes matrices (cf. Chapter 3, §5) arise generally from (linear) *network flow problems* in various branches of physics. Thus the 5-point difference approximation to $-\nabla \cdot (p\nabla u) + q(x,y)u = f(x,y)$ of Chapter 3, (4.2), corresponds to a rectangular D.C. network, whose nodes are the mesh-points and whose conducting wires are the mesh-segments. In this *network analogy*, $U_{i,j}$ is the voltage at the terminal $k = (i,j)$; $R_{i,j} = L_{i+1,j} = c_{k,l} = c_{l,k}$ in (4.2a) of Chapter 3 is the conductance of the wire connecting node $k = (i,j)$ to node $l = (i+1,j)$; $q_{i,j}\alpha_{i,j}$ is the external current flowing into node (i,j); and so on.

In particular, this analogy can be applied to obtain (symmetric) diagonally dominant Stieltjes matrices having Property A in a *general* domain overlaid with a rectangular mesh. Thus it can be used as a substitute for Eq. (4.2) of Chapter 3 when this would give 'irregular stencils' (see our paper in Birkhoff-Schoenstadt [84]).

Analog devices. Using this analogy, one can build rectangular networks of electrical conductors for solving the difference equation $\nabla_h^2 U = 0$ or (4.2)–(4.2a) of Chapter 3 by analogy. Similarly, an electrolytic tank or telegraphic 'teledeltos' paper can be used as an analog computer to solve the DE $\nabla^2 u = 0$ with limited accuracy. However, such 'analog' devices are far less accurate than the digital computer methods described in this book.

Mechanical analogy. Likewise, a mechanical analogue of the Poisson ΔE is provided by a square network of intersecting taut strings under

[32] See R. H. MacNeal, Quart. Appl. Math. 11 (1953), 295–310.

constant tension T, supporting loads $w_k = gm_k$ at the 'mesh-points' where the strings intersect. The linearized 'strain energy' is

$$\frac{1}{2}\mathbf{u}^T A\mathbf{u} = \frac{1}{2}\langle \mathbf{u}, A\mathbf{u}\rangle = \frac{1}{2}T\mathbf{D}_h\langle \mathbf{u},\mathbf{u}\rangle,$$

and the gravitational energy is

$$g\mathbf{w}^T\mathbf{u} = g\textstyle\sum_k w_k u_k = g\langle \mathbf{w},\mathbf{u}\rangle = -\langle \mathbf{b},\mathbf{u}\rangle = -\mathbf{b}^T\mathbf{u}.$$

Static equilibrium occurs when the *total* energy

$$\mathbf{J}[\mathbf{u}] = \frac{1}{2}\langle \mathbf{u}, A\mathbf{u}\rangle - \mathbf{b}^T\mathbf{u}$$

is a minimum, i.e., when $\delta\mathbf{J} = 0$, or, equivalently, $A\mathbf{u} = \mathbf{b}$.

Southwell's relaxation methods were inspired by this classical mechanical analogy. He interpreted the Poisson DE $-\nabla^2 u = f(x,y)$ as "governing the small transverse displacement of a transversely loaded membrane" (Southwell [46, §38]), and $-\nabla_h^2 U_{i,j} = f_{i,j}$ as associated with the corresponding net.

Minimum principles. In equilibrium, the displacements $U_{i,j}$ of the mesh-points minimize the potential energy. The mathematical problem is, therefore, to find those $U_{i,j}$ that *minimize* $\mathbf{J}[\mathbf{u}]$, where the discrete 'energy norm' squared, $\mathbf{u}^T A\mathbf{u}$, is twice the strain energy of the net.

Hand relaxation. Before 1950, a system such as $-\nabla_h^2 U_{i,j} = f_{i,j}$ was solved by 'hand'. An initial guess $U_{i,j}^{(0)}$ of $U_{i,j}$ was made and the residuals $r_{i,j} = h^2(f_{i,j} + \nabla_h^2 U_{i,j}^{(0)})$ were computed at each mesh-point. In one simple version, the mesh-point (i,j) with largest residual was located and the estimated value of $U_{i,j}$ was altered ('relaxed') to make $r_{i,j}$ equal to zero. This changed the residuals at the four adjacent mesh-points by $r_{i,j}/4$, making it easy to locate the largest residual of the new approximation and to repeat the process. With practice, one could achieve skill in 'relaxing', or partially relaxing, several neighboring residuals (a 'block' of residuals)[33] at one step, thus accelerating the process.

Many improvements in 'relaxation methods' by Southwell [40, 46] and his collaborators, including not only block relaxation but also 'overrelaxation' and local mesh refinement, made hand relaxation the most effective procedure for solving elliptic problems in the late 1930's and 1940's. However, the development of SOR in the late 1940's, and of electronic computers with adequate storage in the early 1950's, made hand relaxation obsolete by 1955.

---

[33] Block relaxation was discussed by G. H. Shortley and R. Weller, J. Appl. Phys. 9 (1938), 334–344.

10. Convergence criteria. We now sharpen, systematize, and generalize the discussion of the convergence of one-step stationary iterative methods given in §§7–8. We first observe that

$$\mathbf{u}^{(m+1)} = G\mathbf{u}^{(m)} + \mathbf{k} \tag{10.1}$$

has the same solution as $A\mathbf{u} = \mathbf{b}$ if and only if, for some nonsingular M,

$$G = I - M^{-1}A \qquad \text{and} \qquad \mathbf{k} = M^{-1}\mathbf{b}. \tag{10.1'}$$

For, it leaves the solution $\mathbf{u}$ unchanged if and only if $(I-G)\mathbf{u} = \mathbf{k}$, and therefore, if and only if $M(I-G)\mathbf{u} = M\mathbf{k}$ for some (hence every) nonsingular M.

Residuals; error. For any (real or complex) 'iteration matrix' G and initial 'trial vector' $\mathbf{u}^{(0)}$, define $\mathbf{u}^{(m)}$ recursively by (10.1), the vector *residual* $\mathbf{r}^{(m)}$ by

$$\mathbf{r}^{(m)} = \mathbf{b} - A\mathbf{u}^{(m)}, \tag{10.2}$$

and the more easily computed vector *pseudo-residual* (see [HY, p. 61]) $\mathbf{s}^{(m)}$ by

$$\mathbf{s}^{(m)} = \mathbf{u}^{(m+1)} - \mathbf{u}^{(m)} = \mathbf{k} - (I-G)\mathbf{u}^{(m)}. \tag{10.2'}$$

These vectors are related by $\mathbf{r}^{(m)} = M\mathbf{s}^{(m)}$, where $M = A(I-G)^{-1}$. For Jacobi iteration, $\mathbf{r}^{(m)} = D\mathbf{s}^{(m)}$; for Gauss-Seidel, $\mathbf{r}^{(m)} = (D-E)\mathbf{s}^{(m)}$; and so on.

The effects of iterating the one-step method (10.1) are as follows. First

$$\begin{aligned}\mathbf{s}^{(m+1)} &= \mathbf{u}^{(m+2)} - \mathbf{u}^{(m+1)} = G\mathbf{u}^{(m+1)} + \mathbf{k} - G\mathbf{u}^{(m)} - \mathbf{k} \\ &= G(\mathbf{u}^{(m+1)} - \mathbf{u}^{(m)}) = G\mathbf{s}^{(m)},\end{aligned}$$

whence

$$\mathbf{s}^{(m)} = G^m\mathbf{s}^{(0)} \quad \text{and} \quad \mathbf{r}^{(m)} = MG^mM^{-1}\mathbf{r}^{(0)}.$$

Finally, the *error vector* $\mathbf{e}^{(m)} = \mathbf{u}^{(m)} - \mathbf{u}$ satisfies

$$\mathbf{e}^{(m+1)} = \mathbf{u}^{(m+1)} - \mathbf{u} = G\mathbf{u}^{(m)} + \mathbf{k} - G\mathbf{u} - \mathbf{k} = G(\mathbf{u}^{(m)} - \mathbf{u}) = G\mathbf{e}^{(m)},$$

and hence $\mathbf{e}^{(m)} = G^m\mathbf{e}^{(0)}$. This proves

LEMMA 4. *For any one-step iterative method, the errors* $\mathbf{e}^{(m)}$, *residuals* $\mathbf{r}^{(m)}$, *and pseudo-residuals* $\mathbf{s}^{(m)}$ *satisfy*

$$\mathbf{e}^{(m)} = G^m\mathbf{e}^{(0)}, \qquad \mathbf{r}^{(m)} = MG^mM^{-1}\mathbf{r}^{(0)}, \qquad \mathbf{s}^{(m)} = G^m\mathbf{s}^{(0)}, \tag{10.3}$$

where $M = A(I-G)^{-1}$.

Convergence; spectral radius. On any finite-dimensional vector space, all choices of norm give topologically *equivalent* definitions of convergence, because any two norms differ by a bounded factor:

$$\frac{1}{K}||\mathbf{u}||_\mu \leqslant ||\mathbf{u}||_\nu \leqslant K||\mathbf{u}||_\mu,$$

for some $K < \infty$ independent of $\mathbf{u}$. Hence the following definition and theorems are independent of the choice of norm.

DEFINITION. An iterative process (10.1) *converges* to $\mathbf{u}$ in the norm $||\cdot||$ when $||\mathbf{u}^{(m)} - \mathbf{u}|| \to 0$ as $m \to \infty$ for any initial trial vector $\mathbf{u}^{(0)}$.

It follows by continuity that any limit vector $\mathbf{u}$ satisfies $\mathbf{u} = G\mathbf{u} + \mathbf{k}$. Hence for any nonsingular M, $A\mathbf{u} = \mathbf{b} = M\mathbf{k}$ for $A = M(I - G)$.

By Lemma 4, a one-step iterative method (10.1) converges to the solution $\mathbf{u}$ for arbitrary $\mathbf{u}^{(0)}$ if and only if $G^m \to 0$, the zero matrix. A classic necessary and sufficient condition for $G^m \to 0$ can be stated in terms of the spectral radius $\rho(G)$, defined as follows.[34]

DEFINITION. The *spectral radius* $\rho(A)$ of a square (real or complex) matrix A is the largest of the magnitudes of its eigenvalues. In symbols:

$$\rho(A) = \max\{|\lambda| \text{ such that } \det(A - \lambda I) = 0\}. \tag{10.4}$$

If A is symmetric, then its spectral radius is equal to its 'Euclidean norm' or 'l_2-norm':

$$\rho(A) = ||A||_2 \equiv \max_{||\mathbf{v}||_2 = 1}\{||A\mathbf{v}||_2\}. \tag{10.4'}$$

THEOREM 3. *A scheme* (10.1) *is convergent if and only if* $\rho(G) < 1$. *In words, it is convergent if and only if the spectral radius of* G *is less than one. It converges to the solution of* $A\mathbf{u} = \mathbf{b}$ *if and only if* (10.1′) *holds.*

Sketch of proof.[35] By a fixed change of basis altering nonzero norms in a uniformly bounded ratio, we can write G in the *Jordan canonical form* $J = P^{-1}GP$ as a direct sum of diagonal blocks, $J_1, \cdots, J_k$:

$$J = \begin{bmatrix} J_1 & & & \\ & J_2 & O & \\ & & \cdot & \\ & O & \cdot & \\ & & & J_k \end{bmatrix}, \quad J_i = \begin{bmatrix} \lambda_i & 1 & & & \\ & \lambda_i & 1 & O & \\ & & \cdot & \cdot & \\ & O & & \lambda_i & 1 \\ & & & & \lambda_i \end{bmatrix} = \lambda_i I + E, \tag{10.5}$$

where each square diagonal block $J_i = \lambda_i I + E$ is itself bidiagonal. Moreover, J is unique up to permutations of the blocks. If J_i is $q \times q$, $q = q(i) > 0$, then its powers are given by the polynomial

[34] Young [54, p. 54]; [V, p. 13, pp. 61–68]; and [Y, pp. 77–78, pp. 84–89].

[35] See [V, §1.3], [Y, §2.4], or Birkhoff-Mac Lane [77, §10.10] for detailed explanations and proofs.

$$(10.6)\qquad \begin{aligned} J_i^{\,m} &= \lambda_i^{\,m} I + \binom{m}{1}\lambda_i^{\,m-1}E + \cdots + \binom{m}{p-1}\lambda_i^{\,m-p+1}E^{p-1},\\ p &= \min\{m, q-1\}, \qquad \binom{m}{p} = \frac{m!}{p!\,(m-p)!}. \end{aligned}$$

Evidently, if $\rho(G) > 1$, then $G^m\mathbf{e}^{(0)} \to \infty$ for almost all initial errors $\mathbf{e}^{(0)}$. It follows that $\|G^m\| \to 0$ as $m \to \infty$ if and only if $\rho(G) < 1$.

Moreover, if $\rho = \rho(G) < 1$, then asymptotically,

$$(10.7)\qquad \|J^m\|/\rho^{m-q+1} = O(1), \qquad m \to \infty.$$

A more careful consideration of (10.7) gives the following *quantitative* result about the rate of convergence.

THEOREM 4. *For any one-step iterative method* (10.1) *and any norm* $\|\cdot\|$, *the norms of the error* $\mathbf{e}^{(m)}$, *the residual* $\mathbf{r}^{(m)}$, *and the pseudo-residual* $\mathbf{s}^{(m)}$, *all satisfy*

$$(10.8)\qquad \log\|\cdot\| \leqslant \beta + m\log\rho(G) + \gamma\log m,$$

for suitable finite constants β *and* γ. *For almost all* $\mathbf{u}^{(0)}$,

$$(\log\|\cdot\|)/m \sim \log\rho(G).$$

We omit the proof. The exceptional cases arise when $\mathbf{u}^{(0)} - \mathbf{u}$ ($\mathbf{u}$ the exact solution) is in the invariant subspace spanned by generalized eigenvectors having eigenvalues with modulus less than the spectral radius.

The rate of convergence of an iteration scheme can be measured in terms of its 'average' or its 'asymptotic' rate of convergence.

COROLLARY. *For almost all* $\mathbf{u}^{(0)}$, *in any norm,*

$$(10.9)\qquad \lim_{m\to\infty}\left[\frac{\|\mathbf{e}^{(m)}\|}{\|\mathbf{e}^{(0)}\|}\right]^{1/m} = \rho(G).$$

Hence $\|\mathbf{e}^{(m)}\| \leqslant k\rho^m(G)$ *and* $\log\|\mathbf{e}^{(m)}\| \leqslant a + m\log\rho(G)$ *asymptotically.*

Because of this fact, $-\log\rho(G)$ is called the *asymptotic rate of convergence* for G.

Stein-Rosenberg theorem. A theorem relating the spectral radii of the Jacobi and Gauss-Seidel methods, due to Stein and Rosenberg,[36] asserts that if the Jacobi iteration matrix B is *nonnegative* and $\rho(B) < 1$, then $\rho(\mathbf{L}_1) < \rho(B)$. Therefore, if the Jacobi method converges, the Gauss-Seidel method converges asymptotically faster; cf. [V, §3.3] or [Y, pp. 120–127 and 133]. Likewise, if $\rho(B) > 1$, then $\rho(\mathbf{L}_1) > \rho(B)$. Here B need not be symmetrizable nor have Property A for the Stein-Rosenberg

[36] P. Stein and R. L. Rosenberg, J. Lond. Math. Soc. 23 (1948), 111–118.

theorem to apply. These results have been generalized by J. J. Buoni and R. S. Varga.[37]

Asymptotically, for the 5-point discretization of the linear source problem, the Gauss-Seidel method converges exactly *twice* as rapidly as the Jacobi method in either the 'natural' or the 'red-black' ordering; see §12.

Optimum Richardson. We now determine the (real) ω giving the optimum (fastest asymptotic) rate of convergence of Richardson's method:

$$\mathbf{u}^{(m+1)} = G\mathbf{u}^{(m)} + \mathbf{k}, \quad \text{with} \quad G = I - \omega A \quad \text{and} \quad \mathbf{k} = \omega b,$$

knowing only the least and greatest eigenvalues of A.

THEOREM 5. *Let the eigenvalues of the matrix A be real and positive, and let $\alpha > 0$ be the least and $\beta = \kappa\alpha$ be the greatest eigenvalue of A. Then for*

$$\omega = \omega_{\text{opt}} = 2/(\alpha + \beta), \tag{10.10}$$

Richardson's method converges and

$$\rho(G) = \max\{|1-\omega\alpha|, |1-\omega\beta|\} = \frac{\kappa - 1}{\kappa + 1} < 1; \tag{10.10'}$$

moreover, any value of $\omega \neq \omega_{\text{opt}}$ gives a slower (asymptotic) rate of convergence.

Proof. For any real[38] value ω, the eigenvalues of $G = I - \omega A$ are between $\mu_1 = 1 - \omega\alpha$ and $\mu_2 = 1 - \omega\beta$; thus $\rho(G) = \max\{|\mu_1|, |\mu_2|\}$. This maximum is minimized when $1 - \omega\alpha = \omega\beta - 1$, and thus when ω is given by (10.10). With this ω, $1 - \omega\alpha = (\beta - \alpha)/(\beta + \alpha) < 1$.

11. Estimating $\rho(G)$. In §§7–10, we have referred freely to such quantities as $\rho(A), \lambda_{\min}(A) = \alpha, \rho(B)$, and $\rho(G)$, as if they were known numbers. In practice, these are seldom known *a priori*, and their accurate determination (e.g., using EISPACK; cf. §6) would be prohibitively expensive for a 1500×1500 matrix.

Fortunately, there are a few 'model problems' for which the entire spectrum is known. Hence so are the $\rho(B)$ for the Jacobi iteration matrix, and the optimum ω_{opt} for Richardson's method, from which their rates of convergence for any ω can be computed (and compared). We now turn our attention to some such problems.

Example 5. Eigenfunctions for $-\nabla_h^2$ which are zero on the boundary of the unit square are $\phi_{i,j}(x_k, y_l) = \sin(ik\pi h)\sin(jl\pi h)$ (see Chapter 3, §8,

[37] J. J. Buoni and R. S. Varga in R. Ansorge et al, eds., *Numerical Mathematics*, Birkhäuser, 1979, pp. 65–75.

[38] Complex ω give larger spectral radii.

Example 8) and the eigenvalues for $i, j = 1, \cdots, I-1$ are

$$\lambda_{i,j} = \frac{4}{h^2}[\sin^2(i\pi h/2) + \sin^2(j\pi h/2)]. \tag{11.1}$$

They range from

$$\lambda_{\min} = \frac{8}{h^2}\sin^2(\pi h/2) = \alpha \quad \text{to} \quad \lambda_{\max} = \frac{8}{h^2}\cos^2(\pi h/2) = \beta,$$

whence the condition number κ (Chapter 3, §8) is

$$\kappa(\nabla_h^2) = \frac{\lambda_{\max}}{\lambda_{\min}} = \cot^2(\pi h/2) = \frac{4}{\pi^2 h^2}[1 - \pi^2 h^2/6 + O(h^4)].$$

It follows from Theorem 5 that, for Richardson's method, the optimum ω is $\omega_{\text{opt}} = 2/(\alpha+\beta) = h^2/4$, and

$$\rho(I - \omega_{\text{opt}} A) = 1 - \pi^2 h^2/2 + O(h^4). \tag{11.2}$$

Thus its asymptotic rate of convergence is

$$-\log \rho(I - \omega_{\text{opt}} A) \sim \pi^2 h^2/2. \tag{11.2'}$$

THEOREM 6. *For $-\nabla_h^2 U = f$ or for $\nabla_h^4 U = f$ with given h in a unit hypercube $[0,1]^n$, the rate of convergence of the optimum Richardson method is independent of the dimension number n. Thus in particular, it is the same in the unit cube as in the unit square.*

Proof. For $-\nabla_h^2$ in an n-cube with a cubic mesh, the extreme eigenvalues are, as in Example 5,

$$\lambda_{\min} = \frac{4n}{h^2}\sin^2(\pi h/2) \quad \text{and} \quad \lambda_{\max} = \frac{4n}{h^2}\cos^2(\pi h/2),$$

with $\kappa(\nabla_h^2) = \cot^2(\pi h/2)$; hence by Theorem 5 the asymptotic rate of convergence is independent of n. For ∇_h^4, we have likewise

$$\lambda_{\min} = \left[\frac{4n}{h^2}\sin^2(\pi h/2)\right]^2 \quad \text{and} \quad \lambda_{\max} = \left[\frac{4n}{h^2}\cos^2(\pi h/2)\right]^2,$$

and $\kappa(\nabla_h^4) = \cot^4(\pi h/2)$.

Since $1/\omega_{\text{opt}}$ is equal to the (constant) diagonal entries of the matrix A (in the natural ordering) for $-\nabla_h^2 U = f$, optimum Richardson is the same as the Jacobi method. Hence we have the following.

COROLLARY. *For $-\nabla_h^2 U = f$ as in Theorem 6, the rate of convergence of the Jacobi method is independent of the dimension number n.*

Some values of the condition number of ∇_h^2 and ∇_h^4 on the unit square are given in Table 1.

General domains. For the discrete Laplace operator $-\nabla_h^2$, with Dirichlet boundary conditions, least and greatest (positive real) eigenvalues $\lambda_{\min}$ and $\lambda_{\max}$ (and hence the condition number $\kappa(A)$) can be estimated for

TABLE 1. *Condition numbers.*

$1/h$	10	20	50	100	200
$\kappa(\nabla_h^2)$	3.99×10^1	1.61×10^2	1.01×10^3	4.05×10^3	1.62×10^4
$\kappa(\nabla_h^4)$	1.59×10^3	2.61×10^4	1.03×10^6	1.64×10^7	2.63×10^8

general domains by combining the *domain monotonicity* of the $\lambda_j(\Omega)$ for the Helmholtz equation[39] with an *isoperimetric inequality* stemming from Steiner, Pólya and Szegö.[40]

As pointed out by Garabedian [56], the error $e_{i,j}^{(m)} = U_{i,j} - U_{i,j}^{(m)}$ in SOR gives an approximation to $v(ms,ih,jh)$ where v is the solution of the damped wave equation (see §15). Thus for small h, the error satisfies $\|\mathbf{e}^{(m)}\| \approx K\, e^{mh\,\mathrm{Re}\,[\lambda]}$, where $\lambda = -\mu \pm i\nu$ is related to the overrelaxation parameter ω by the relations

$$\lambda = -2\sigma \pm \sqrt{4\sigma^2 - 2k^2}, \qquad \mu = 2(2-\omega)/\omega h, \qquad \nu = \sqrt{2k^2 - \mu^2}.$$

A solution $v = e^{\lambda s}\phi(x,y)$ tends to zero as $s \to \infty$ if and only if $\mu = -\max\{\mathrm{Re}\,[\lambda]\} > 0$. Its rate of decay is governed by μ and hence by the λ associated with the minimum eigenvalue k_0^2 of $-\nabla^2\phi = k^2\phi$, $\phi = 0$ on Γ. With $A[\Omega]$ denoting the area of Ω, we thus have $k_0^2 \leqslant j_{0,1}{}^2\pi/A[\Omega]$, where $j_{0,1} \approx 2.405$ is the smallest zero of the Bessel function J_0 (see Chapter 1, §5). The maximum μ is obtained with $\omega = 2/(1+k_0h/\sqrt{2})$. This leads to Garabedian's recommendation, to choose the SOR overrelaxation parameter as

$$\omega = \frac{2}{1 + j_{0,1}h\sqrt{\pi/2A[\Omega]}} \approx \frac{2}{1 + 3.014h/\sqrt{A[\Omega]}}. \tag{11.3}$$

Varga [V, pp. 291–296] gives sharper estimates which involve the smallest eigenvalue k_0^2. These use the result of Weinberger[41] that if $k_h{}^2$ is the smallest eigenvalue of the 5-point operator, $-\nabla_h^2\phi_h = k_h{}^2\phi$, $\phi_h = 0$ on Γ_h, then $k_0^2 \leqslant k_h{}^2/(1 - 3h^2k_h{}^2)$.

Finally, in diffusion problems with absorption, when the diffusion length is only a few mesh-lengths (e.g., when $h^2q/p > 0.1$, say, for the DE $-\nabla(p\nabla u) + qu = f$), and more generally, when D 'strongly dominates' $E+F$ in (8.5), the spectral radius can be estimated from this fact alone.

---

[39] This was already known by 1900; see F. Pockels, *Über die partielle Differentialgleichung* $\Delta u + k^2u = 0$, Leipzig 1891. For domain monotonicity, see Chapter 6, §8.

[40] Pólya-Szegö [51]; see also Chapter 6, §8.

[41] H. F. Weinberger, Comm. Pure Appl. Math. 9 (1956), 613–623. See also Weinberger [74, Chap. 6, §3].

12. Property A. As we observed in Chapter 3, §5, the matrix A arising from the standard 5-point difference approximation to the linear source problem on any rectangular mesh has 'Property A'. That is, its indices $i, j = 1, \cdots, N$ can be divided into two nonvoid complementary subsets, such that $a_{i,j} = 0$ unless $i = j$ or i and j belong to complementary subsets.

If we permute the indices of any such matrix so that all those of the first subset come first, the resulting matrix $P^{-1}AP$ (P a permutation matrix) clearly has the form

$$P^{-1}AP = \begin{bmatrix} D_1 & -S_2 \\ -S_1 & D_2 \end{bmatrix} = D' - S, \quad D' = \begin{bmatrix} D_1 & 0 \\ 0 & D_2 \end{bmatrix}, \quad S = \begin{bmatrix} 0 & S_2 \\ S_1 & 0 \end{bmatrix}, \tag{12.1}$$

where D_1 and D_2 are diagonal square matrices whose entries are those of A, rearranged. This amounts to using a 'red-black' ordering (§3) of the unknowns and equations.

For any such red-black ordering, the Jacobi iteration matrix $B = I - (D')^{-1}P^{-1}AP = (D')^{-1}S$ has the form

$$B = \begin{bmatrix} 0 & B_2 \\ B_1 & 0 \end{bmatrix}. \tag{12.2}$$

Matrices of the form (12.2) are called 'weakly cyclic of index two'; we will call them *weakly* 2*-cyclic* for short.[42]

In §13 we will show that, with a red-black ordering, the eigenvalues of the SOR iteration matrix are directly related to those of the Jacobi iteration matrix. This is a consequence of the following basic and beautiful result, derived in Young's Ph.D. Thesis [50]:[43]

THEOREM 7. *Let B be weakly 2-cyclic and in the form of* (12.2). *If μ is an eigenvalue of B, then so is $-\mu$; if $(\mathbf{u},\mathbf{v})^T$ is an eigenvector of B, then so is $(\mathbf{u},-\mathbf{v})^T$.*

Proof. Since similarity transformations leave the spectrum invariant, we can consider B in the form (12.2). If $(\mathbf{u},\mathbf{v})^T$ denotes an eigenvector of the bipartite matrix B with eigenvalue μ, then by definition

$$\begin{bmatrix} 0 & B_2 \\ B_1 & 0 \end{bmatrix} \begin{bmatrix} \mathbf{u} \\ \mathbf{v} \end{bmatrix} = \mu \begin{bmatrix} \mathbf{u} \\ \mathbf{v} \end{bmatrix}, \tag{12.3}$$

and so $B_2\mathbf{v} = \mu\mathbf{u}$, $B_1\mathbf{u} = \mu\mathbf{v}$. Hence

$$\begin{bmatrix} 0 & B_2 \\ B_1 & 0 \end{bmatrix} \begin{bmatrix} \mathbf{u} \\ -\mathbf{v} \end{bmatrix} = -\mu \begin{bmatrix} \mathbf{u} \\ -\mathbf{v} \end{bmatrix},$$

and $-\mu$ is also an eigenvalue of B with eigenvector $(\mathbf{u},-\mathbf{v})^T$.

---

[42] See [V, p. 54] for historical remarks. For the more general concept of 'p-cyclic' matrices, see [V, §4.1] and Chapter 5, §2, below.

[43] See also G. Frobenius, S.-B. Preuss Akad. Wiss., Berlin, (1912), 456–477; V. Romanovsky, Acta Math. 66 (1936), 147–251. See [V, pp. 129–130] for historical comments.

There are two cases. If $\mu = 0$, then $B_2\mathbf{v} = B_1\mathbf{u} = \mathbf{0}$, whence $(\mathbf{u},\mathbf{0})^T$ and $(\mathbf{0},\mathbf{v})^T$ are both in the null space of B. If $\mu \neq 0$, then B has a *linearly independent* pair of eigenvectors with eigenvalues μ and $-\mu$.

LEMMA 5. *Let $(\mathbf{u},\mathbf{v})^T$ be an eigenvector of the Jacobi iteration matrix B in* (12.2) *with nonzero eigenvalue μ. Then $(\mathbf{u},\mathbf{0})^T$ is an eigenvector of the Gauss-Seidel iteration matrix G with eigenvalue* 0, *while $(\mathbf{u},\mu\mathbf{v})^T$ is an eigenvector of G with eigenvalue μ^2. Moreover, the null space of G contains the null space of B.*

Proof. For any α,

$$\begin{bmatrix} 0 & B_2 \\ 0 & 0 \end{bmatrix} \begin{bmatrix} \mathbf{u} \\ \alpha\mathbf{v} \end{bmatrix} = \alpha \begin{bmatrix} B_2\mathbf{v} \\ \mathbf{0} \end{bmatrix} = \alpha\mu \begin{bmatrix} \mathbf{u} \\ \mathbf{0} \end{bmatrix}. \tag{12.4}$$

Therefore, if $\mathbf{w} = (\mathbf{u},\mathbf{0})^T$ (the case $\alpha = 0$), then, with U, L, and the Gauss-Seidel iteration matrix G defined by

$$U = \begin{bmatrix} 0 & B_2 \\ 0 & 0 \end{bmatrix}, \qquad L = \begin{bmatrix} 0 & 0 \\ B_1 & 0 \end{bmatrix}, \quad \text{and} \quad G = (I - L)^{-1}U,$$

we have $G\mathbf{w} = (I-L)^{-1}\mathbf{0} = \mathbf{0}$, which proves the first statement. Likewise, if $\mathbf{w} = (\mathbf{u},\mu\mathbf{v})^T$, then $U\mathbf{w} = {}^2(\mathbf{u},\mathbf{0})^T$ by (12.4). Since

$$(I-L)\begin{bmatrix} \mathbf{u} \\ \mu\mathbf{v} \end{bmatrix} = \begin{bmatrix} I & 0 \\ -B_1 & I \end{bmatrix} \begin{bmatrix} \mathbf{u} \\ \mu\mathbf{v} \end{bmatrix} = \begin{bmatrix} \mathbf{u} \\ -B_1\mathbf{u} + \mu\mathbf{v} \end{bmatrix} = \begin{bmatrix} \mathbf{u} \\ \mathbf{0} \end{bmatrix},$$

it follows that $G\mathbf{w} = (I-L)^{-1}U\mathbf{w} = \mu^2\mathbf{w}$, which proves the second statement.

Finally, if $(\mathbf{u},\mathbf{v})^T$ is an eigenvector of B with eigenvalue 0, then trivially $U\mathbf{w} = \mathbf{0}$ with $\mathbf{w} = (\mathbf{u},\mathbf{v})^T$, and so $(\mathbf{u},\mathbf{v})^T$ is in the null space of G. This proves the last statement of the lemma.

COROLLARY. *Under the hypotheses of Theorem 7, the Gauss-Seidel method converges exactly twice as fast as the Jacobi method, asymptotically.*

The following result of Forsythe and Straus[44] shows that the diagonal scaling built into the Jacobi method is beneficial.

THEOREM 8. *Let A be a symmetric positive definite matrix with Property A and condition number $\kappa(A) = \lambda_{\max}/\lambda_{\min}$, and let $D = \operatorname{diag}(A)$ and $A' = D^{-1/2}AD^{-1/2}$. Then for any diagonal matrix P, $\kappa(A') \leqslant \kappa(PAP)$. In particular, $\kappa(A') \leqslant \kappa(A)$.*

Proof. By Remark 2 of §8, Richardson's method for solving $(I-B)\mathbf{u} = D^{-1}\mathbf{b}$, with $\omega = 1$ is the same as the Jacobi method if $D = dI$.

By Lemma 1 of §7, the eigenvalues of B are real. If A also has Property A, then by Theorem 7

---

[44] G. E. Forsythe and E. G. Straus, Proc. Am. Math. Soc. 6 (1955), 340–345; [Y, pp. 214–215].

(12.5) $$\mu_{\max}(B) = -\mu_{\min}(B) = 1-\epsilon.$$

Since A is by hypothesis symmetric positive definite, it then follows from the Ostrowski-Reich theorem (§8) that SOR converges for $\omega = 1$ (i.e., Gauss-Seidel converges). In particular, it converges for any red-black ordering and Lemma 5 then shows that the spectral radius of B is less than unity, whence $0 < \epsilon \leqslant 1$ in (12.5). Therefore,

$$\kappa(I-B) = \frac{2-\epsilon}{\epsilon} \quad \text{and} \quad \epsilon = \frac{2}{\kappa(I-B)+1}.$$

Consequently, because $A' = D^{-1/2}AD^{-1/2} = I - D^{1/2}BD^{-1/2}$, the spectral radius, $\rho(B) = 1-\epsilon$, of B is

$$\rho(B) = \frac{\kappa(I-B)-1}{\kappa(I-B)+1} = \frac{\kappa(I-D^{1/2}BD^{-1/2})-1}{\kappa(I-D^{1/2}BD^{-1/2})+1} = \frac{\kappa(A')-1}{\kappa(A')+1}.$$

But then, since $f(x) = (x-1)/(x+1)$ is increasing for $x \geqslant 0$, it follows from Theorems 5 and 8 that

$$\rho(B) = \frac{\kappa(A')-1}{\kappa(A')+1} \leqslant \frac{\kappa(A)-1}{\kappa(A)+1} = \rho(I-\omega_{\text{opt}}A) \leqslant \rho(I-\omega A),$$

which completes the proof.

COROLLARY. *If A is symmetric positive definite with Property A, then the Jacobi method converges asymptotically at least as fast as Richardson's method with $\omega = \omega_{\text{opt}}$.*

A simple example. The following 2×2 example shows that the Jacobi method can converge much more rapidly than optimum Richardson, for positive definite symmetric matrices with Property A:

$$A = \begin{bmatrix} M^2 & M-\epsilon \\ M-\epsilon & 1 \end{bmatrix}, \qquad B = D^{-1/2}\begin{bmatrix} 0 & 1-\epsilon/M \\ 1-\epsilon/M & 0 \end{bmatrix}D^{-1/2};$$

consequently $\rho(B) = 1-\epsilon/M$. On the other hand,

$$|A-\lambda I| = \lambda^2 - (M^2+1)\lambda - (1-2\epsilon M).$$

Hence asymptotically, for small ϵ and large M,

$$\lambda_i \approx \frac{1}{2}(M^2+1)\left[1 \pm \sqrt{1-8\epsilon/M^2}\right].$$

Therefore, $\lambda_1 \approx M^2$, $\lambda_2 \approx 2\epsilon/M$, $\kappa = \lambda_1/\lambda_2 \approx M^3/2\epsilon$, and $\rho(G)$ for the optimum ω satisfies

$$\rho(G) = \frac{\kappa-1}{\kappa+1} \approx 1 - \frac{2}{\kappa} \approx 1 - \frac{4\epsilon}{M^3}.$$

Example 6. Consider the 9-point approximation $\mathbf{N}_h$ to the Laplacian given in (10.2) of Chapter 3. The associated matrix does not have Property A. However, in the natural ordering, its Jacobi iteration matrix has eigenvalues which differ by $O(h^2)$ from those of the corresponding matrix for the 5-point approximation ∇_h^2. Therefore, the Jacobi method converges, and by the Stein-Rosenberg theorem (§10), the Gauss-Seidel method converges asymptotically faster. Also, by the Ostrowski-Reich theorem (§8), SOR converges for $0 < \omega < 2$.

Example 7. Consider the 13-point approximation ∇_h^4 to the biharmonic operator and conditions $u = \nabla^2 u = 0$ on the boundary of a unit square. In the natural ordering, its Jacobi iteration matrix has spectral radius $-1 + (64/20)\cos^4(\pi h/2) = 2.2 + O(h^2)$ and so the Jacobi method diverges. Since the matrix corresponding to ∇_h^4 is positive definite, optimum Richardson converges, while by the Ostrowski-Reich theorem, SOR converges for $0 < \omega < 2$. In particular Gauss-Seidel ($\omega = 1$) converges. (The Stein-Rosenberg theorem does not apply because the Jacobi iteration matrix has some negative entries, but its extensions by Buoni and Varga[37] do.)

Gauss-Seidel iteration. For matrices with Property A in a red-black ordering, one can write $\mathbf{u}$ in partitioned form as $(\mathbf{v},\mathbf{w})^T$ where $\mathbf{v}$ and $\mathbf{w}$ are the vectors whose components are those of $\mathbf{u}$ at 'red' and 'black' points, respectively. Then, after multiplication by D^{-1}, the Gauss-Seidel method (8.5) becomes [V, §5.4]:

$$\begin{bmatrix} I & 0 \\ -B_1 & I \end{bmatrix} \begin{bmatrix} \mathbf{v}^{(m+1)} \\ \mathbf{w}^{(m+1)} \end{bmatrix} = \begin{bmatrix} 0 & B_2 \\ 0 & 0 \end{bmatrix} \begin{bmatrix} \mathbf{v}^{(m)} \\ \mathbf{w}^{(m)} \end{bmatrix} + \begin{bmatrix} \mathbf{k}_1 \\ \mathbf{k}_2 \end{bmatrix}.$$

Writing this as a pair of vector equations, one gets

$$\mathbf{v}^{(m+1)} = B_2 \mathbf{w}^{(m)} + \mathbf{k}_1, \qquad \mathbf{w}^{(m+1)} = B_1 \mathbf{v}^{(m+1)} + \mathbf{k}_2.$$

One can then iterate

$$\mathbf{w}^{(m+1)} = B_1(B_2 \mathbf{w}^{(m)}) + (B_1 \mathbf{k}_1 + \mathbf{k}_2).$$

13. Point SOR. We now treat the *successive overrelaxation* (SOR) method (see p. 123). This one-step method differs from the Gauss-Seidel method in that it multiplies each successive component of a Gauss-Seidel pseudo-residual $u_i^{(m+1)} - u_i^{(m)}$ by a parameter ω called the *overrelaxation factor*. The components of $\mathbf{u}^{(m+1)}$ are computed by

$$(13.1) \quad u_i^{(m+1)} = u_i^{(m)} + \omega \left[\sum_{j=1}^{i-1} b_{i,j} u_j^{(m+1)} + \sum_{j=i+1}^{N} b_{i,j} u_j^{(m)} + k_i - u_i^{(m)} \right],$$

with $b_{i,j}$ and k_i as in (8.2′). For $\omega = 1$, SOR reduces to the Gauss-Seidel method.

For example, with the standard 5-point difference approximation to the Poisson equation $-\nabla^2 u = f$ on a square mesh, with spacing h in the natural ordering, the values of the approximations $U_{i,j}$ of $u(x_i, y_j)$ are computed by iterating

$$U_{i,j}^{(m+1)} = (1-\omega)\,U_{i,j}^{(m)} + \frac{\omega}{4}\,[U_{i-1,j}^{(m+1)} + U_{i,j-1}^{(m+1)} + U_{i+1,j}^{(m)} + U_{i,j+1}^{(m)} + h^2 f_{i,j}]. \tag{13.1'}$$

In general, for any $A = D - E - F$, with $D = \operatorname{diag}(A)$ nonsingular and E and F strictly lower and upper triangular, respectively, the SOR method applied over a complete cycle of components is equivalent to

$$(D-\omega E)\,\mathbf{u}^{(m+1)} = [(1-\omega)D + \omega F]\,\mathbf{u}^{(m)} + \omega\mathbf{b}. \tag{13.2}$$

If we set $L = D^{-1}E$ and $U = D^{-1}F$, this becomes

$$\mathbf{u}^{(m+1)} = \mathbf{L}_\omega \mathbf{u}^{(m)} + \mathbf{k}, \tag{13.3}$$

as in (8.10)–(8.10′), with

$$\mathbf{L}_\omega = (I-\omega L)^{-1}[(1-\omega)I + \omega U], \qquad \mathbf{k} = \omega(I-\omega L)^{-1}D^{-1}\mathbf{b}. \tag{13.3'}$$

Whereas formula (13.1′) assumes the natural ordering of mesh-points (i,j), formulas (13.2)–(13.3) hold for *any* (cyclic) ordering of mesh-points, if improved values $u_i^{(m+1)}$ are used as soon as available. Moreover, formula (13.3) can also be written in terms of the residual $\mathbf{r}^{(m)} = \mathbf{b} - A\mathbf{u}^{(m)}$ or pseudo-residual $\mathbf{s}^{(m)} = \mathbf{u}^{(m+1)} - \mathbf{u}^{(m)}$ as

$$\mathbf{u}^{(m+1)} = \mathbf{u}^{(m)} + \omega(I-\omega L)^{-1}D^{-1}\mathbf{r}^{(m)} = \mathbf{u}^{(m)} + \mathbf{s}^{(m)}.$$

The error $\mathbf{e}^{(m)} = \mathbf{u} - \mathbf{u}^{(m)}$ is transformed under (13.3) by the formula

$$\mathbf{e}^{(m+1)} = \mathbf{L}_\omega \mathbf{e}^{(m)} = \mathbf{L}_\omega^m \mathbf{e}^{(0)}.$$

Young's identity. For any positive definite symmetric A with Property A, Young derived an identity which, for any red-black ordering and any ω, relates the spectrum of $\mathbf{L}_\omega$ to that of the underlying Jacobi iteration matrix B.

THEOREM 9. *Let A be symmetric positive definite and have Property A. Then under any red-black ordering, the eigenvalues λ of $\mathbf{L}_\omega$ are related to corresponding eigenvalues μ of B by Young's identity:*[45]

$$(\lambda + \omega - 1)^2 = \lambda\omega^2\mu^2. \tag{13.4}$$

[45] In his Ph.D. Thesis, Young [50] treated other orderings for which (13.4) holds; see §14. In MTAC 7 (1953), 152–159, A. M. Ostrowski independently derived (13.4) for 2×2 matrices. Frankel [50] analyzed SOR for the special case of the Poisson equation and obtained the optimal ω for the 5-point approximation in the natural ordering.

Proof. We have, much as in (12.3),

$$(13.5)\quad D^{-1}A=\begin{bmatrix} I & -B_2\\ -B_1 & I\end{bmatrix},\quad B=\begin{bmatrix} 0 & B_2\\ B_1 & 0\end{bmatrix},\quad L=\begin{bmatrix} 0 & 0\\ B_1 & 0\end{bmatrix},\quad U=\begin{bmatrix} 0 & B_2\\ 0 & 0\end{bmatrix},$$

where the diagonal blocks are square matrices.

Now let $\mathbf{w}=(\mathbf{r},\mathbf{s})^T$ denote an eigenvector of $\mathbf{L}_\omega$ with eigenvalue λ. Then

$$(I-\omega L)^{-1}[(\omega-1)I+\omega U]\,\mathbf{w}=\lambda\,\mathbf{w},$$

or equivalently

$$[(\omega-1)I+\omega U]\mathbf{w}=\lambda\,(I-\omega L)\mathbf{w}.$$

For L and U as in (13.5), we have also

$$\begin{bmatrix} (\omega-1)I & \omega B_2\\ 0 & (\omega-1)I\end{bmatrix}\begin{bmatrix}\mathbf{r}\\ \mathbf{s}\end{bmatrix}=\lambda\begin{bmatrix} I & 0\\ -\omega B_1 & I\end{bmatrix}\begin{bmatrix}\mathbf{r}\\ \mathbf{s}\end{bmatrix},$$

whence

$$(13.6)\qquad \omega B_2\mathbf{s}=(\lambda+\omega-1)\mathbf{r},\qquad \omega\lambda B_1\mathbf{r}=(\lambda+\omega-1)\mathbf{s}.$$

Unless $\lambda\omega=0$, this implies

$$(13.7)\qquad \begin{bmatrix} 0 & B_2\\ B_1 & 0\end{bmatrix}\begin{bmatrix}\mathbf{r}\\ \gamma^{-1}\mathbf{s}\end{bmatrix}=\begin{bmatrix} B_2(\gamma^{-1}\mathbf{s})\\ B_1\mathbf{r}\end{bmatrix}=\frac{(\lambda+\omega-1)}{\gamma\omega}\begin{bmatrix}\mathbf{r}\\ \gamma^{-1}\mathbf{s}\end{bmatrix},\quad \gamma=\lambda^{1/2}.$$

By Theorem 7 of §12, both μ and $-\mu$ are eigenvalues of B. Comparing the first expression of (13.7) with that of (12.3), we see that

$$(13.8)\qquad \frac{(\lambda+\omega-1)}{\omega\lambda^{1/2}}=\pm\mu$$

is an eigenvalue of B. From this (13.4) follows.

In case $\lambda=0$, (13.6) reduces to

$$(13.9)\qquad \omega B_2\mathbf{s}=(\omega-1)\mathbf{r},\qquad \mathbf{0}=(\omega-1)\mathbf{s}.$$

For $(\mathbf{r},\mathbf{s})^T$ to be an eigenvector, one of $\mathbf{r}$, $\mathbf{s}$ must be nonzero. If $\mathbf{s}=\mathbf{0}$, then the first equation in (13.9) shows that $\omega=1$; if $\mathbf{s}\neq\mathbf{0}$, then the second shows that $\omega=1$. Thus $\lambda=0$ implies $\omega=1$, which is the Gauss-Seidel case treated in §12. In this case (13.4) holds trivially, completing the proof.

The following important result is also due to Young [54].

THEOREM 10. *In Theorem 9, let B with $\rho=\rho(B)<1$ have a real spectrum, so that $\mu\in[-\rho,\rho]$. Let*

$$\omega_b \equiv \frac{2}{\rho^2[1-\sqrt{1-\rho^2}]} = \frac{2}{1+\sqrt{1-\rho^2}}. \tag{13.10}$$

If $\omega_b \leqslant \omega < 2$, *then every eigenvalue* λ *of* $\mathbf{L}_\omega$ *has modulus* $\omega - 1$. *If* $0 < \omega < \omega_b$, *then* $\rho(\mathbf{L}_{\omega_b}) < \rho(\mathbf{L}_\omega) < 1$.

Proof. If $\mu = 0$, then, by Theorem 9, the corresponding λ is $1-\omega$. For given $\omega \neq 0$ and $\pm\mu$, the corresponding λ's are obtained from (13.8) as

$$\lambda_+, \lambda_- = [\mu\omega \pm \sqrt{\mu^2\omega^2 - 4(\omega-1)}]^2. \tag{13.11}$$

Thus, the eigenvalues λ occur in complex conjugate pairs when $\omega_b < \omega < 2$; and by (13.4)

$$\lambda^2 + [2(\omega-1) - \omega^2\mu^2]\lambda + (\omega-1)^2 = 0.$$

This shows that the product of the roots λ_+, λ_- satisfies

$$\lambda_+\lambda_- = (\omega-1)^2 = |\lambda_+|^2 = |\lambda_-|^2.$$

When the radicand in (13.8) is positive, so that the eigenvalues λ_+, λ_- are real, the larger eigenvalue is an increasing function of μ which attains its maximum when $\mu = \rho$. For $0 < \omega < \omega_b$, the eigenvalues λ_+, λ_- corresponding to $\pm\rho(B)$ are real and nonnegative; the larger is given by

$$4\lambda_+ = [\omega\rho + \sqrt{\omega^2\rho^2 - 4(1-\omega)}]^2; \tag{13.12}$$

the right side is a strictly decreasing function of ω, $0 \leqslant \omega \leqslant \omega_b$, which varies between 1 and $\omega_b - 1$.

COROLLARY. *In Theorem* 10, *the SOR method converges for* $0 < \omega < 2$; *the spectral radius of* $\mathbf{L}_\omega$ *is minimized with* $\omega = \omega_b$ *and* $\rho(\mathbf{L}_{\omega_b}) = \omega_b - 1$.

DEFINITION. For B with a real spectrum and spectral radius $\rho = \rho(B)$, $0 \leqslant \rho < 1$, the *optimum omega* $\omega = \omega_b$ is defined by Eq. (13.10).

The most interesting range (in the sense that accelerating Gauss-Seidel is most crucial) is $\rho = 1-\epsilon$ ($\epsilon > 0$ small). In this range, $\omega_b = 2 - 2\sqrt{2\epsilon} + O(\epsilon^{3/2})$ and

$$\rho(\mathbf{L}_{\omega_b}) = \omega_b - 1 = \frac{1-\sqrt{1-\rho^2}}{1+\sqrt{1-\rho^2}} = 1 - 2\sqrt{2\epsilon} + O(\epsilon^{3/2}).$$

If A is symmetric positive definite, let $\kappa = \kappa(A')$ denote the ratio of the largest to the smallest eigenvalue of $A' = D^{-1/2}AD^{-1/2}$, i.e., its (Euclidean and spectral) *condition number*. Then, in terms of κ,

$$\rho(B) = 1 - \epsilon = \frac{\kappa-1}{\kappa+1} \quad \text{and} \quad \epsilon = \frac{2}{\kappa+1},$$

whence

$$\rho(\mathbf{L}_{\omega_b}) \sim 1 - 4/\sqrt{\kappa}. \tag{13.13}$$

Example of Young [50]. The effect of the *ordering of unknowns* on the spectral radius is illustrated by $-\nabla_h^2 U = f$ on a 2×2 set of interior mesh-points. We consider three orderings of the mesh-points: 'natural', 'red-black', and 'clockwise'. These orderings are shown in Fig. 5.

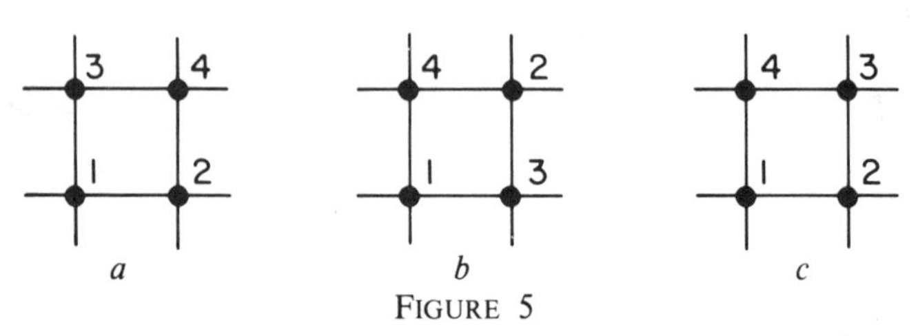

FIGURE 5

The Jacobi iteration matrices for these are:

$$\frac{1}{4}\begin{bmatrix} 0 & 1 & 1 & 0 \\ 1 & 0 & 0 & 1 \\ 1 & 0 & 0 & 1 \\ 0 & 1 & 1 & 0 \end{bmatrix}, \quad \frac{1}{4}\begin{bmatrix} 0 & 0 & 1 & 1 \\ 0 & 0 & 1 & 1 \\ 1 & 1 & 0 & 0 \\ 1 & 1 & 0 & 0 \end{bmatrix}, \quad \frac{1}{4}\begin{bmatrix} 0 & 1 & 0 & 1 \\ 1 & 0 & 1 & 0 \\ 0 & 1 & 0 & 1 \\ 1 & 0 & 1 & 0 \end{bmatrix},$$

for the 'natural', the 'red-black', and the 'clockwise' orderings, respectively. Each has spectrum $\{-1/2, 0, 0, 1/2\}$.

The resulting Gauss-Seidel iteration matrices ($\mathbf{L}_\omega$ with $\omega = 1$) are:

$$\frac{1}{8}\begin{bmatrix} 0 & 8 & 8 & 0 \\ 0 & 2 & 2 & 8 \\ 0 & 2 & 2 & 8 \\ 0 & 1 & 1 & 1 \end{bmatrix}, \quad \frac{1}{8}\begin{bmatrix} 0 & 0 & 2 & 2 \\ 0 & 0 & 2 & 2 \\ 0 & 0 & 1 & 1 \\ 0 & 0 & 1 & 1 \end{bmatrix}, \quad \frac{1}{256}\begin{bmatrix} 0 & 64 & 0 & 64 \\ 0 & 16 & 64 & 16 \\ 0 & 4 & 16 & 68 \\ 0 & 17 & 4 & 33 \end{bmatrix},$$

for the 'natural', the 'red-black', and the 'clockwise' orderings, respectively. The first two of these have spectrum $\{0, 0, 0, 1/4\}$ with spectral radius 1/4, exactly one-half that of the corresponding Jacobi matrix. The 'clockwise' Gauss-Seidel iteration matrix has spectrum $\{0, 0.277, -0.011 \pm 0.118i\}$.

The reason why the third Gauss-Seidel iteration matrix has a different spectrum, which does *not* satisfy Young's identity (13.4), is that the determinant of the *Jacobi* iteration matrix B to which it corresponds has a nonzero term

$$b_{12} b_{23} b_{34} b_{41} \neq 0,$$

having a different number of factors $b_{12} b_{23} b_{34}$ above the main diagonal than below it. In contrast, each nonzero term of the determinants of the other two Jacobi iteration matrices has just as many factors above as below the main diagonal. A fuller explanation of this 'consistent ordering' condition, which is discussed at greater length in [V, pp. 121–122] and [Y, Chap. 5], will be given in the next section.

14. Property Y; numerical examples. We now consider the effect of the node ordering on the spectral radius of SOR iteration. The iteration matrices for the Jacobi and SOR methods are $B = I - D^{-1}A$ and $\mathbf{L}_\omega = (I - \omega L)^{-1}([1-\omega]I + \omega U)$, respectively. Although the spectrum of B is invariant under any permutation of nodes, $B \mapsto P^{-1}BP$, that of $\mathbf{L}_\omega$ is not. However, for matrices A arising from the 5-point discretization of any linear source problem, the spectrum of $\mathbf{L}_\omega$ for any red-black ordering *is* the same as for the natural ordering (by rows), the natural ordering by columns, and for many other node orderings. Young [50] proved this by using a 'consistent ordering' property satisfied by all these orderings. We will use instead a slightly more general Property Y, defined as follows.[46]

DEFINITION. The *directed graph* $G(A)$ of the $N \times N$ matrix $A = \|a_{i,j}\|$ has the vertices (nodes) $k = 1, \cdots, N$, with $\overline{ij}$ an *arc* if and only if $i \neq j$ and $a_{i,j}$ is nonzero. We will say that A has *Property Y* when, in every simple cycle $\gamma = \overline{k_0 k_1}, \overline{k_1 k_2}, \cdots, \overline{k_m k_0}$ of $G(A)$, $k_n > k_{n+1}$ exactly as often as $k_n < k_{n+1}$ $(n+1 \bmod m)$.

LEMMA. *Let $A = \|a_{i,j}\|$ have Property Y, and let $a_{i,j} \neq 0$ imply $a_{j,i} \neq 0$. Then A has Property A.*

Sketch of proof. By hypothesis, the directed graph of A is symmetric; hence $G(A)$ is the direct sum of connected components (actually, in most applications, it is connected). In each connected component, Property Y implies that every cycle has even length. Hence for each k, either every path from node 1 to node k has *even* length or every such path has *odd* length. Calling k 'red' in the first case and 'black' in the second, we see that $G(B) = G(D-A)$ is bipartite (if A has Property Y).

Examples. It is easy to show that any tridiagonal matrix has Property Y. Also, if A has Property A, then its red-black ordering has Property Y. Similarly, the natural ordering (by rows or columns) of the standard 5-point approximation to a second-order linear source problem, and ordering by diagonals, all lead to matrices which have Property Y.

Because $k_n < k_{n+1}$ occurs when the corresponding matrix entry is above the main diagonal of A and $k_n > k_{n+1}$ when it is below it, a matrix has Property Y precisely when each nonzero term in its determinant has just as many factors above as below the main diagonal. There follows

THEOREM 11. *If A has Property Y, and its Jacobi iteration matrix $B = I - D^{-1}A$ has lower and upper triangular components L and U, respectively, then for any $\alpha \neq 0$ the matrix $B(\alpha) = \alpha L + \alpha^{-1}U$ has the same characteristic polynomial as B.*[47]

[46] Where we say that a matrix has Property Y, Young calls it "consistently ordered". The graph-theoretic definition of Property Y is due to Varga; see [V, §2.4, §4.1, and pp. 121–122].

[47] Varga defines a 2-cyclic matrix A to be *consistently ordered* if the spectrum of $B(\alpha)$ is the same as B for all $\alpha \neq 0$; see [V, p. 101].

Proof. Each term $c(\beta)=\prod_k c_{k,\beta_k}$ in the expansion of the determinant of $C=B(\alpha)-\mu I$ arises from a permutation β which is a product of simple cycles γ. Each cycle, by definition of Property Y, contains just as many entries of αL as of $\alpha^{-1}U$, thus in each nonzero term $c(\beta)$, α and α^{-1} occur as factors equally often. Since each nonzero term is independent of α, so is the sum

$$\det[B(\alpha)-\mu I]=\sum_\beta c(\beta)=\det[B-\mu I],$$

which completes the proof.

The next theorem generalizes Theorem 9 from matrices indexed by a red-black ordering to matrices having Property Y.

THEOREM 12.[48] *For any matrix A having Property Y, the eigenvalues λ of $\mathbf{L}_\omega$ are related to the eigenvalues μ of $B=I-D^{-1}A$ by Young's identity*

$$(\lambda+\omega-1)^2=\lambda\mu^2\omega^2. \tag{14.1}$$

Proof. We first obtain a relation between the characteristic polynomials of $\mathbf{L}_\omega$ and B. Since L is strictly lower triangular, $\det[I-\omega L]=1$ and the characteristic polynomial of $\mathbf{L}_\omega$ is

$$\begin{aligned}P_\omega(\lambda)&=\det[\mathbf{L}_\omega-\lambda I]=\det[(I-\omega L)^{-1}\{(1-\omega)I+\omega U\}-\lambda I]\\&=\det[\lambda\omega L+\omega U-(\lambda+\omega-1)I].\end{aligned}$$

If $\omega=0$, then $\mathbf{L}_\omega=I$. If $\lambda=0$ is an eigenvalue of $\mathbf{L}_\omega$, then $P_\omega(0)=(1-\omega)^N=0$; thus $\mathbf{L}_\omega$ is nonsingular if $\omega\neq 1$ (i.e., except for Gauss-Seidel iteration).

If $\lambda\omega\neq 0$, then

$$P_\omega(\lambda)=(\omega\sqrt{\lambda})^N\det\left[\sqrt{\lambda}L+\frac{2}{\sqrt{\lambda}}U-\frac{\lambda+\omega-1}{\omega\sqrt{\lambda}}I\right]. \tag{14.2}$$

Set $\alpha=\sqrt{\lambda}\neq 0$ in Theorem 11; then the characteristic polynomial of $B=I-D^{-1}A$ is

$$Q(\mu)=\det[L+U-\mu I]. \tag{14.2'}$$

Comparison of (14.2) with (14.2') then gives (14.1) for $\lambda\omega\neq 0$, which completes the proof.

It follows from Theorem 12 and Theorem 6 of §11, that the rate of convergence of SOR for solving $-\nabla_h^2U=f$, when the matrix has Property Y, is independent of the dimension number n.

Number of iterations. To illustrate the effectiveness of the SOR method, we consider the number m of iterations required to reduce the initial error

---

[48] This result, originally due to Young [50], [54], has been generalized by Varga to consistently ordered p-cyclic matrices; see Chapter 5, §2.

by a factor of one thousand, for the Model Problem ($-\nabla_h^2 u = f$ in the unit square). If ρ is the spectral radius of the iteration matrix, then $R = -\log\rho$ is the asymptotic *rate of convergence*.

For the Jacobi, optimum Richardson, and Gauss-Seidel methods, $m = 6.908/R$ (since $\log 1000 = 6.908$). For the Model Problem, the spectral radii of the Jacobi and the optimum Richardson methods are equal and, by Lemma 5 of §12, are twice the spectral radius of the Gauss-Seidel method; thus m for them is twice that for Gauss-Seidel. For SOR with 'optimum ω', m can be estimated from $m\rho^{m-1} = 1/1000$.[49] Table 2 lists ρ and m for the Model Problem and various values of h.

TABLE 2.

$1/h$	10	20	50	100	200
	Spectral radius ρ				
Gauss-Seidel	.90451	.97553	.99606	.99901	.99975
SOR	.52786	.72945	.88184	.93909	.96907
	m required for 1000-fold error reduction				
Gauss-Seidel	69	279	1749	6998	27995
SOR	17	35	92	195	413

Next we consider the effect of not using the 'optimum' $\omega = \omega_b$, in the special case $h = 1/20$, $\rho(B) = .98769$, $\omega_b = 1.72945$. Table 3 records, for an initial error given by random numbers of magnitude $O(1)$, and a series of exponents p, the (experimentally observed) number m of iterations required to reduce the l_2-norm of the error by a factor of 10^p.

Table 3 shows two things. First, it is much better to *overestimate* ω_b than to underestimate it. And second, a value of ω *slightly* larger than ω_b will usually reduce the l_2-norm of the error by a factor of 10^p a little faster than ω_b itself (see [Y, p. 227]).

Finally, we consider the number of multiplications per unknown required to reduce the error to h^2, the order of the discretization error. Using 6 multiplications per unknown per iteration[50] and m iterations, where $m\rho^{m-1} = h^2$, and the spectral radii and the h listed in Table 2, then SOR uses 90, 186, 600, 1410, and 3234 multiplications, respectively, per unknown. This compares with 81, 361, 2401, 9801, and 39601 for

[49] The Jordan canonical form of the iteration matrix has a 2×2 block J_p (see §10) associated with $\lambda = \omega_b - 1 = \rho(\mathbf{L}_{\omega_b})$ [HY, Theorem 9-3.1, p. 218].

[50] The Model Problem can be done with 2 multiplications and 6 additions.

TABLE 3. *Number of SOR iterations to reduce error to* 10^{-p}.

ω	$\rho(\mathbf{L}_\omega)$	$p=1$	2	3	4	5	6	7	8	9	10
1.67945	.84863	16	31	45	59	73	87	101	115	129	143
1.71945	.79032	14	26	35	45	55	65	74	84	94	104
$\omega_b=1.72945$	.72945	13	25	32	42	49	57	65	72	79	87
1.73945	.73945	13	24	30	38	44	52	61	67	76	84
1.74945	.74945	13	23	30	40	47	51	62	71	80	86
1.77945	.77945	12	23	33	40	51	59	69	79	87	99

band elimination. On the other hand, Gauss-Seidel uses many more than either; using $\rho^m = h^2$, we obtain 254287 for $h = 1/200$.

15. Frankel's method. Several other iterative methods have nearly the same rate of convergence as optimum SOR. One of these is the 'second-order Richardson' method invented (and named) by S. Frankel [50];[51] we shall call this *Frankel's method*.

Frankel, having noticed "the formal equivalence between the (first-order) Richardson method (see §8) and the solution of the time-dependent equation $u_t = L[u]$," looked for and found a "modified iteration method . . . equivalent to . . . "

$$u_{tt} + \gamma u_t = \mathbf{L}[u],$$

the equation of damped linear oscillations. Guided by this physical analogy, Frankel proposed an iterative scheme to be analyzed below.

In the same vein, Paul Garabedian [56] pointed out that for the *Laplace* equation on a domain Ω, the SOR method applied to $\nabla_h^2 U_{i,j} = 0$ in the natural ordering can be interpreted as giving an approximation $U_{i,j}^{(m)} \approx v$ at time $t = mh$ for v a solution of the (damped) wave equation

$$v_{ss} + 4\sigma v_s = 2(v_{xx} + v_{yy}), \quad s = t + x/2 + y/2, \ \sigma = (2-\omega)/\omega h. \tag{15.1}$$

Solutions of the form $v = e^{\lambda s}\phi(x,y)$, with $\lambda = -\mu \pm i\nu$ *complex*, correspond to the *real* eigenvalues k^2 of the reduced wave equation $\nabla^2\phi + k^2\phi = 0$, through the formulas

$$\lambda = -2\sigma \pm \sqrt{4\sigma^2 - 2k^2}, \qquad \mu = 2(2-\omega)/\omega h, \qquad \nu = \sqrt{2k^2 - \mu^2}.$$

[51] See also James D. Riley, MTAC 8 (1954), 125–131; also see E. L. Wachspress, P. M. Stone, and C. E. Lee, Proc. of Second U. N. Internat. Conf. on Peaceful Uses of Atomic Energy, U. N. Geneva, Vol. 16, pp. 483–488.

The smallest eigenvalue, $k^2 = k_0^2$, limits the rate of decay; this rate is maximized when $\mu^2 = 2k_0^2$.

Garabedian also showed that SOR applied to the 9-point approximation $\mathbf{N}_h U_{i,j} = 0$ (Chapter 3, §10) to the Laplace equation can be interpreted as giving an approximation to the solution of the hyperbolic equation

$$2v_{xt} + 3v_{yt} + 5\sigma v_t = 3(v_{xx} + v_{yy}). \tag{15.2}$$

This leads to Garabedian's estimate of ω_b:

$$\omega = \frac{2}{1 + 3.014h/\sqrt{A[\Omega]}}. \tag{15.3}$$

Frankel's method for solving $A\mathbf{u} = \mathbf{b}$ uses the *two-step stationary* iterative scheme

$$\mathbf{u}^{(m+1)} = \mathbf{u}^{(m)} + \sigma[\mathbf{b} - A\mathbf{u}^{(m)}] + \eta[\mathbf{u}^{(m)} - \mathbf{u}^{(m-1)}], \tag{15.4}$$

where σ *and* η are iteration parameters, for which see (15.7)–(15.7″).

This is a two-step method of 'simultaneous displacements' which expresses $\mathbf{u}^{(m+1)}$ in terms of $\mathbf{u}^{(m)}$ and $\mathbf{u}^{(m-1)}$; new values, $u_i^{(m+1)}$, are *not* used as soon as available. It has the great potential advantage of being well-suited to parallel computers. Moreover, it does *not* require the matrix to have 'Property A' or to be consistently ordered (i.e., to have Property Y).

It has the disadvantage of requiring the storage of two vectors $\mathbf{u}^{(k)}$ ($u_i^{(m+1)}$ overwrites $u_i^{(m-1)}$ on a sequential computer), whereas SOR requires only one. Thus for a 5-point symmetric difference approximation, which requires one to store the coefficients and right side, the Frankel method requires about 20% more storage than SOR; it also requires about twice as many iterations as optimum SOR. For all these reasons, it is much less efficient on sequential computers.

Under the iteration of (15.4), the error $\mathbf{e}^{(m)} = \mathbf{u}^{(m)} - \mathbf{u}$ is transformed by the corresponding *homogeneous* formula

$$\mathbf{e}^{(m+1)} = \mathbf{e}^{(m)} - \sigma A\mathbf{e}^{(m)} + \eta[\mathbf{e}^{(m)} - \mathbf{e}^{(m-1)}].$$

The residual $\mathbf{r}^{(m)} = \mathbf{b} - A\mathbf{u}^{(m)}$ and pseudo-residual $\mathbf{s}^{(m)} = \mathbf{u}^{(m+1)} - \mathbf{u}^{(m)}$ then satisfy the same equation:

$$\mathbf{r}^{(m+1)} = \mathbf{r}^{(m)} - \sigma A\mathbf{r}^{(m)} + \eta[\mathbf{r}^{(m)} - \mathbf{r}^{(m-1)}],$$
$$\mathbf{s}^{(m+1)} = \mathbf{s}^{(m)} - \sigma A\mathbf{s}^{(m)} + \eta[\mathbf{s}^{(m)} - \mathbf{s}^{(m-1)}].$$

Let A have a basis of *real* eigenvectors, $\boldsymbol{\phi}_1, \cdots, \boldsymbol{\phi}_N$, with corresponding real eigenvalues $\lambda_1, \cdots, \lambda_N$, in the interval $[\alpha, \beta]$, with $0 < \alpha \leqslant \beta = \kappa\alpha$. The error $\mathbf{e}^{(m)}$ after the m-th iteration is therefore a linear combination $a_1^{(m)}\boldsymbol{\phi}_1 + \cdots + a_N^{(m)}\boldsymbol{\phi}_N$, whose coefficients $a_j^{(m)}$ satisfy

$$a_j^{(m+1)} - (1 - \sigma\lambda_j + \eta) a_m^{(m)} + \eta a_j^{(m-1)} = 0, \tag{15.5}$$

with $a_j^{(0)}, a_j^{(1)}$ the coefficients of the (unknown) initial errors: $\mathbf{e}^{(m)} = \mathbf{u}^{(m)} - \mathbf{u}$, $m = 0, 1$. Consequently, if the roots ν_{j+}, ν_{j-} of

$$\nu^2 - (1 - \sigma\lambda_j + \eta)\nu + \eta = 0 \tag{15.6}$$

are distinct, then the coefficients are

$$a_j^{(m)} = b_j(\nu_{j+})^m + c_j(\nu_{j-})^m.$$

If $\nu_{j-} = \nu_{j+}$, then

$$a_j^{(m)} = (b_j + c_j m)(\nu_{j+})^m.$$

As with SOR, the objective is to minimize the spectral radius ρ:

$$\rho = \max |\nu| < 1.$$

By (15.6), the product of the roots, $\nu_{j+}\nu_{j-}$, is η. If they are real and distinct, then the magnitude of one of them must exceed η. Consequently, the maximum spectral radius is obtained by choosing σ and η to make the complex roots (all of the same magnitude $|\eta|^{1/2} = \rho$) just fit the range $[\alpha, \beta]$, $\beta = \kappa\alpha$, of the (real) eigenvalues of A. This requires

$$1 - \sigma\alpha + \eta = 2\sqrt{\eta}, \qquad 1 - \sigma\kappa\alpha + \eta = -2\sqrt{\eta}.$$

Thus, $(\kappa - 1)(\eta + 1) = 2(\kappa + 1)\sqrt{\eta}$, and so with $\mu = (\kappa - 1)/(\kappa + 1)$, η satisfies $\eta - 2\sqrt{\eta}/\mu + 1 = 0$. The smaller solution of this equation is

$$\rho = \sqrt{\eta} = \frac{1 - \sqrt{1 - \mu^2}}{\mu} = \frac{\mu}{1 + \sqrt{1 - \mu^2}}.$$

Hence

$$\rho^2 = \eta = \frac{1 - \sqrt{1 - \mu^2}}{1 + \sqrt{1 - \mu^2}} = \omega - 1, \tag{15.7}$$

$$\rho = \sqrt{\omega - 1}, \qquad \omega = \frac{2}{1 + \sqrt{1 - \mu^2}}, \qquad \mu = \frac{\kappa - 1}{\kappa + 1} \tag{15.7$'$}$$

and

$$\sigma = \frac{4\rho}{(\kappa - 1)\alpha} = \frac{4\rho}{\alpha + \beta}. \tag{15.7$''$}$$

Comparing (15.7′) with (13.10), one sees that if μ is equal to the spectral radius of the Jacobi iteration matrix B of A, then $\omega = \omega_b$, the optimal parameter for SOR. Therefore by (15.7′) and the corollary of Theorem 10, the spectral radius of the optimal Frankel method is the *square root* of the spectral radius of the optimal SOR method. By (13.13), the asymptotic rate of convergence of the Frankel method is $-\log\rho \approx 2/\sqrt{\kappa}$, half that of optimum SOR.

Examples. For the 5-point approximation of the 'Model Problem' of Example 5, we have $-\log\rho \approx \pi h$ for the Frankel method. For the 9-point approximation of Chapter 3, §10, whose matrix does not have Property A, we have $-\log\rho \approx \pi\sqrt{3/2}\,h$. Its asymptotic rate of convergence is about 22% less than for the 5-point approximation.

The matrix associated with the 13-point difference approximation, ∇_h^4, to the biharmonic operator does not have Property A. For the clamped plate problem on a unit square and for the natural ordering, $\kappa = \cot^4 \pi h/2$ (see §11). Substitution into (15.7′) and simplification shows that

$$\rho = \cos^2 \pi h/2 - \sin^2 \pi h/2 = 1 - \pi^2 h^2/2 + O(h^4),$$

so that the spectral radius and the asymptotic rate of convergence of the optimum Frankel method for ∇_h^4 are the same as for the Jacobi scheme for $-\nabla_h^2$. Thus the spectral radius, and the number of iterations to reduce the initial error by a factor of 1000, are both twice those tabulated for the Gauss-Seidel method in Table 2 of §14.

With $U_{i,j}$ denoting the solution of $-\nabla_h^2 U = f$ at the mesh-point (ih, jh), one can evaluate the components of the iterates of (15.4) for each i, j as follows:

$$r_{i,j}^{(m)} = f_{i,j} - (4U_{i,j}^{(m)} - U_{i+1,j}^{(m)} - U_{i-1,j}^{(m)} - U_{i,j-1}^{(m)} - U_{i,j+1}^{(m)})/h^2$$

$$s_{i,j}^{(m)} = \sigma r_{i,j}^{(m)} + \eta(U_{i,j}^{(m)} - U_{i,j}^{(m-1)}), \qquad U_{i,j}^{(m+1)} = U_{i,j}^{(m)} + s_{i,j}^{(m)}.$$

The value $U_{i,j}^{(m+1)}$ can overwrite $U_{i,j}^{(m-1)}$, because this latter value is not needed after this calculation, i.e., storage for only two copies of each unknown is needed.

Norms of the residual $\mathbf{r}$ or the pseudo-residual $\mathbf{s}$ can be used to estimate the condition number κ and the iteration error (see Chapter 5, §7).

Some other equations equivalent to $A\mathbf{u} = \mathbf{b}$ can be used without changing the spectral radius, for example $(I - B)\mathbf{u} = \mathbf{k}$, where, with D the diagonal of A, $I - B = D^{-1}A$ and $\mathbf{k} = D^{-1}\mathbf{b}$ so that B is the Jacobi iteration matrix. In (15.4) $[\mathbf{b} - A\mathbf{u}^{(m)}]$ is replaced by $[\mathbf{k} - (I - B)\mathbf{u}^{(m)}]$. For other choices, see [Y, Chap. 16].

Vector and parallel machines. The algorithms discussed in this book, like the Fortran language, were designed for *sequential* performance on a machine having just one 'arithmetic unit'. However, *vector* machines having many 'chips' capable of performing arithmetic operations on streams of values ('vectors'), and designed to be able to compute *inner products* much faster, are becoming increasingly available. Moreover, *parallel* machines capable of performing many arithmetic operations simultaneously are currently under design and some are available commercially.

For a discussion of the architecture of vector (pipeline) and parallel computers, as well as numerical techniques for vector and parallel computation, see R. W. Hockney and C. R. Jesshope, *Parallel Computers*, Adams Hilger, 1981; K. Hwang and F. A. Briggs, *Computer Architecture and Parallel Processing*, McGraw-Hill, 1984. See also Part I of Birkhoff-Schoenstadt [84]. Specific applications and vector algorithms, such as FFT and cyclic reduction, are given in papers in *Parallel Computations*, G. Rodrigue (ed.), Academic Press, 1982.

Whereas the Gauss-Seidel and SOR algorithms of 'successive displacements' can be carried out on vector machines just as efficiently[52] as methods of 'simultaneous displacements' like Jacobi, Richardson, or Frankel, the latter are clearly better suited to parallel machines. Since Frankel's method converges in many fewer iterations than Jacobi's or Richardson's, it seems the best suited of the methods discussed in this chapter for solving linear source problems on vector or parallel machines.

[52] See H. S. Price and K. H. Coats, Trans. SPE of AIME 257 (1974), 295–308; D. J. Hunt in *Infotech State of the Art Report: Suppercomputers*, vol 2, C. R. Jesshope and R. W. Hockney (eds.), (1979), 205–219; J. W. Watts, *Proc. SPE-AIME* 54*th Annual Fall Technical Conf. and Exhibition* (1979).

Chapter 5

Accelerating Convergence

1. Introduction. In Chapter 4 we described various 'direct' and 'iterative' methods for solving large sparse linear systems $A\mathbf{u} = \mathbf{b}$ arising from elliptic problems. All of them are asymptotically more efficient (have less 'computational complexity') than the simple band-elimination method described in Chapter 3 (and available in LINPACK). The present chapter is concerned with further improvements in efficiency, that have been found to be especially effective for solving linear source problems.

Four of the five iterative methods considered in Chapter 4 (Jacobi, Gauss-Seidel, Richardson, and SOR) were *stationary one-step* schemes of the form $\mathbf{u}^{(m+1)} = G\mathbf{u}^{(m)} + \mathbf{k}$, for an appropriate 'iteration matrix' G; cf. Chapter 4, §8.

We will explain in §2 how to speed up SOR by combining it with 'block decomposition'. This replaces the *purely* iterative methods described in Chapter 4, §§8, 7, and 12, by a 'hybrid' method of the more general form

$$M\mathbf{u}^{(m+1)} = N\mathbf{u}^{(m)} + \mathbf{b}, \qquad N = MG, \quad \mathbf{b} = M\mathbf{k}, \tag{1.1}$$

where M is *block* diagonal with square, often tridiagonal, blocks on the main diagonal. Extensions of SOR theory to matrices not having Property Y will also be discussed in §2, and 'symmetric' SOR (SSOR) as well.

We will then take up, in §3, the *conjugate gradient* (CG) method of Hestenes and Stiefel. This applies to $A\mathbf{u} = \mathbf{b}$, whenever A is *positive definite symmetric*, and more generally whenever A is *similar* to such a matrix — i.e., whenever A has a basis of real eigenvectors with positive eigenvalues. Whereas Gauss-Seidel and SOR are stationary methods of 'successive' displacements, the CG method is a *non*stationary method of 'simultaneous' displacements.

In this last respect, it is like the (stationary) Jacobi, Richardson, and Frankel schemes of Chapter 4, §7, §8, and 15. And like them, it gives rise to a sequence of approximate solutions of the form

$$\begin{aligned} \mathbf{u}^{(m)} &= \mathbf{u}^{(0)} - c_{m,1}\mathbf{r}^{(0)} - c_{m,2}A\mathbf{r}^{(0)} - \cdots - c_{m,m}A^{m-1}\mathbf{r}^{(0)} \\ &= \mathbf{u} - A^{-1}p_m(A)\mathbf{r}^{(0)}, \qquad \mathbf{r}^{(0)} = \mathbf{b} - A\mathbf{u}^{(0)}, \end{aligned} \tag{1.2}$$

where $p_m(A)$ is a polynomial of degree m satisfying $p_m(0) = 1$. The *residual* of the m-th approximation is then $\mathbf{r}^{(m)} = p_m(A)\mathbf{r}^{(0)}$. The (linear) subspace of *all* vectors of the form $\Sigma_{j=0}^{m} A^j \mathbf{r}^{(0)}$ is called the m-th *Krylov subspace* K_m of A and $\mathbf{r}^{(0)}$. The CG method *minimizes the energy norm of the error* $\mathbf{e}^{(m+1)}$ *in the translated Krylov subspace* $\mathbf{u}^{(0)} + K_m$.

In §4 we take up the Chebyshev semi-iterative (CSI) method. This is also a polynomial method of the form (1.2); it has a remarkable *minimax* property. Namely, given the class $\mathbf{A}$ of all matrices whose spectrum is in a given interval $[\alpha, \kappa\alpha]$, $\alpha > 0$, each m-th 'residual polynomial' $p_m(A)$ minimizes the maximum possible spectral radius: $\max_{A \in \mathbf{A}} \rho(p_m(A))$. More precisely, the choice $p(A) = p_m(A)$ minimizes $\max_{A \in \mathbf{A}} \rho(p(A))$ in the class of all polynomials of degree m satisfying $p(0) = 1$. (This residual polynomial is different from that of the CG method.)

This minimax property is achieved by making $p_m(x)$ be a scaled translate of $T_m(x)$, the Chebyshev polynomial of degree m. Since the Chebyshev polynomials satisfy a three-term recurrence formula,[1] so do the successive CSI approximations $\mathbf{u}^{(m)}$. Specifically,

$$\mathbf{u}^{(m+1)} = \gamma_m \omega(\mathbf{b} - A\mathbf{u}^{(m)}) + \gamma_m \mathbf{u}^{(m)} + (1-\gamma_m)\mathbf{u}^{(m-1)}, \tag{1.3}$$

where $\omega = 2/(\alpha+\beta)$, as for the optimum (stationary) Richardson method. Like Frankel's method (Chapter 4, §15), CSI is thus a *two-step* scheme; but unlike Richardson's and Frankel's methods, it is not stationary.

In §5 and §6 we take up the current 'methods of choice' for solving large linear source problems. There we first describe the 'cyclic' CSI method (CCSI), which combines the best features of CSI and optimum SOR. We next treat the *Chebyshev acceleration* of SSOR which, if $\rho(LU) \leqslant 1/4 + O(h^2)$, converges more rapidly than optimum SOR by an order of magnitude (like $O(\kappa^{-1/4})$ instead of $O(\kappa^{-1/2})$), thereby attaining a given accuracy in many *fewer* iterations. In §6 we go on to *conjugate gradient acceleration*, which can be appreciably more efficient for some classes of problems.

Next, in §7, we discuss the complicated and difficult question of adaptive parameter selection. This can greatly speed up the numerical solution when large classes of related elliptic problems are being treated. We also discuss stopping criteria.

We devote §8 and §9 to the alternating direction implicit (ADI) methods of Peaceman, Rachford, Douglas, and others. Like CG acceleration, these also converge very rapidly in favorable cases, but their most important use is in conjunction with *parabolic* ADI, to solve time-dependent problems.

---

[1] Lanczos [52] used the three-term recurrence relation. Algorithms based on Chebyshev polynomials which do not use it are very sensitive to rounding errors; see D. M. Young, J. Math. Phys. 32 (1954), 243–255, and [Y, p. 365].

Finally, in §10, we discuss the multigrid method and mesh refinement.

ITPACK. Most of the methods described in §§2–7 can be implemented using ITPACK,[2] a transportable package of Fortran subroutines developed at the University of Texas for solving linear systems iteratively. Besides SOR, ITPACK provides modules which accelerate Jacobi and SSOR by conjugate gradient or Chebyshev acceleration. Parameters, such as ω for SOR, are obtained adaptively from repeatedly improved estimates of the smallest and largest eigenvalues.

2. Improvements in SOR. In Chapter 4, §13 and §14, we analyzed the rate of convergence of the SOR method for solving $A\mathbf{u}=\mathbf{b}$, assuming that the matrix A had Property Y, and that the spectral radius $\rho(B)=\mu_1$ of the associated Jacobi iteration matrix $B=I-D^{-1}A$ was known. From $\rho(B)$, the 'optimum omega', ω_b, and the asymptotic rate of convergence can then be calculated easily using (13.10) of Chapter 4, as

$$\omega_b=\frac{2}{1+\sqrt{1-\rho(B)^2}} \quad \text{and} \quad \rho(\mathbf{L}_{\omega_b})=\omega_b-1. \tag{2.1}$$

In order to get these results as quickly as possible, we omitted several important related topics.

Block reduction.[3] These omitted topics include 'line', 'multiline', and 'block' SOR, which effectively combine SOR with direct methods, by using the following notion of 'block decomposition' to eliminate groups of unknowns.

Clearly, any partition of the unknowns u_k of $A\mathbf{u}=\mathbf{b}$ into n successive disjoint groups $\mathbf{u}_i$ decomposes the matrix A into n^2 *blocks*. In symbols, we have

$$A\mathbf{u}=\begin{bmatrix} A_{11} & A_{12} & \cdot & A_{1n} \\ A_{21} & A_{22} & \cdot & A_{2n} \\ \cdot & \cdot & \cdot & \cdot \\ A_{n1} & A_{n2} & \cdot & A_{nn} \end{bmatrix}\begin{bmatrix} \mathbf{u}_1 \\ \mathbf{u}_2 \\ \cdot \\ \mathbf{u}_n \end{bmatrix}=\begin{bmatrix} \mathbf{b}_1 \\ \mathbf{b}_2 \\ \cdot \\ \mathbf{b}_n \end{bmatrix}=\mathbf{b}, \tag{2.2}$$

where the diagonal blocks A_{ii} are square, and the components of $\mathbf{u}_i$ are the values of the unknowns in the i-th group. The block SOR method is given by[4]

[2] D. M. Young and D. R. Kincaid, in Schultz [81, pp. 163–185]; D. R. Kincaid, J. R. Respess, D. M. Young, and R. G. Grimes, Algorithm 586, TOMS 8 (1982), 302–322. ITPACK is available from IMSL; it is also included in ELLPACK.

[3] See [V, §6.4], Hageman-Varga [64], and the references given there. Early relevant papers include J. Schroeder, ZaMM 34 (1954), 241–253; R. J. Arms, L. D. Gates and B. Zondek, J. SIAM 4 (1956), 220–229; J. Heller, ibid. 8 (1960), 150–173. For historical comments, see [V, p. 207].

[4] For $\omega=1$, this is the block Gauss-Seidel method, and if $\mathbf{u}_j^{(m+1)}$ is replaced with $\mathbf{u}_j^{(m)}$, it becomes the block Jacobi method.

$$(2.3)\qquad A_{ii}\mathbf{u}_i^{(m+1)} = A_{ii}\mathbf{u}_i^{(m)} + \omega\left[\mathbf{b}_i - \sum_{j=1}^{i-1} A_{ij}\mathbf{u}_j^{(m+1)} - \sum_{j=i}^{n} A_{ij}\mathbf{u}_j^{(m)}\right],$$

$i = 1, \cdots, n$. For it to be effective, the systems $A_{ii}\mathbf{x} = \mathbf{y}$ must be easy to solve.

We write $A = D - E - F = D(I - B)$, where D is nonsingular *block* diagonal with diagonal blocks A_{ii}, and E and F are strictly upper and lower block triangular, respectively. The matrix $B = I - D^{-1}A$ is the block Jacobi iteration matrix, and the block SOR iteration matrix is

$$\mathbf{L}_\omega = (I - \omega L)^{-1}([1-\omega]I + \omega U), \qquad L = D^{-1}E, \quad U = D^{-1}F.$$

In the definition of Property A given in Chapter 3, §5, if the entries a_{ij} are replaced with the blocks A_{ij}, one gets the definition of *block Property A*, and similarly for *block consistently ordered* and *block Property Y* (Chapter 4, §14). When A has block Property Y, $\rho(\mathbf{L}_\omega)$ is minimized by the $\omega = \omega_b$ given in (2.1).

Line SOR. From the standard 5-point approximation to a linear source problem and the natural ordering, one gets the *line SOR* method by partitioning the unknowns so that the components of $\mathbf{u}_i$ are the unknown nodal values on the i-th row of the mesh. This makes (2.2) a block tridiagonal system. Since the diagonal blocks A_{ii} are themselves tridiagonal, each i-th system in (2.3) can be solved quickly. Moreover, Cuthill and Varga[5] have constructed a method of solution which, for positive definite tridiagonal A_{ii}, requires little more computation than the point SOR iterative method.

More generally, on a rectangular mesh, *k-line SOR* partitions the matrix A into blocks associated with the nodes on disjoint sets of k adjacent lines.

For the Model Problem of $-\nabla_h^2 \mathrm{U} = \mathbf{f}$ on the unit square, the spectral radius for optimum k-line SOR is asymptotically, for small h,[6]

$$\rho\left(\mathbf{L}_{\omega_b}^{(k-line)}\right) \sim 1 - 2\pi h\sqrt{2k};$$

for optimum point SOR, $\rho(\mathbf{L}_{\omega_b}) \sim 1 - 2\pi h$. Thus for $k = 1$, the asymptotic rate of convergence for line SOR is about 40% greater than for point SOR and the amount of computation is about the same. Likewise, the use of 2-line SOR nearly doubles the rate of convergence. But since the computational effort increases linearly with k, little is gained by using much larger values of k.

---

[5] E. Cuthill and R. S. Varga, J. Assn. Comp. Mach. 6 (1959), 236–244.

[6] S. V. Parter, Numer. Math. 1 (1959), 240–252; ibid. 3 (1961), 305–319; see also [V, §6.5] and [Y, §§14.4–14.5].

Similarly, for the 9-point discretization of ∇^2 having a 3×3 stencil, line SOR gives a block tridiagonal matrix whose diagonal blocks are also tridiagonal. In contrast, for *point* SOR, the matrix does not have point Property A. To make the matrix block tridiagonal for the biharmonic equation, and more generally for any 5×5 stencil, it suffices to take $k=2$. For $\nabla_h^4=\nabla_h^2\nabla_h^2$, the 13-point approximation to the biharmonic operator on an $N\times N$ rectangular array of mesh-points, the band half-width of the $2N\times 2N$ diagonal blocks can be reduced from $N+1$ to five.[7]

Stieltjes matrices. The asymptotic rate of convergence of optimum SOR is the same up to a factor of about 2 for all Stieltjes matrices having a given condition number, whether or not they have Property Y; this was first shown by Kahan [58] and Varga [59]. In particular, if A is a Stieltjes matrix, the lack of Property Y does not dramatically decrease the rate of convergence of SOR.[8] Specifically, Kahan and Varga showed that Young's optimum overrelaxation factor ω_b is still 'good'. Their results can be summarized as follows; for proofs and generalizations, see [V, §§4.4–4.5] and [Y, §12.2].

THEOREM 1. *Let A be a Stieltjes matrix. Then, for the ω_b of* (2.1),

$$\omega_b-1\leqslant\rho(\mathbf{L}_{\omega_b})\leqslant\sqrt{\omega_b-1}\quad\text{and}\quad\omega_b-1\leqslant\min_\omega\{\rho(\mathbf{L}_\omega)\}<\sqrt{\omega_b-1}.$$

Moreover, $\omega_b-1=\rho(\mathbf{L}_{\omega_b})$ if and only if A has Property Y.

For $\rho(B)=1-\epsilon$, where ϵ is small, the asymptotic rate of convergence $-\log\rho(\mathbf{L}_{\omega_b})$ of SOR therefore satisfies

$$\sqrt{2\epsilon}=-\frac{1}{2}\log(\omega_b-1)\leqslant-\log\rho(\mathbf{L}_{\omega_b})\leqslant\log(\omega_b-1)=2\sqrt{2\epsilon}.$$

Consequently, since $\rho(\mathbf{L}_\omega)\geqslant\omega_b-1$ for *any* relaxation factor ω, this ω_b is a *good* relaxation factor.

We also note that, since the determinant of $\mathbf{L}_\omega$ is equal to the product of its eigenvalues and

$$|\det[\mathbf{L}_\omega]|=|\det[(I-\omega L)^{-1}([1-\omega]I+\omega U)]|=|(1-\omega)|^N\leqslant\rho(\mathbf{L}_\omega)^N,$$

we have the lower bound on the spectral radius of $\mathbf{L}_\omega$:

$$|\omega-1|\leqslant\rho(\mathbf{L}_\omega). \tag{2.4}$$

Symmetric SOR (SSOR). An important variant of SOR is Sheldon's 'symmetric' SOR (or SSOR), in which sweeps are alternately made in the forward and backward directions of the ordering.[9] Although SSOR is less

[7] S. V. Parter, Numer. Math. 1 (1959), 240–252; J. ACM 8 (1961), 359–365.

[8] See Varga [59]; see also C. G. Broyden, Numer. Math. 12 (1968), 47–56.

[9] J. W. Sheldon, Math. Tables Aids Comput. 9 (1955), 101–112; J. Assoc. Comput. Mach. 6 (1959), 495–504.

efficient than SOR for symmetric positive definite matrices A, it has the advantage of having *real* eigenvalues. As a result, it can be accelerated by the Chebyshev semi-iterative method (§5) and by the conjugate gradient method (§6). Thus, for the linear source problem, accelerated SSOR achieves convergence in $O(h^{-1/2})$ iterations, as compared with the $O(h^{-1})$ iterations required by ordinary SOR.[10]

For $A\mathbf{u}=\mathbf{b}$, with $A=D-E-F$, $D=\operatorname{diag}(A)$ and E and F strictly lower and upper triangular, respectively, the SSOR iterative method is defined in terms of $L=D^{-1}E$, $U=D^{-1}F$, by

$$(2.5)\qquad \begin{aligned}(I-\omega L)\mathbf{u}^{(m+1/2)}&=([1-\omega]I+\omega U)\mathbf{u}^{(m)}+\omega\mathbf{b},\\ (I-\omega U)\mathbf{u}^{(m+1)}&=([1-\omega]I+\omega L)\mathbf{u}^{(m+1/2)}+\omega\mathbf{b}.\end{aligned}$$

It has a *symmetrizable* iteration matrix $\mathbf{S}_\omega$:

$$(2.6)\qquad \mathbf{S}_\omega=(I-\omega U)^{-1}([1-\omega]I+\omega L)(I-\omega L)^{-1}([1-\omega]I+\omega U).$$

The rate of convergence of SSOR is relatively insensitive to the exact choice of ω, and so the optimum SSOR parameter, ω_0, can be approximated effectively.[11] In particular, if the 'SSOR condition'

$$(2.7)\qquad \rho(LU)\leqslant 1/4$$

is satisfied, then the parameter ω_1 defined by

$$(2.8)\qquad \omega_1=\frac{2}{1+\sqrt{2[1-\rho(B)]}}$$

gives a nearly optimum rate of convergence. Specifically [Y, Theorem 3.1, p. 464],

$$(2.9)\qquad \rho(\mathbf{S}_{\omega_0})\leqslant\rho(\mathbf{S}_{\omega_1})\leqslant\frac{1-\sqrt{[1-\rho(B)]/2}}{1+\sqrt{[1-\rho(B)]/2}}.$$

In particular, (2.7) is satisfied for $-\nabla_h^2U_{i,j}=f_{i,j}$ in the natural ordering (but *not* in the red-black ordering, where the optimum ω is unity and $\rho(\mathbf{S}_1)=\rho(B)^2$) as well as for the standard 5-point discretization (Chapter 3, (4.2)–(4.2a)) of the axially symmetric Poisson equation.

p-cyclic matrices. A square matrix A is called (weakly) *p-cyclic* when, for some permutation matrix P, $P^{-1}AP=B$ has the form

$$(2.10)\qquad B=\begin{bmatrix}0&0&\cdot&0&S_1\\ S_2&0&\cdot&0&0\\ 0&S_3&\cdot&0&0\\ \cdot&\cdot&\cdot&\cdot&\cdot\\ 0&0&\cdot&S_p&0\end{bmatrix}.$$

[10] See Young [77] and [Y, §§15.1 and 15.4]. See also L. W. Ehrlich, J. SIAM 12 (1964), 807–826; G. J. Habetler and E. L. Wachspress, Math. Comp. 15 (1961), 356–362; O. Axelsson in Barker [77, pp. 1–51]; D. M. Young, J. Approx. Theory 5 (1972), 137–148.

[11] See Kincaid-Young [79, p. 367] and the references they cite.

Although it is the case $p=2$ that arises most frequently in applications (e.g., the Jacobi iteration matrix $B=D^{-1}A-I$, obtained for the linear source problem, is 2-cyclic for a red-black ordering), p-cyclic matrices with $p>2$ arise in the 'outer iterations' of the multigroup diffusion equations (Chapter 1, §8).[12]

Varga [V, §4.2] has generalized many basic results about 2-cyclic matrices to p-cyclic matrices. In particular, if μ is an eigenvalue of B in (2.10), then so is $\mu\, e^{2\pi ir/p}$, $r=1,2,\cdots,p-1$. For Gauss-Seidel iteration, $\rho(\mathbf{L}_1)=[\rho(B)]^p$, so the asymptotic rate of convergence is p times that for Jacobi iteration. More generally, the eigenvalues λ of $\mathbf{L}_\omega$ are related to those of B by Varga's identity

$$(\lambda+\omega-1)^p=\lambda^{p-1}\omega^p\mu^p. \tag{2.11}$$

Moreover [V, Theorem 4.5], the spectral radius of SOR is again $1-O(h)$ with the optimum overrelaxation parameter. For recent results about the application of SSOR to p-cyclic systems, see Varga in Birkhoff-Schoenstadt [84, pp. 198–201].

3. Conjugate gradient method.[13] In this section we discuss the *conjugate gradient* (CG) method of Hestenes-Stiefel [52]; in the next, we will take up the related Chebyshev semi-iterative (CSI) method. These methods can be used to solve $A\mathbf{u}=\mathbf{b}$ whenever A is symmetric and positive definite. Moreover, each method has a remarkable *optimality* property, which is associated with the sequence of *Krylov subspaces* $K_0\subset K_1\subset\cdots\subset K_m\subset\cdots$ already mentioned in §1.

This important concept is most easily defined when $\mathbf{u}^{(0)}=\mathbf{0}=A\mathbf{u}^{(0)}$. In this case, the *m-th Krylov subspace* K_m is the *linear* subspace spanned by[14] $\mathbf{b}$, $A\mathbf{b}$, $A^2\mathbf{b}$, $\cdots$, $A^m\mathbf{b}$. Since the *initial residual* $\mathbf{r}^{(0)}=\mathbf{b}-A\mathbf{u}^{(0)}$ is then $\mathbf{b}$, K_m is also spanned by $\mathbf{r}^{(0)}$, $A\mathbf{r}^{(0)}$, $\cdots$, $A^m\mathbf{r}^{(0)}$. Moreover, by setting $\mathbf{v}=\mathbf{u}-\mathbf{u}^{(0)}$ and $\mathbf{v}^{(m)}=\mathbf{u}^{(m)}-\mathbf{u}^{(0)}$, we can reduce the problem of solving $A\mathbf{u}=\mathbf{b}$ to that of solving $A\mathbf{v}=A\mathbf{u}-A\mathbf{u}^{(0)}=\mathbf{b}-A\mathbf{u}^{(0)}=\mathbf{r}^{(0)}$, with $\mathbf{v}^{(0)}=\mathbf{0}$. For general $\mathbf{u}^{(0)}$, $\mathbf{u}^{(m+1)}$ is in the translated Krylov subspace $K_m+\mathbf{u}^{(0)}$ obtained by translating K_m through $\mathbf{u}^{(0)}$.

---

[12] See also Wachspress [66, p. 83 and §8.3 and §9.4].

[13] See Theorems 5.1 and 6.1 in Hestenes-Stiefel [52]. Discussions can also be found in H. A. Antosiewicz and W. C. Rheinboldt in Todd [62, pp. 501–512]; D. K. Faddeev and V. N. Faddeeva, *Computational Methods of Linear Algebra*, Freeman and Co., 1963, Chap. 6; Hestenes [80, Chap. 9, esp. pp. 234–235)]; also, Engeli-Ginsburg-Rutishauser-Stiefel [59]. An Algol program for the conjugate gradient method, due to Th. Ginsburg, is included in Wilkinson-Reinsch [71, pp. 5–69]. For the conjugate residual method, see R. Chandra, S. C. Eisenstat, and M. H. Schultz in R. Vichnevetsky, ed., *Advances in Computer Methods for Partial Differential Equations*, 2 (1977), pp. 13–19, IMACS.

[14] See A. S. Householder, *The Theory of Matrices in Numerical Analysis* Blaisdell, 1964, p. 18.

Residual polynomial. We consider schemes for which each m-th approximate solution $\mathbf{u}^{(m)}$ is given by

$$\mathbf{u}^{(m)} = \mathbf{u}^{(0)} - \sum_{k=1}^{m} c_{m,k} A^{k-1}\mathbf{r}^{(0)} \in K_{m-1} + \mathbf{u}^{(0)}. \tag{3.1}$$

Then, because $\mathbf{r}^{(m)} = \mathbf{b} - A\mathbf{u}^{(m)}$, we have

$$\mathbf{r}^{(m)} = [I + \sum_{k=1}^{m} c_{m,k} A^{k}]\mathbf{r}^{(0)} = p_m(A)\mathbf{r}^{(0)} \in K_m. \tag{3.2}$$

Here $p_m(A)$ is called the *residual polynomial*; evidently $p_m(0) = 1$. Because the error $\mathbf{e}^{(m)} = \mathbf{u}^{(m)} - \mathbf{u}$ is equal to $-A^{-1}\mathbf{r}^{(m)}$, we also have

$$\mathbf{e}^{(m)} = p_m(A)\mathbf{e}^{(0)} \in K_{m-1} + \mathbf{e}^{(0)}, \qquad p_m(0) = 1. \tag{3.2'}$$

Error reduction. For any positive definite symmetric A, with eigenvalues $\lambda_i \in [\alpha, \beta]$, $\beta = \kappa\alpha$, and corresponding orthonormal eigenvectors $\boldsymbol{\phi}_i$, the initial residual is $\mathbf{r}^{(0)} = \Sigma_i \eta_i \boldsymbol{\phi}_i$, for some set of coefficients η_i. It follows from (3.2) that $p_m(\lambda_i)$ gives the factor by which the i-th component $\eta_i \boldsymbol{\phi}_i$ of the initial residual has been reduced at the m-th step, and similarly for the error in (3.2′).

For estimating the reduction in the overall error, clearly some *norm* must be selected as measuring the 'magnitude' of the error. Most convenient for this purpose is the *energy* (or $A^{1/2}$) norm, defined as the square root of $E_m = \mathbf{e}^{(m)T} A \mathbf{e}^{(m)}$. Since $\mathbf{e}^{(m)} = A^{-1}\mathbf{r}^{(m)}$, E_m is given by

$$E_m = \|\mathbf{e}^{(m)}\|^2_{A^{1/2}} = \mathbf{r}^{(m)T} A^{-1} \mathbf{r}^{(m)} = \sum_{i=1}^{N} \frac{p_m(\lambda_i)^2}{\lambda_i} \eta_i^2. \tag{3.3}$$

Following Stiefel [54, p. 387], we will call E_m the 'error measure'.

LEMMA. *For any initial 'trial vector',* $\mathbf{u}^{(0)}$, *the error measure* E_m *is minimized at each step if and only if the residuals* $\mathbf{r}^{(m)} = \mathbf{b} - A\mathbf{u}^{(m)}$ *are orthogonal*: $\mathbf{r}^{(m)T}\mathbf{r}^{(j)} = 0$, $j = 1, \cdots, m-1$.

Proof. The error measure E_m in (3.3) is minimized if and only if

$$\frac{\partial E_m}{\partial c_{m,k}} = \sum_{i=1}^{N} p_m(\lambda_i)\lambda_i^{k} \eta_i^2 = 0, \qquad k = 0, \cdots, m-1.$$

The result then follows because $\mathbf{r}^{(m)T}\mathbf{r}^{(k)} = \Sigma_i\, p_m(\lambda_i) p_k(\lambda_i) \eta_i^2$ and because $p_k(\lambda)$ is a polynomial of degree at most k.

Three-term recurrence relation. Any such set of orthogonal polynomials $p_m(\lambda)$ satisfies a three-term recurrence relation, hence so does the set of residuals $\mathbf{r}^{(m)} = p_m(A)\mathbf{r}^{(0)}$. Because $p_m(0) = 1$, we can write this as

$$\mathbf{r}^{(m+1)} = -\gamma_m \delta_m A \mathbf{r}^{(m)} + \gamma_m \mathbf{r}^{(m)} + (1 - \gamma_m)\mathbf{r}^{(m-1)}, \tag{3.4}$$

with $\gamma_0 = 1$. Forming inner products with $\mathbf{r}^{(m)}$ and $\mathbf{r}^{(m-1)}$, we determine from (3.4) that

$$(3.5)\qquad \delta_m=\frac{\langle \mathbf{r}^{(m)},\mathbf{r}^{(m)}\rangle}{\langle \mathbf{r}^{(m)},A\mathbf{r}^{(m)}\rangle},\qquad \gamma_m=\frac{\langle \mathbf{r}^{(m-1)},\mathbf{r}^{(m-1)}\rangle}{\delta_m\langle \mathbf{r}^{(m-1)},A\mathbf{r}^{(m)}\rangle+\langle \mathbf{r}^{(m-1)},\mathbf{r}^{(m-1)}\rangle}.$$

Clearly, the coefficients of the recurrence relation can be computed *without* knowing the spectrum of A *or* the components η_i of the initial residual.

Because $\mathbf{r}^{(m)}=\mathbf{b}-A\mathbf{u}^{(m)}$, this proves that the CG method *minimizes the energy norm of the error*. It does this at the cost of computing inner products to evaluate δ_m and γ_m at each step.

THEOREM 2. *For any initial approximation* $\mathbf{u}^{(0)}$, *the iterates computed by*

$$(3.6)\qquad \mathbf{u}^{(m+1)}=\gamma_m\delta_m\mathbf{r}^{(m)}+\gamma_m\mathbf{u}^{(m)}+(1-\gamma_m)\mathbf{u}^{(m-1)},$$

where $\mathbf{r}^{(m)}=\mathbf{b}-A\mathbf{u}^{(m)}$ *and* δ_m, γ_m *are given by* (3.5), *minimize the energy norm of* $\mathbf{e}^{(m+1)}=\mathbf{u}^{(m+1)}-\mathbf{u}$ *in the affine subspace* $K_m+\mathbf{u}^{(0)}$. *In exact arithmetic, this norm decreases monotonically until, for some* $m\leqslant N$, $\mathbf{e}^{(m)}=\mathbf{0}$.

In addition to the residuals of the CG method being orthogonal, the differences of successive iterates form an A-orthogonal sequence.

COROLLARY. *The differences* $\mathbf{d}^{(m)}=\mathbf{u}^{(m+1)}-\mathbf{u}^{(m)}$ *of the iterates in* (3.6) *are A-orthogonal*: $\mathbf{d}^{(m)T}A\mathbf{d}^{(k)}=0$, $k=0,\cdots,m-1$.

Proof. Because of the normalization $p_k(0)=1$, $[p_{k+1}(x)-p_k(x)]/x$ is a polynomial of degree k. Consequently, $\mathbf{d}^{(m)}$ and the differences $\boldsymbol{\rho}^{(m)}=\mathbf{r}^{(m+1)}-\mathbf{r}^{(m)}$ of the residuals satisfy

$$(3.7)\qquad \mathbf{d}^{(m)T}A\mathbf{d}^{(k)}=\mathbf{d}^{(m)T}\boldsymbol{\rho}^{(k)}=\boldsymbol{\rho}^{(m)T}A^{-1}\boldsymbol{\rho}^{(k)}=0,\qquad k=0,\cdots,m-1.$$

Note that whereas the Gauss-Seidel and SOR algorithms are stationary methods of 'successive' displacements, the CG method is a *non*stationary method of 'simultaneous' displacements (see Chapter 4, §8).

Operation count. For the standard 5-point approximation to the linear source problem, A has at most five nonzero entries per row. Taking into account the symmetry of A, one can compute $\mathbf{u}^{(m+1)}$ from the previous two iterates in about 11 multiplications and 15 additions per unknown. For SOR, since $D^{-1}A=I-B$ is not symmetric, each unknown requires about 5 multiplications and 6 additions per iteration.

One-step algorithm. To get the original form of the conjugate gradient algorithm (Hestenes-Stiefel [52]), set $a_m=\gamma_m\delta_m$ and $\mathbf{d}^{(m)}=a_m\mathbf{p}^{(m)}$. Then (3.6) can be rewritten as $\mathbf{p}^{(m)}=\mathbf{r}^{(m)}+b_m\mathbf{p}^{(m-1)}$, where $b_m=(\gamma_m-1)a_{m-1}/a_m$. Because the $\mathbf{p}^{(m)}$ are A-orthogonal, we have

$$\begin{aligned}\mathbf{p}^{(m)T}A\mathbf{p}^{(m)}&=\mathbf{p}^{(m)T}A\mathbf{r}^{(m)}=\mathbf{d}^{(m)T}A\mathbf{r}^{(m)}/a_m\\&=[-\mathbf{r}^{(m+1)}+\mathbf{r}^{(m)}]^T\mathbf{r}^{(m)}/a_m=\mathbf{r}^{(m)T}\mathbf{r}^{(m)}/a_m.\end{aligned}$$

CONJUGATE GRADIENT ALGORITHM. For any $\mathbf{u}^{(0)}$, set $\mathbf{r}^{(0)}=\mathbf{p}^{(0)}=\mathbf{b}-A\mathbf{u}^{(0)}$ and compute cyclically for $m=0,1,2,\cdots$,

$$a_m=\frac{\langle \mathbf{r}^{(m)},\mathbf{r}^{(m)}\rangle}{\langle \mathbf{p}^{(m)},A\mathbf{p}^{(m)}\rangle},\qquad \mathbf{u}^{(m+1)}=\mathbf{u}^{(m)}+a_m\mathbf{p}^{(m)},\tag{3.8}$$

$$\mathbf{r}^{(m+1)}=\mathbf{b}-A\mathbf{u}^{(m+1)}=\mathbf{r}^{(m)}-a_mA\mathbf{p}^{(m)},\tag{3.8a}$$

$$b_{m+1}=-\frac{\langle \mathbf{r}^{(m+1)},A\mathbf{p}^{(m)}\rangle}{\langle \mathbf{p}^{(m)},A\mathbf{p}^{(m)}\rangle},\qquad \mathbf{p}^{(m+1)}=\mathbf{r}^{(m+1)}+b_{m+1}\mathbf{p}^{(m)}.\tag{3.8b}$$

The CG 'direct' method. Originally, the conjugate gradient method was proposed as a *direct* method, more efficacious than Gauss elimination for solving $A\mathbf{x}=\mathbf{b}$ when A was positive definite, and symmetric or symmetrizable. It was pointed out that, in the absence of roundoff error, it would give the solution in ν or fewer steps. Indeed, if A has only ν distinct eigenvalues $\mu_1,\cdots,\mu_\nu$, then $p_\nu(\lambda)=\prod_{k=1}^{\nu}(\lambda-\mu_i)$, and $\|\mathbf{e}^{(\nu)}\|_{A^{1/2}}^2=0$ in (3.3) for $\mathbf{e}^{(0)}$ being *any* eigenvector and thus for *every initial* $\mathbf{u}^{(0)}$. Hence we have the following result.

THEOREM 3. *If the number of distinct eigenvalues of A is ν, then the error $\mathbf{e}^{(\nu)}$ of the CG method after ν iterations is entirely due to roundoff.*

Thus for $-\nabla_h^2$ in the Model Problem, with $n^2=N$ interior mesh-points, the eigenvalues of ∇_h^2 are sums $\lambda_j+\lambda_k$, $j,k=1,\cdots,n$. Hence the conjugate gradient method would converge in at most $(n^2+n)/2$ steps if there were no roundoff error. More generally, the same is true for any elliptic difference operation of the form $\mathbf{L}_h\otimes\mathbf{I}_h+\mathbf{I}_h\otimes\mathbf{L}_h$ which approximates

$$-[p(x)u_x]_x-[p(y)u_y]_y+[a(x)+a(y)]u.$$

4. The CSI and CCSI methods. During the late 1950's, block SOR with near-optimal ω was the 'method of choice' for solving linear source problems, although elliptic ADI methods (see §§8 and 9) were also widely used, especially in petroleum reservoir calculations. Since then, several major advances have occurred.

First came the CCSI method of Golub-Varga [61], which we will describe at the end of this section. Then came a recognition of the potentialities of the conjugate gradient and Chebyshev algorithms for *accelerating* the convergence (see §5) of 'basic' iterative methods (Jacobi, Richardson, Gauss-Seidel, SSOR, etc.). Finally, these were coordinated with various matrix 'splittings' (Chapter 4, §8) and other 'preconditioning' techniques. A few such combinations will be discussed in §6.

The *Chebyshev semi-iterative* (CSI) method for solving $A\mathbf{u}=\mathbf{b}$, to be analyzed next, is a *polynomial* method in the following sense. Each m-th *approximate solution* $\mathbf{u}^{(m)}$ is given by (3.1), i.e., by the formula

$$\text{(4.1)} \qquad \mathbf{u}^{(m)} = \mathbf{u}^{(0)} - \sum_{k=1}^{m} c_{m,k} A^{k-1}\mathbf{r}^{(0)} = \mathbf{u} - A^{-1}p_m(A)\mathbf{r}^{(0)},$$

where the polynomials $p_m(A) = I + \sum_{k=1}^{m} c_{m,k} A^k$ depend only on A, and the $c_{m,k}$ are *independent* of $\mathbf{u}^{(0)}$. The residual $\mathbf{r}^{(m)} = \mathbf{b} - A\mathbf{u}^{(m)}$ is thus in K_m, the m-th Krylov subspace of A, making formulas (3.2)–(3.2′) applicable.

Clearly, any 'stationary' one-step iterative method $\mathbf{u}^{(m)} = G\mathbf{u}^{(m-1)} + \mathbf{k}$, such as the Jacobi or Richardson method of Chapter 4, §8, is a polynomial method, with $p_m(A) = G^m = (I - A)^m$. Frankel's method (Chapter 4, §15) is also a polynomial method. However, the CG method of §3 is *not* a polynomial method, because the $p_m(A)$ *depend on* $\mathbf{u}^{(0)}$.

For the CSI method, the polynomials p_m in (4.1), like those for optimum Richardson or optimum SOR, are determined by the *spectral interval* $[\alpha, \beta]$ known (or assumed) to contain all the eigenvalues of A.[15] For solving $(I - B)\mathbf{u} = \mathbf{k}$, with $B = D^{-1}A$, $\mathbf{k} = D^{-1}\mathbf{b}$, the spectrum of $A' = D^{-1/2}AD^{-1/2}$ plays a similar role.

Specifically, if all we know about the eigenvalues of the matrix A (or A') is that they lie in the interval $[\alpha, \beta]$, then all we know about the spectral radius of any polynomial $p_m(A)$ with $p_m(0) = 1$ is that

$$\text{(4.2)} \qquad \rho(p_m(A)) \leqslant \max_{\lambda \in [\alpha,\beta]} |p_m(\lambda)| = \rho^*(p_m(A)).$$

The number $\rho^*(p_m(A))$ defined by (4.2) is called the *virtual spectral radius* of $p_m(A)$ [HY, p. 42]. If A is *symmetric* with $\alpha > 0$, then $\rho^*(p_m(A))$ is the largest *guaranteed* factor by which the Euclidean norm of the residual is multiplied. In CSI, p_m is chosen to *minimize* the virtual spectral radius of $p_m(A)$ (or $p_m(A')$).

But the preceding *minimax* problem in polynomial algebra has a classic *known solution*.[16] Namely, the Chebyshev polynomial $T_m(z)$, defined by the equation $T_m(\cos\theta) = \cos m\theta$, is the polynomial of degree m with $T_m(1) = 1$ which has the smallest maximum absolute value on $[-1,1]$. Consequently, the polynomial $p_m(x)$ defined by

$$\text{(4.3)} \qquad p_m(x) = \frac{T_m(\sigma - 2x/[\beta - \alpha])}{T_m(\sigma)}, \qquad \sigma = \frac{\beta + \alpha}{\beta - \alpha} = \frac{\kappa + 1}{\kappa - 1},$$

---

[15] CSI can also be used when A has complex eigenvalues; see [HY, §12.2].

[16] See, for example, Lanczos's summary of properties of Chebyshev polynomials in *Tables of Chebyshev Polynomials $S_n(x)$ and $C_n(x)$*, Nat. Bur. Standards Applied Math. Ser. 9 (1952), pp. V–XXVI. See also Birkhoff-Rota [78], Chap. 11, §§6–7, and p. 308, Theorem 8. Early uses of Chebyshev polynomials in the solution of linear systems and eigenvalue problems include C. Lanczos, J. Res. Nat. Bur. of Standards 45 (1950), 255–282; D. Flanders and G. Shortley, J. Appl. Phys. 21 (1950), 1326–1332; and G. Shortley, J. Appl. Phys. 24 (1953), 392–396. See [V, p. 159] and [Y, pp. 385–386] for other references.

satisfies $p_m(0) = 1$ and, because $y = -(\sigma - 2x/[\beta-\alpha])$ maps $[\alpha,\beta]$ onto $[-1,1]$, it also has the following minimax property:

$$\max_{\alpha \leqslant x \leqslant \beta} |p_m(x)| = \frac{1}{T_m(\sigma)} \leqslant \max_{\alpha \leqslant x \leqslant \beta} |q_m(x)| \tag{4.4}$$

for any polynomial $q_m(x)$ of degree m which satisfies $q_m(0) = 1$. Furthermore, the spectral radius of $p_m(A)$ satisfies

$$\rho(p_m(A)) \leqslant 1/T_m(\sigma). \tag{4.4'}$$

The Chebyshev polynomials satisfy the three-term recurrence relation

$$T_0(z) = 1, \qquad T_1(z) = z, \qquad T_{m+1}(z) = 2zT_m(z) - T_{m-1}(z). \tag{4.5}$$

Consequently, by (4.3), $p_0(x) = 1$, $p_1(x) = 1 - 2x/(\beta+\alpha)$, and for $m > 1$,

$$p_{m+1}(x) = 2\frac{\beta+\alpha-2x}{\beta-\alpha}\frac{T_m(\sigma)}{T_{m+1}(\sigma)}p_m(x) - \frac{T_{m-1}(\sigma)}{T_{m+1}(\sigma)}p_{m-1}(x). \tag{4.6}$$

Replacing x in (4.6) with A and multiplying on the right by $\mathbf{e}^{(0)}$, we express the error $\mathbf{e}^{(m+1)} = p_{m+1}(A)\mathbf{e}^{(0)}$ in terms of $\mathbf{e}^{(m)}$ and $\mathbf{e}^{(m-1)}$. Substituting $\mathbf{u}^{(m)} - \mathbf{u}$ for $\mathbf{e}^{(m)}$ and rearranging, we get

$$\begin{aligned}\mathbf{u}^{(m+1)} &= 2\sigma\frac{T_m(\sigma)}{T_{m+1}(\sigma)}\mathbf{u}^{(m)} - \frac{T_{m-1}(\sigma)}{T_{m+1}(\sigma)}\mathbf{u}^{(m-1)} \\ &\quad + \frac{4}{\beta-\alpha}\frac{T_m(\sigma)}{T_{m+1}(\sigma)}A(\mathbf{u}-\mathbf{u}^{(m)}) + \left[\frac{T_{m+1}(\sigma) - 2\sigma T_m(\sigma) + T_{m-1}(\sigma)}{T_{m+1}(\sigma)}\right]\mathbf{u}.\end{aligned}$$

Because of (4.5) with $z = \sigma$, the last term vanishes. Deleting it and replacing $-T_{m-1}(\sigma)\mathbf{u}^{(m-1)}$ by $T_{m+1}(\sigma)\mathbf{u}^{(m-1)} - 2\sigma T_m(\sigma)\mathbf{u}^{(m-1)}$, and $A\mathbf{u}$ by $\mathbf{b}$, we get the *Chebyshev semi-iterative algorithm*.

CHEBYSHEV SEMI-ITERATIVE ALGORITHM. *Let A have real eigenvalues and spectrum in $[\alpha,\beta] = [\alpha,\kappa\alpha]$, $0 < \alpha$. For any $\mathbf{u}^{(0)}$, set*

$$\omega = \frac{2}{\alpha+\beta}, \qquad \sigma = \frac{\kappa+1}{\kappa-1}, \qquad \gamma_0 = 2, \tag{4.7}$$

$$\mathbf{u}^{(1)} = \mathbf{u}^{(0)} + \omega(\mathbf{b} - A\mathbf{u}^{(0)}), \tag{4.7a}$$

and compute cyclically for $m = 1, 2, 3, \cdots$,

$$\gamma_m = 2\sigma\frac{T_m(\sigma)}{T_{m+1}(\sigma)} = \frac{1}{1-\gamma_{m-1}/4\sigma^2}, \tag{4.7b}$$

$$\mathbf{u}^{(m+1)} = \gamma_m\omega(\mathbf{b} - A\mathbf{u}^{(m)}) + \gamma_m\mathbf{u}^{(m)} + (1-\gamma_m)\mathbf{u}^{(m-1)}. \tag{4.7c}$$

THEOREM 4. *In $A\mathbf{u} = \mathbf{b}$, let A be any matrix with real eigenvalues and spectrum in $[\alpha,\beta]$, $0 < \alpha < \beta = \kappa\alpha$. Then the spectral radius of the residual polynomial $p_m(A)$ in (4.3) of the Chebyshev semi-iterative method satisfies $\rho(p_m(A)) \leqslant 1/T_m(\sigma)$.*

Since $T_m(z)$ assumes its maximum magnitude at $z = \pm 1$, equality holds in Theorem 4 if A has either α or β as an eigenvalue.

Applications. If A is positive definite symmetric, or symmetrizable with positive real eigenvalues, and all we know is that its spectrum is contained in $[\alpha,\beta]$, $0 < \alpha \leqslant \beta$, CSI *minimizes* the maximum possible *spectral radius* of $p_m(A)$. For A symmetric, this minimizes the maximum possible ratio $\|\mathbf{r}^{(m)}\|/\|\mathbf{r}^{(0)}\|$, thus giving the largest possible *guaranteed reduction* in the *Euclidean* norm of the residual in m steps; because of (3.2′), it also minimizes the maximum possible ratio $\|\mathbf{e}^{(m)}\|/\|\mathbf{e}^{(0)}\|$.

Since the asymptotic rate of convergence is independent of the norm (Chapter 4, §10), CSI therefore has the same maximum possible asymptotic rate of convergence in *any* norm, given only the information that the spectrum of A is in $[\alpha,\beta]$.

However, as Varga [57] pointed out, when A has block Property Y, optimum SOR is often preferable to CSI with $A = I - G$, where G is the Gauss-Seidel iteration matrix, the 'optimality' of CSI notwithstanding. This is because optimum SOR, being a *one-step* method, requires less storage than CSI [V, p. 143]. We will consider some of these facts in more detail in §5, as well as the Chebyshev acceleration of SSOR whose rate of convergence is an order of magnitude greater than SOR.

Rate of convergence. By the general results of Chapter 4, §10, the asymptotic rate of convergence of the CSI method is

$$R_\infty = \lim_{m\to\infty} [-\log \rho(p_m(A))]^{1/m} \geqslant \lim_{m\to\infty} \log [T_m(\sigma)]^{1/m},$$

where the inequality follows from (4.4′). We now prove the following related theorem.

THEOREM 5. *If the spectrum of A is in $[\alpha,\kappa\alpha]$, $\alpha > 0$, then the asymptotic rate of convergence of the CSI method is at least* $2/\sqrt{\kappa} + 2/3\kappa^{3/2} + O(\kappa^{-5/2})$.

Proof. For fixed $z = \sigma = (\kappa+1)/(\kappa-1)$, the general solution of the recurrence relation in (4.5) is $T_m(\sigma) = as^m + bs^{-m}$ with

$$s = \sigma + \sqrt{\sigma^2 - 1} = \frac{\sqrt{\kappa}+1}{\sqrt{\kappa}-1}. \tag{4.8}$$

Using the initial conditions in (4.5), we obtain

$$T_m(\sigma) = s^m(1 + s^{-2m})/2. \tag{4.8′}$$

Because $\kappa > 1$, $s > 1$ and $T_m(\sigma)/s^m \to 1/2$; therefore we have

$$R_\infty \geqslant \log s = \log\left[\frac{\sqrt{\kappa}+1}{\sqrt{\kappa}-1}\right] = \frac{2}{\sqrt{\kappa}} + \frac{2}{3\kappa^{3/2}} + O(1/\kappa^{5/2}). \tag{4.9}$$

Note that if γ_k in (4.7c) were set equal to unity for all k, then (4.7)–(4.7c) becomes the optimum Richardson method with asymptotic rate of convergence $2/\kappa + O(\kappa^{-3})$ (see Chapter 4, §10, Theorem 5). Thus the CSI method can be considered as a scheme for *accelerating* the rate of convergence of the Richardson method from $O(\kappa^{-1})$ to $O(\kappa^{-1/2})$.

The CCSI method. In the 1960's, the most effective technique for solving the equations of the 5-point approximation of the linear source problem was the following *cyclic Chebyshev semi-iterative* (CCSI) method of Golub-Varga [61]. This method applies to the discretized linear source problem of Chapter 3, (4.2), when a red-black ordering is used with the CSI method. We then have

$$\mathbf{u}^{(m+1)} = B\mathbf{u}^{(m)} + \mathbf{k}, \qquad B = I - D^{-1}A = \begin{bmatrix} 0 & B_2 \\ B_1 & 0 \end{bmatrix}. \tag{4.10}$$

We can apply the *cyclic reduction* method of Chapter 4, §3, to this CSI scheme to reduce the operation count, noting that in Theorem 7 of Chapter 4, §12, the extreme eigenvalues of B are $\pm\rho$, $\rho = \rho(B)$. Thus $\alpha = 1 - \rho$, $\beta = 1 + \rho$, and $\omega = 1$, in (4.7)–(4.7c). With $\mathbf{u} = (\mathbf{u}_1, \mathbf{u}_2)^T$ and $\mathbf{k} = (\mathbf{k}_1, \mathbf{k}_2)^T$ partitioned the same way, (4.7a)–(4.7c) now give

$$\begin{aligned} \mathbf{u}_1^{(1)} &= B_2\mathbf{u}_2^{(0)} + \mathbf{k}_1, \qquad \mathbf{u}_2^{(1)} = B_1\mathbf{u}_1^{(0)} + \mathbf{k}_2, \\ \mathbf{u}_1^{(m+1)} &= \gamma_m(B_2\mathbf{u}_2^{(m)} + \mathbf{k}_1) + (1-\gamma_m)\,\mathbf{u}_1^{(m-1)}, \\ \mathbf{u}_2^{(m+1)} &= \gamma_m(B_1\mathbf{u}_1^{(m)} + \mathbf{k}_2) + (1-\gamma_m)\,\mathbf{u}_2^{(m-1)}. \end{aligned} \tag{4.11}$$

Note that since

$$\kappa = \frac{1+\rho}{1-\rho} \quad \text{and} \quad s = \frac{\sqrt{\kappa}+1}{\sqrt{\kappa}-1} = \left[\frac{1+\sqrt{1-\rho^2}}{1-\sqrt{1-\rho^2}}\right]^{1/2} = (\omega_b - 1)^{-1/2},$$

where ω_b is the optimum SOR parameter, it follows from Theorem 5 that the asymptotic rate of convergence is at least half that of optimum SOR. But, clearly, only *half* of these vectors need be computed; it is sufficient to compute just the odd iterates of $\mathbf{u}_1$ and the even iterates of $\mathbf{u}_2$. This is the *cyclic Chebyshev semi-iterative* (CCSI) method.

Since only half of the vectors are computed, and the spectral radius of this scheme for generating $(\mathbf{u}_1^{(2m+1)}, \mathbf{u}_2^{(2m)})^T$ is asymptotically at most $\omega_b - 1$, its asymptotic rate of convergence is at least twice that for ordinary CSI. Thus we have the following result [V, Theorem 5.4, p. 152].

THEOREM 6. *If $B = I - A$ is weakly 2-cyclic and if A is similar to a symmetric positive definite matrix, then the CCSI method has at least the asymptotic rate of convergence of optimum SOR.*

5. Chebyshev acceleration. Originally envisaged by Stiefel [56] as a scheme for solving (say) Fredholm integral equations,[17] his three-term Chebyshev method (CSI) can also be used to *accelerate the convergence* of

[17] For 'full' or 'dense' matrices, elimination methods may be more efficient.

any other 'one-step' iterative scheme[18] of the form

(5.1) $$\mathbf{u}^{(m+1)} = G\mathbf{u}^{(m)} + \mathbf{k},$$

as long as the spectrum of $I - G$ is real and positive. In (4.7c), one replaces A with $I - G$ and $\mathbf{b}$ with $\mathbf{k}$. This gives

(5.2) $$\mathbf{u}^{(m+1)} = \gamma_m \omega(\mathbf{k} - \mathbf{u}^{(m)} + G\mathbf{u}^{(m)}) + \gamma_m \mathbf{u}^{(m)} + (1 - \gamma_m)\mathbf{u}^{(m-1)},$$

where, as in (4.7) and (4.7b), $\gamma_0 = 2$ and

(5.2') $$\omega = \frac{2}{\alpha + \beta}, \qquad \sigma = \frac{\kappa + 1}{\kappa - 1}, \qquad \gamma_m = \frac{1}{1 - \gamma_{m-1}/4\sigma^2}.$$

Indeed, because of the min-max property of the Chebyshev methods, the CSI method (5.2) will in general converge faster than (5.1). Moreover, CSI converges — even if (5.1) diverges because $\rho(G) > 1$. However, the eigenvalues of optimum SOR are on a circle with center $\mathbf{0}$, and hence CSI *cannot* be used to accelerate its convergence.

For example, the CCSI method of (4.11) can be viewed as CSI acceleration of the Jacobi method, combined with cyclic reduction. Gauss-Seidel iteration can also be accelerated by CSI, as is explained below.

CSI acceleration of Gauss-Seidel. Since Gauss-Seidel iteration has real eigenvalues, CSI acceleration is also effective for it. More precisely, instead of using the Jacobi method as in (4.10), we can use the Gauss-Seidel scheme of Chapter 4, §12:

(5.3) $$\begin{aligned} \mathbf{u}_1^{(m+1)} &= B_2 \mathbf{u}_2^{(m)} + \mathbf{k}_1, \\ \mathbf{u}_2^{(m+1)} &= B_1 \mathbf{u}_1^{(m+1)} + \mathbf{k}_2. \end{aligned}$$

Equivalently, we have

(5.4) $$\mathbf{u}_1^{(m+1)} = B_2 B_1 \mathbf{u}_1^{(m)} + \mathbf{y}_1, \qquad \mathbf{y}_1 = B_2 \mathbf{k}_2 + \mathbf{k}_1,$$

(5.4') $$\mathbf{u}_2^{(m+1)} = B_1 B_2 \mathbf{u}_2^{(m)} + \mathbf{y}_2, \qquad \mathbf{y}_2 = B_1 \mathbf{k}_1 + \mathbf{k}_2;$$

(5.4)–(5.4') is called the *cyclic reduction* of (4.10) [V, p. 155] and (5.4') is called the *reduced system* [HY, p. 165].

When $\mathbf{u}_2^{(m+1)}$ has been obtained to sufficient accuracy, one computes $B_2 \mathbf{u}_2^{(m+1)} + \mathbf{k}_1$ as the estimate of $\mathbf{u}_1$. When accelerated by CSI, this becomes the *reduced system semi-iterative method* (RSSI) [HY, p. 167].

The next result is due to Varga [57]; see also [V, p. 143].

THEOREM 7. *Let A have Property Y and real positive eigenvalues. Then Chebyshev acceleration of the Gauss-Seidel method for solving $A\mathbf{u} = \mathbf{b}$ has at least the asymptotic rate of convergence as optimum SOR.*

---

[18] See E. L. Wachspress, Report KAPL-1333, Knolls Atomic Power Laboratory, 1955; D. M. Young, Proc. Sixth Symp. Appl. Math., J. H. Curtiss ed., Am. Math. Soc. (1956), 283–298; Varga [57]; J. W. Sheldon, J. ACM 6 (1959), 494–505, and Ralston-Wilf [60, pp. 144–156].

Proof. The asymptotic rate of convergence of optimum SOR is

$$-\log(\omega_b - 1) = \log\left[\frac{1+\sqrt{1-\rho(B)^2}}{1-\sqrt{1-\rho(B)^2}}\right]. \tag{5.5}$$

For $\omega = 1$, Young's identity (Theorem 12, Chapter 4, §14) reduces to $\lambda^2 - \lambda\mu^2 = 0$, so the eigenvalues of the Gauss-Seidel iteration matrix, $G = \mathbf{L}_1$, are between 0 and $\rho(B)^2$; those of $I - G$ are then between $\alpha = 1 - \rho(B)^2$ and $\beta = \kappa\alpha = 1$. Thus the asymptotic rate of convergence of CSI is at least $\log[(\sqrt{\kappa}+1)/(\sqrt{\kappa}-1)]$ by (4.9). Substitution of $\kappa = 1/[1-\rho(B)^2]$ and simplification gives the right side of (5.5).

Caution. Although the asymptotic rate of convergence of the Chebyshev acceleration of Gauss-Seidel is the same as that of SOR, convergence can be very slow when the *natural* ordering is used. Faster convergence can be obtained with red-black orderings.[19]

SSOR. The rate of convergence of SSOR (§2) can be speeded up by Chebyshev acceleration if the 'SSOR condition' $\rho(LU) \leqslant 1/4$ of (2.7) is satisfied. The standard 5-point approximation to the Poisson DE, and its axially symmetric analogue in cylindrical coordinates,

$$u_{rr} + \frac{1}{r}u_r + u_{zz} = f(r,z),$$

satisfy this SSOR condition if a uniform mesh and the natural ordering are used.

More generally, Young[20] has shown that $\rho(LU) \leqslant 1/4 + O(h^2)$ on a uniform mesh, whenever the coefficient functions of a linear source problem are in $C^2(\overline{\Omega})$, while Varga has recently proved even sharper results. Under these circumstances, CSI acceleration of SSOR converges an order of magnitude more rapidly than CCSI. This is the essential content of the following theorem.

THEOREM 8. *Let $A = I - L - U = I - B$ be symmetric positive definite, where L and U are the strictly lower and upper triangular parts of A. Set $\rho(B) = (\kappa - 1)/(\kappa + 1) < 1$. Let $\mathbf{S}_{\omega_1}$ be the SSOR iteration matrix with ω_1 as in* (2.8). *If $\rho(LU) \leqslant 1/4 + O(h^2)$, then the asymptotic rates of convergence of SSOR and its CSI acceleration are at least*

$$2/\kappa^{1/2} + O(\kappa^{-1}) \quad \text{and} \quad \sqrt{8}/\kappa^{1/4} + O(\kappa^{-3/4}), \tag{5.6}$$

respectively.

---

[19] See G. J. Tee, Comput. J. 6 (1963), 250–263; G. A. Miles, K. L. Stewart, and G. J. Tee, Comput. J. 6 (1964), 352–355; and [Y, pp. 374 and 386].

[20] Young [77]. See also [Y, Chap. 15], especially §15.4 for the Model Problem and §15.5 for the linear source problem.

Proof. By (2.8) and (2.9), the spectral radius of $\mathbf{S}_{\omega_1}$ satisfies

$$\rho(\mathbf{S}_{\omega_1}) \leqslant \frac{1-\sqrt{1/(\kappa+1)}}{1+\sqrt{1/(\kappa+1)}} = 1 - \frac{2}{\sqrt{\kappa}} + O(\kappa^{-1}),$$

which gives the first result. The eigenvalues of $\mathbf{S}_{\omega_1}$ are between $a = 0$ and $b = \rho(\mathbf{S}_{\omega_1})$, and those of $I - \mathbf{S}_{\omega_1}$ are between $\alpha = 1 - \rho(\mathbf{S}_{\omega_1})$ and $\beta = 1$; hence we have

$$\sigma = \frac{\beta+\alpha}{\beta-\alpha} = \frac{2-b}{b} = 1 + 4/\sqrt{\kappa} + O(1/\kappa).$$

The asymptotic rate of convergence of the CSI acceleration of SSOR is then given by (4.8)–(4.9) as

$$\log(\sigma + \sqrt{\sigma^2 - 1}) = \sqrt{8}/\kappa^{1/4} + O(1/\kappa^{3/4}),$$

which gives the second result.

For the Model Problem (i.e., the 5-point approximation $-\nabla_h^2 U = f$ to the Poisson equation on the unit square with a square mesh with spacing h), the asymptotic rates of convergence are πh, $2\pi h$, and $2\sqrt{\pi h}$, for SSOR, SOR, and CSI acceleration of SSOR, respectively. Thus, for $h = 1/80$, CSI acceleration of SSOR requires about five times *fewer* iterations than optimum SOR to reduce the error by a given factor.

For convenient reference, we display in Table 1 the asymptotic rates of convergence of some one-step iterative methods and of their acceleration by CSI.

TABLE 1.

Asymptotic Rates of Convergence; $A = I - E - E^T$; $\kappa = \text{cond}(A)$.

Method	One-step	CSI acceleration
Jacobi	--	$2/\kappa^{1/2}$
Richardson	$2/\kappa$	$2/\kappa^{1/2}$
SSOR, if $\rho(LU) \leqslant 1/4$ and $\omega = \omega_1$	$2/\kappa^{1/2}$	$2^{3/2}/\kappa^{1/4}$
A with Property Y		
Jacobi	$2/\kappa$	$2/\kappa^{1/2}$
Gauss-Seidel	$4/\kappa$	$4/\kappa^{1/2}$
Optimum SOR	$4/\kappa^{1/2}$	--

Table 2 lists results from numerical experiments with ITPACK as contained in ELLPACK. These were performed on the VAX 11/780 with floating point accelerator. The spectral interval was determined adaptively and the iterations were terminated when the (estimated) error was less than 10^{-4} (see §7).

TABLE 2. *Adaptive Chebyshev acceleration, Model Problem.*

		Number of iterations; (time/N) $\times 10^4$; time $\times h^\nu \times 10^5$		
h	N	Jacobi, $\nu = 3$	Reduced System, $\nu = 3$	SSOR, $\nu = 2.5$
1/8	49	31; 57; 55	16; 50; 45	12; 51; 138
1/16	225	62; 103; 57	31; 61; 34	16; 55; 120
1/32	961	135; 206; 60	69; 108; 32	22; 69; 115
1/64	3961	202; 302; 46	108; 151; 23	27; 83; 100

Computation. For solving $A\mathbf{u} = \mathbf{b}$, the vector equation for the 5-point approximation to the linear source problem with N unknowns, CSI requires about $7N$ storage cells; $3N$ are used for the vectors $\mathbf{u}^{(m-1)}$, $\mathbf{u}^{(m)}$, and $\mathbf{k}$, and the rest store the nonzero entries in, say, $D^{-1}A = I - L - U$.[21]

When CSI is used to accelerate the Jacobi method, the four nonzero diagonals of $G = B = I - D^{-1}A$ (or two diagonals of its better conditioned symmetric form $I - D^{-1/2}AD^{-1/2}$) are stored. For other one-step schemes, such as Gauss-Seidel, SSOR, and so on, the entries of the iteration matrix G are not explicitly computed.

For example, consider acceleration of Gauss-Seidel iteration; this can be written in terms of the pseudo-residual, $\mathbf{s}^{(m)} = \mathbf{k} - (I - G)\mathbf{u}^{(m)}$, of the Gauss-Seidel method as

$$\mathbf{u}^{(m+1)} = \gamma_m \omega \mathbf{s}^{(m)} + \gamma_m \mathbf{u}^{(m)} + (1 - \gamma_m)\mathbf{u}^{(m-1)}.$$

Components of $\mathbf{v} = \mathbf{u}^{(m)} + \mathbf{s}^{(m)}$ are computed successively, as in Chapter 4, (8.2):

$$\mathbf{v}_i = \sum_{j<i} L_{i,j}\mathbf{v}_j + \sum_{j>i} U_{i,j}(\mathbf{u}^{(m)})_j + \mathbf{k}_i;$$

one then computes

$$(\mathbf{u}^{(m+1)})_i = \gamma_m \omega [\mathbf{v}_i - (\mathbf{u}^{(m)})_i] + \gamma_m (\mathbf{u}^{(m)})_i + (1 - \gamma_m)(\mathbf{u}^{(m-1)})_i.$$

Because the i-th components of $\mathbf{u}^{(m)}$ and $\mathbf{u}^{(m-1)}$ are no longer needed, the storage cells for them can be used for $\mathbf{v}_i$ and $(\mathbf{u}^{(m+1)})_i$, respectively.

6. Conjugate gradient acceleration. The conjugate gradient method was originally proposed as a general 'machine' method for solving $A\mathbf{u} = \mathbf{b}$ and was stated (Hestenes-Stiefel [52, p. 409]) to have the following advantages over Gaussian elimination: (i) it requires less storage if A is

[21] The identity matrix is not stored. If the symmetric form, $D^{-1/2}AD^{-1/2}$ is used, then only about $5N$ storage cells are required. If N is so large that the amount of storage exceeds the size of the fast memory, then one could process several lines of nodal values by the 'concurrent iteration procedure'; see [HY, pp. 270–273].

sparse, since there is no fill-in; (ii) used as a *direct* method, it gives the solution in N steps, up to roundoff error; (iii) the error decreases (in the energy norm) at each step; and (iv) it can also be used as an *iterative* method, to give adequate accuracy after fewer than N steps. Moreover, used as an iterative method, it has the advantage over the SOR, CSI, and ADI (see §8) methods of *not requiring any information about the spectrum of* A *or of any associated 'iteration matrix'* such as the Jacobi iteration matrix $B = (I - D^{-1}A)$, and it *adapts to advantageous choices of the initial approximation* $\mathbf{u}^{(0)}$.[22]

The CG method was never used much as a 'stand-alone' method. This was partly because it required much more storage and was harder to program (in machine language) than SOR, and partly because its roundoff errors (when used as a direct method) were appreciable and not self-correcting (see [FW, p. 214]). However, for the last decade it has been used effectively to *accelerate* the convergence of a wide variety of one-step iterative methods. Its effectiveness for this purpose was pointed out first by Reid [71], later by O. Axelsson in Barker [74], and explained more fully by Concus, Golub, and O'Leary in Bunch-Rose [76, pp. 309–332].[23] Our exposition below will be based mainly on [HY, Chap. 7], to which we refer the reader for further details.

For any splitting $A = M - N$ (see Chapter 4, (8.9)), where A and M are nonsingular, $A\mathbf{u} = \mathbf{b}$ is equivalent to $M\mathbf{u} = N\mathbf{u} + \mathbf{b}$. This suggests solving $A\mathbf{u} = \mathbf{b}$ by iterating the stationary one-step scheme

$$M\mathbf{u}^{(m+1)} = N\mathbf{u}^{(m)} + \mathbf{b}, \tag{6.1}$$

where M is chosen so that $M\mathbf{x} = \mathbf{y}$ is easy to solve and so that $\rho(M^{-1}N)$ is small. The acceleration of (6.1) by the conjugate gradient method is called the 'generalized conjugate gradient method'. It often converges very quickly, as illustrated by the next example.

Example 1.[24] Consider the Poisson equation on the L-shaped region $[0,2] \times [0,1] \cup [0,1] \times [1,2]$, as in Chapter 3, §1, with Dirichlet boundary conditions. For $h = 1/(n+1)$, with n even, and the natural ordering, the matrix for $\nabla_h^2 U_{i,j} = f_{i,j}$ has the form

$$A = \begin{bmatrix} M_1 & -N_1 \\ -N_1{}^T & M_2 \end{bmatrix}, \qquad M_i = \begin{bmatrix} T_i & D_i & 0 & \cdot \\ D_i & T_i & D_i & \cdot \\ 0 & D_i & T_i & \cdot \\ \cdot & \cdot & \cdot & \cdot \end{bmatrix}, \qquad N_1 = \begin{bmatrix} 0 & 0 \\ S & 0 \end{bmatrix},$$

[22] For the effect of the smoothness of the initial error on the rate of convergence of CG, see W. Gautschi and R. E. Lynch, ZaMP 33 (1983), 24–35.

[23] See also M. R. Hestenes in Proc. Sixth Symp. Appl. Math., J. H. Curtiss ed., Am. Math. Soc. (1956), 83–102; J. W. Daniel, SIAM J. Numer. Anal. 4 (1967), 10–26.

[24] This is a slight modification of an example of P. Concus, G. H. Golub, and D. P. O'Leary in Bunch-Rose [76, p. 328].

where M_i is block tridiagonal with tridiagonal diagonal blocks T_i and diagonal off-diagonal blocks D_i. The matrices T_2, D_2 are are $n \times n$, and T_1, D_1 are $(2n-1)\times(2n-1)$, and the matrix N_1 is zero except for the $(2n-1)\times n$ submatrix S. With the splitting

$$A = M - N, \qquad M = \begin{bmatrix} M_1 & 0 \\ 0 & M_2 \end{bmatrix}, \qquad N = \begin{bmatrix} 0 & N_1 \\ N_1{}^T & 0 \end{bmatrix},$$

and with $\mathbf{u}^{(m)} = (\mathbf{v}_1^{(m)}, \mathbf{v}_2^{(m)})^T$, $\mathbf{b} = (\mathbf{b}_1, \mathbf{b}_2)^T$, one can take an arbitrary $\mathbf{v}_1^{(0)}$ and set $\mathbf{v}_2^{(0)} = M_2^{-1}(N_1{}^T\mathbf{v}_1^{(0)} - \mathbf{b}_2)$. Using FFT, the computation of $\mathbf{u}^{(1)}, \mathbf{u}^{(2)}, \cdots$, then becomes very fast. The splitting matrix M is positive definite. The iteration matrix is 2-cyclic:

$$G = \begin{bmatrix} 0 & M_1{}^{-1}N_1 \\ M_2{}^{-1}N_1{}^T & 0 \end{bmatrix}.$$

Since $M_1^{-1}N_1$ and $M_2^{-1}N_1{}^T$ have only n nonzero columns, the rank of G is at most $2n$. Consequently, by Theorem 3, CG acceleration with exact arithmetic would give the solution in at most $2n$ steps.

Preconditioning. In some cases, instead of solving $A\mathbf{u} = \mathbf{b}$, it is more economical to solve $WA\mathbf{u} = W\mathbf{b}$, where the 'preconditioning' matrix W is a symmetrization matrix for $I - G$, so that $W(I-G)W^{-1}$ is positive definite (see the articles by T. L. Jordan, O. Axelsson, A. Bayliss, C. I. Goldstein, E. Turkel, and the references they give, in Birkhoff-Schoenstadt [84]). Given a one-step iterative scheme such as (6.1), we then have

$$\text{(6.2)} \quad \mathbf{u}^{(m+1)} = G\mathbf{u}^{(m)} + \mathbf{k}, \quad \text{with} \quad G = M^{-1}N \quad \text{and} \quad \mathbf{k} = M^{-1}\mathbf{b}.$$

Writing $\mathbf{v} = W\mathbf{u}$, we can solve $C\mathbf{v} = \mathbf{c}$ where

$$C = W(I-G)W^{-1} = (W^T)^{-1}W^TW(I-G)W^{-1} = (W^T)^{-1}AW^{-1},$$

and $\mathbf{c} = W\mathbf{k}$. The CG method applied to $C\mathbf{v} = \mathbf{c}$ minimizes the energy norm of the error, i.e.,

$$\langle(\mathbf{v} - \mathbf{v}^{(m)}), C(\mathbf{v} - \mathbf{v}^{(m)})\rangle = (\mathbf{v} - \mathbf{v}^{(m)})C^T(\mathbf{v} - \mathbf{v}^{(m)})$$

is minimized.

In particular, if A and $M = W^TW$ are symmetric positive definite, and W is known, we can use that W. This is the case for the Richardson, Jacobi, and SSOR methods. As in (2.6) and Chapter 4, §8, the relevant symmetrization matrices are $W = \omega^{-1/2}I$, $W = D^{1/2}$, and

$$W = (2-\omega)^{-1/2}(D - \omega E)D^{-1/2},$$

respectively, where $-E$ is the strictly lower triangular part of $A = D - E - E^T$. Hence these methods can be speeded up, often substantially, by CG acceleration. The most effective use of CG acceleration for elliptic problems seems to be when it is used to accelerate the SSOR method.[25]

CG acceleration. The CG acceleration scheme is obtained from the two-step CG algorithm (3.5)–(3.6) by the substitutions

$$\text{(6.3)} \quad \mathbf{u}^{(m)} \mapsto W\mathbf{u}^{(m)}, \qquad A \mapsto W(I-G)W^{-1}, \qquad \mathbf{r}^{(m)} \mapsto W\mathbf{s}^{(m)},$$

and then multiplying the recurrence relation by W^{-1}, where $W^T W = M$.

With $A = M - N = M(I-G)$, one chooses $\mathbf{u}^{(0)}$ and applies the one-step scheme (6.1) once to get

$$\mathbf{u}^{(1)} = \mathbf{u}^{(0)} + [\mathbf{k} - (I-G)\mathbf{u}^{(0)}] = \mathbf{u}^{(0)} - \mathbf{s}^{(0)}, \qquad \sigma_0 = \langle \mathbf{s}^{(0)}, W\mathbf{s}^{(0)} \rangle,$$

and $\eta_0 = \langle \mathbf{s}^{(0)}, W(I-G)\mathbf{s}^{(0)} \rangle$. Then for $m = 1, 2, \cdots$, one solves

$$\text{(6.4)} \qquad M\mathbf{s}^{(m)} = \mathbf{k} - (I-G)\mathbf{u}^{(m)} = N\mathbf{s}^{(m-1)}$$

for $\mathbf{s}^{(m)}$ and computes

$$\text{(6.4a)} \qquad \sigma_m = \langle \mathbf{s}^{(m)}, M\mathbf{s}^{(m)} \rangle, \qquad \eta_m = \langle \mathbf{s}^{(m-1)}, M(I-G)\mathbf{s}^{(m)} \rangle,$$

$$\text{(6.4b)} \qquad \delta_m = \frac{\sigma_m}{\langle \mathbf{s}^{(m)}, M(I-G)\mathbf{s}^{(m)} \rangle}, \qquad \gamma_m = \frac{\sigma_{m-1}}{\delta_m \eta_m + \sigma_{m-1}},$$

$$\text{(6.4c)} \qquad \mathbf{u}^{(m+1)} = \gamma_m \delta_m \mathbf{s}^{(m)} + \gamma_m \mathbf{u}^{(m)} + (1-\gamma_m)\mathbf{u}^{(m-1)}.$$

In exact arithmetic, this would terminate when $\mathbf{s}^{(m)} = \mathbf{0}$; in practice, one stops when a norm of $\mathbf{s}^{(m)}$ is sufficiently small.

Theorems 2 and 3 apply directly after making the changes (6.3) and $\mathbf{e}^{(m)} \mapsto W\mathbf{e}^{(m)}$. At each step the error $\mathbf{e}^{(m)} = \mathbf{u} - \mathbf{u}^{(m)}$ is minimized in the norm

$$\text{(6.5)} \qquad \|\mathbf{e}^{(m)}\| = \langle \mathbf{e}^{(m)}, M(I-G)\mathbf{e}^{(m)} \rangle^{1/2}.$$

Table 3 lists results for conjugate gradient acceleration for the same problem and stopping criterion as in Table 2.

TABLE 3. *Conjugate gradient acceleration, Model Problem.*

		Number of iterations; $(\text{time}/N) \times 10^4$; $\text{time} \times h^{\nu} \times 10^5$		
h	N	Jacobi, $\nu = 3$	Reduced System, $\nu = 3$	SSOR, $\nu = 2.5$
1/8	49	9; 31; 29	4; 20; 20	6; 47; 127
1/16	225	20; 41; 22	11; 26; 14	10; 53; 117
1/32	961	41; 72; 21	21; 36; 11	14; 70; 116
1/64	3961	85; 139; 21	41; 66; 10	22; 100; 121

[25] ITPACK contains procedures for CG acceleration of Jacobi, SSOR, and the 'reduced system' which uses the equations of cyclic reduction (see §5).

7. Adaptive parameter selection; stopping criteria.[26] We have discussed the rates of convergence of SOR, CSI, CCSI, etc., as if the relevant spectral radius $\rho(G)$ and spectral intervals $[\alpha,\beta]$ were precisely *known*. However, this is rarely the case in practice (cf. Chapter 4, §11) when problems of an unfamiliar kind are run. It is often wise to make the first few iterations of SOR with conservative 'trial' values of ω, such as $\omega = 1$ or $\omega = 1.5$, and to infer better estimates of $\rho(G)$, ω_b, etc., from the observed rates of convergence.

Adaptive Chebyshev. We first consider the adaptive Chebyshev acceleration of a one-step stationary method of the form $\mathbf{u}^{(m+1)} = G\mathbf{u}^{(m)} + \mathbf{k}$, discussed in §5. Let $A = I - G$ be symmetrizable with positive eigenvalues in the interval $[\alpha,\beta]$, $\beta = \kappa\alpha$; then G is symmetrizable with spectrum in $[a,b]$, where $a = 1-\beta$, $b = 1-\alpha < 1$. Under these circumstances, the following procedure has been found to work well [HY, Chaps. 4–6].

Define $s_m(y;a,b) = p_m(1-y)$ to be the residual polynomial in (4.3), and choose conservative initial estimates a_0, b_0 so that surely $a_0 \leqslant a \leqslant b_0 < b$. Then $s_m(1;a_0,b_0) = 1$, where

$$s_m(y;a_0,b_0) = \frac{T_m(\sigma_0 + 2[y-1]/[b_0-a_0])}{T_m(\sigma_0)}, \qquad \sigma_0 = \frac{2-b_0-a_0}{b_0-a_0}.$$

Consequently, $s_m(y;a_0,b_0)$ has the Chebyshev minimax property on $[a_0,b_0]$.

Now apply (5.2), the Chebyshev acceleration of the one-step scheme (5.1), after replacing σ by σ_0. The function $s_m(y;a_0,b_0)$ is strictly increasing on the interval $b_0 \leqslant y \leqslant b$ (see Fig. 1). Since b is the largest eigenvalue of G, the spectral radius of $s_m(G;a_0,b_0)$ is given by

$$\rho_m = s_m(b;a_0,b_0) = \frac{T_m(\eta_0)}{T_m(\sigma_0)}, \qquad \text{where} \quad \eta_0 = \frac{2b-b_0-a_0}{b_0-a_0}, \tag{7.1}$$

with $b_0 < b < 1$.

By Theorem 3 of Chapter 4, §10, the ratio $P_m = ||\mathbf{s}^{(m)}||/||\mathbf{s}^{(0)}||$ of norms of pseudo-residuals, $\mathbf{s}^{(m)} = G\mathbf{u}^{(m)} + \mathbf{k} - \mathbf{u}^{(m)}$, is an accurate approximation of ρ_m for sufficiently large m. Consequently, by (4.8), (4.8′) and (7.1), we have

$$P_m = \frac{||\mathbf{s}^{(m)}||}{||\mathbf{s}^{(0)}||} \approx \frac{1+r^m(\eta_0)}{2r^{m/2}(\eta_0)} Q_m, \qquad Q_m = \frac{2r^{m/2}(\tau_0)}{1+r^m(\tau_0)}, \tag{7.2}$$

where

$$r(\eta_0) = s^{-2} = \frac{1-\sqrt{1-\eta_0^2}}{1+\sqrt{1-\eta_0^2}}. \tag{7.3}$$

---

[26] See [HY] for complete description of parameter selection and stopping tests; their Appendices A–C give programs for executing the algorithms they describe. See also [V, Chap. 9].

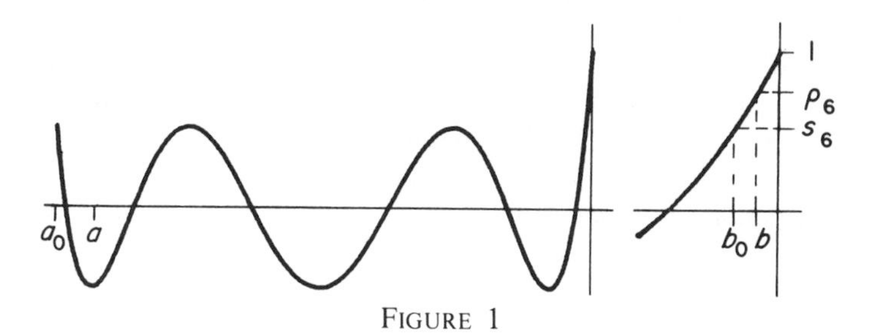

FIGURE 1

Therefore, since $\eta_0 > \sigma_0$, P_m will ultimately exceed Q_m. Hence for sufficiently large m,

$$\rho_m > 1/T_m(\sigma_0) = Q_m.$$

To avoid changing the estimate of b too often, Hageman and Young recommend [HY, pp. 82–86] waiting until $P_m/(Q_m)^F > 1$, where $F = 0.75$ is an empirical 'damping factor'. When $P_m/(Q_m)^F$ exceeds unity, one solves

$$\frac{1+r^m(\eta_0)}{2r^{m/2}(\eta_0)} = \frac{P_m}{Q_m} \tag{7.4}$$

for η_0.[27] Then $\eta_0 = (b_0 - a_0)/(2z - b_0 - a_0)$ is solved for z and this value is taken as b_1, the next approximation of b. The next estimate of σ is taken as $\sigma_1 = (2 - b_1 - a_0)/(b_1 - a_0)$.

The iteration is then continued by using (5.2)–(5.2′) with $\mathbf{u}^{(m+1)}$, the last computed iterate, taken as $\mathbf{u}^{(0)}$, and $\sigma = \sigma_1$. This process is repeated to obtain b_j and σ_j, $j = 2, 3, \cdots$.

Adaptive SOR. One can accelerate the convergence of SOR or block SOR similarly, for A symmetrizable with Property Y (see [V, Ch. 9]). First, choose a trial value $\omega_0 < \omega_b$ (say $\omega_0 = 1$). Then, after a moderate number of iterations, approximate the spectral radius of the SOR iteration matrix $\mathbf{L}_{\omega_0}$ by

$$\left[\frac{\|\mathbf{s}^{(m)}\|}{\|\mathbf{s}^{(0)}\|}\right]^{1/m} = R_{\omega_0}^{(m)} \approx \rho(\mathbf{L}_{\omega_0}). \tag{7.5}$$

Since $\omega_0 < \omega_b$, $\rho(\mathbf{L}_{\omega_0}) = \lambda_0$ where λ_0 is a real eigenvalue of $\mathbf{L}_{\omega_0}$. The eigenvalues of $\mathbf{L}_{\omega_0}$ and those of the Jacobi iteration matrix $B = I - D^{-1}A$ are related by Young's identity (Chapter 4, (14.1)) which gives

---

[27] One obtains $r^m(\eta_0)$ from (7.4), which can be written as a quadratic polynomial in $r^{m/2}(\eta_0)$; (7.3) then gives a linear algebraic equation for $(1-\eta_0^2)^{1/2}$.

$$\rho = \rho(B) = \frac{\lambda_0 + \omega_0 - 1}{\omega_0 \sqrt{\lambda_0}}.$$

Setting $\lambda_0 = R_{\omega_0}^{(m)}$ above, one gets an estimate ρ_1 of ρ and takes $\omega_1 = 2/(1+\sqrt{1-\rho_1})$ as the next approximation of $\omega_b = 2/(1-\sqrt{1-\rho})$.

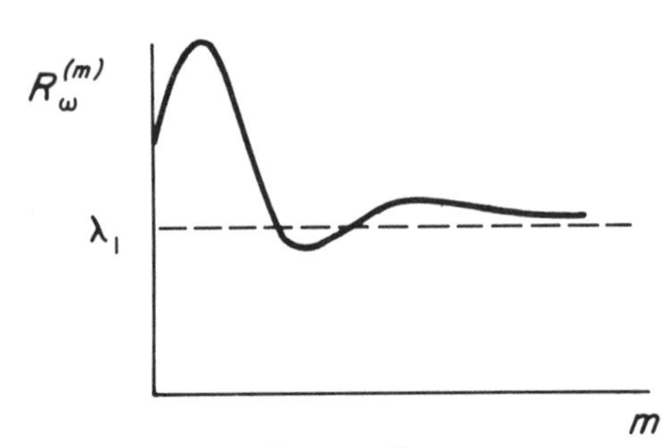

FIGURE 2

Changing omega. To avoid computing $\|\mathbf{s}^{(m)}\|$ too often, and to avoid changing ω_k prematurely, one should verify several conditions before changing it. Figure 2 shows typical behavior of $R_\omega^{(m)}$ as a function of iteration number m for fixed ω, $1 \leqslant \omega < \omega_b$.[28] To avoid the (possible) initial rise in $R_\omega^{(m)}$, Hageman and Young [HY, Chap. 9] recommend doing at least five iterations. Since $\omega - 1$ is a lower bound on $\rho(\mathbf{L}_\omega)$ (see (2.4)), one should also require $R_\omega^{(m)} < (\omega-1)^F$ with $F \in [0.65, 0.8]$ [HY, p. 225]. Another condition is $m(\omega-1)^{m-1} < 1/2$; this is to allow for an eigenvalue μ_i of the Jacobi matrix B which satisfies $\omega_k^2 \mu_i^2 = 4(\omega_k - 1)$; in this case the iteration matrix $\mathbf{L}_\omega$ has a principal vector associated with μ_i which gives a contribution to the pseudo-residual which behaves like $Km(\omega-1)^{m-1}$ [HY, pp. 218–219]. Finally, Hageman and Young require that

$$-10^{-p+1} \leqslant R_\omega^{(m-1)} - R_\omega^{(m)} \leqslant 10^{-p},$$

where p is a 'strategy parameter';[29] this is to be sure that m is large enough so that $R_\omega^{(m)}$ is on the 'tail' of the curve in Fig. 2.

Stopping criteria. All criteria for 'stopping' an iterative or semi-iterative process for solving a discretized linear elliptic problem $A\mathbf{u} = \mathbf{b}$ are somewhat arbitrary. Generally speaking, a one-step iterative process is terminated when its residual $\mathbf{r}^{(m)}$ or pseudo-residual $\mathbf{s}^{(m)}$ has become sufficiently small. Thus, for G as above we have $(I-G)\mathbf{e}^{(m)} = \mathbf{s}^{(m)}$. If $W(I-G)W^{-1}$ is symmetric with eigenvalues in $[\alpha, \beta]$, $0 < \alpha$, then

$$(7.6) \qquad \|W\mathbf{e}^{(m)}\| \leqslant \rho([I-G]^{-1})\, \|W\mathbf{s}^{(m)}\| = \frac{1}{\alpha} \|W\mathbf{s}^{(m)}\|,$$

where $\alpha = 1 - b$ and the norm is the Euclidean norm. Thus a simple and practical stopping criterion is $\|W\mathbf{s}^{(m)}\| \leqslant R\eta$, where R is an estimate of $1/\alpha$ and η is the desired tolerance on $\|W\mathbf{e}\|$. For the Richardson, Jacobi,

[28] We thank Lou Hageman and David Young for allowing us to use this figure which appears in [HY, p. 227].

[29] ITPACK uses $p = 4$ for computers with a macheps 10^{-14}.

and SSOR schemes, W is known (cf. §6); hence this test can be applied to their acceleration by the Chebyshev and CG methods.

The SOR iteration matrix is, in general, not symmetrizable. In adaptive SOR, one uses overrelaxation parameters which are less than the optimum ω_b. The spectral radius is then equal to a real eigenvalue λ_1 and one uses the approximate relationship

$$\|\mathbf{e}^{(m)}\| \approx R\,\|\mathbf{s}^{(m)}\|, \qquad R = \frac{1}{1-\lambda_1}. \tag{7.7}$$

One can take $R = 1/(1-H)$, where $H = \max\{(\omega-1), R_\omega^{(m)}\}$ ($\omega - 1$ is a lower bound on $\rho(\mathbf{L}_\omega)$ by (2.4)) and $R_\omega^{(m)}$ is the estimate (7.5).

In typical linear source problems, $1-b$ and $1/\kappa(A) = O(h^{-2})$ are both in the range 0.001–0.01, so that $\|\mathbf{s}^{(m)}\| < 10^{-5}$ is within one order of magnitude of being a satisfactory stopping criterion.

8. ADI methods. Another ingenious family of semi-iterative methods for solving linear source problems is provided by *alternating direction implicit* (ADI) methods. As in Chapter 3, (4.1), consider the DE

$$-\frac{\partial}{\partial x}A(x,y)\frac{\partial u}{\partial x} - \frac{\partial}{\partial y}C(x,y)\frac{\partial u}{\partial y} + F(x,y)u = G(x,y), \tag{8.1}$$

with 'leakage' boundary conditions $\partial u/\partial n + h(\mathbf{y})u = 0$. As usual, A and C in (8.1) are assumed to be *positive* functions (to make the DE elliptic), while F and h are *nonnegative*.

The standard 5-point difference operator of Chapter 3, (4.2), can be split up into three components, and (4.2) rewritten as

$$(H + V + \Sigma)\mathbf{u} = \mathbf{b}. \tag{8.2}$$

In (8.2), H comes from the 3-point difference approximation to $(Au_x)_x$ (by 'horizontal' divided differences) and V from that to $(Cu_y)_y$. The diagonal component Σ comes from the absorption term Fu, while $\mathbf{b}$ comes from the right side G and the boundary conditions.

Thus, for a uniform rectangular mesh with mesh lengths h and k and $U_{i,j} \approx u(ih, jk)$, we have

$$HU_{i,j} = -a_{i,j}U_{i-1,j} + 2b_{i,j}U_{i,j} - c_{i,j}U_{i+1,j}, \tag{8.3}$$

$$VU_{i,j} = -\alpha_{i,j}U_{i,j-1} + 2\beta_{i,j}U_{i,j} - \gamma_{i,j}U_{i,j+1}, \tag{8.3'}$$

where, much as in Chapter 3, (4.2a),

$$a_{i,j} = kA_{i-1/2,j}/h, \qquad c_{i,j} = kA_{i+1/2,j}/h, \qquad 2b_{i,j} = a_{i,j} + c_{i,j},$$
$$\alpha_{i,j} = hC_{i,j-1/2}/k, \qquad \gamma_{i,j} = hC_{i,j+1/2}/k, \qquad 2\beta_{i,j} = \alpha_{i,j} + \gamma_{i,j}.$$

These choices make H and V *symmetric* matrices acting on the vector of nodal values $U_{i,j} = U(x_i, y_j)$.

The matrix Σ is then a nonnegative diagonal matrix with diagonal entry $hkF(x_i, y_j)$ at (x_i, y_j). Moreover, H and V have positive diagonal entries and negative or zero off-diagonal entries. The matrices H and V are also *diagonally dominant* (see Chapter 4, §7). For any $\theta \geqslant 0$, the same is true *a fortiori* of $H+\theta\Sigma$, $V+\theta\Sigma$, and of $\theta_1 H+\theta_2 V+\theta\Sigma$ if $\theta_1 > 0$, $\theta_2 > 0$; indeed, they are all *Stieltjes matrices* (see Chapter 3, §5). The vector $\mathbf{b}$ is computed by adding to the source term $hkG(x_i, y_j)$ the terms in (8.3)–(8.3′) associated with points on the *boundary* of the domain.

Ordering the mesh-points by rows makes H *tridiagonal*; likewise, ordering them by columns makes V tridiagonal. That is, both H and V are similar to tridiagonal matrices, although one cannot make H and V both tridiagonal simultaneously by any one ordering. The basic idea of ADI methods is to factor each iterative cycle into two steps (half-cycles), the first of which uses an ordering that makes H tridiagonal and the second an ordering that makes V tridiagonal. This can be done as follows.

Equation (8.2) is clearly equivalent, for any diagonal matrices D and E, to each of the two vector equations

$$(H+\Sigma+D)\mathbf{u} = \mathbf{k} - (V-D)\mathbf{u}, \tag{8.4}$$

$$(V+\Sigma+E)\mathbf{u} = \mathbf{k} - (H-E)\mathbf{u}. \tag{8.4′}$$

This was first observed by Peaceman and Rachford[30] for the case $\Sigma = 0$, $D = E = \rho I$ a scalar matrix. In this case, (8.4)–(8.4′) reduce to

$$(H+\rho I)\mathbf{u} = \mathbf{k} - (V-\rho I)\mathbf{u}, \qquad (V+\rho I)\mathbf{u} = \mathbf{k} - (H-\rho I)\mathbf{u}.$$

The generalization to $\Sigma \neq 0$ and arbitrary $D = E$ was made by Wachspress and Habetler [60]; see also G. J. Habetler and E. L. Wachspress, Math. Comp. 15 (1961), 356–362.

ADI parameters. The *Peaceman-Rachford* method for solving (8.2) involves choosing an appropriate sequence of positive numbers ρ_m (ADI parameters), and calculating the sequence of vectors $\mathbf{u}^{(m)}$, $\mathbf{u}^{(m+1/2)}$ defined from the matrices $D_m = E_m = \rho_m I$, by the formulas

$$(H+\Sigma+D_m)\mathbf{u}^{(m+1/2)} = \mathbf{k} - (V-D_m)\mathbf{u}^{(m)}, \tag{8.5}$$

$$(V+\Sigma+E_m)\mathbf{u}^{(m+1)} = \mathbf{k} - (H-E_m)\mathbf{u}^{(m+1/2)}. \tag{8.5′}$$

Since $(H+\Sigma+D)$ and $(V+\Sigma+E)$ are similar under conjugation by permutation matrices to tridiagonal Stieltjes matrices, each of the Eqs. (8.5) and (8.5′) can be solved rapidly.[31] The aim is to choose the initial trial vector $\mathbf{u}^{(0)}$ and the matrices D_1, E_1, D_2, E_2, $\cdots$, so as to make the sequence $\{\mathbf{u}^{(m)}\}$ converge quickly to the solution.

[30] Peaceman-Rachford [55]

[31] For example, by subroutine SPTSL of LINPACK. Of course, band elimination would also be quite efficient.

Douglas-Rachford method. A more flexible variant of the ADI method was suggested by Douglas and Rachford [56, p. 422, (2.3)]. It uses two sets of iteration parameters, ρ_n and ρ'_m. Originally proposed for $\Sigma = 0$, it can be defined for general $\Sigma \geqslant 0$ by setting $D_m = \rho_m I - \Sigma/2$ and $E_m = \rho'_m I - \Sigma/2$ in (8.5) and (8.5′), respectively. This gives

$$(8.6) \qquad (H + \tfrac{1}{2}\Sigma + \rho_m I)\mathbf{u}^{(m+1/2)} = \mathbf{k} - (V - \tfrac{1}{2}\Sigma - \rho_m I)\mathbf{u}^{(m)},$$

and

$$(8.6') \qquad (V + \tfrac{1}{2}\Sigma + \rho'_m I)\mathbf{u}^{(m+1)} = \mathbf{k} - (H - \tfrac{1}{2}\Sigma + \rho'_m I)\mathbf{u}^{(m+1/2)}.$$

The rate of convergence depends strongly on the choice of iteration parameters ρ_m, ρ'_m.

Permutable H, V, Σ. For some 'model problems' and choices of the ρ_m, ADI methods are truly 'fast'; they converge orders of magnitude faster than SOR or CCSI, for example. This is true in particular of the Poisson equation and the modified Helmholtz equation, $-\nabla^2 u + \sigma u = f(x, y)$, in a rectangle. This is because these problems involve *permutable operators*,[32] satisfying

$$(8.7) \qquad HV = VH, \qquad H\Sigma = \Sigma H, \quad \text{and} \quad V\Sigma = \Sigma V.$$

More generally, H and V are permutable whenever the variables x and y are 'separable' for a given problem (cf. Chapter 2, §1).

When (8.7) holds, let the ρ_m be distributed in the intervals containing the (real) eigenvalues of

$$(H + \Sigma + \rho_m I)^{-1}(V - \rho_m I) \quad \text{and} \quad (V + \Sigma + \rho_m I)^{-1}(H - \rho_m I)$$

with equal proportionate spacing.[33] As $h \downarrow 0$, the number of ADI iterations required to reduce the error by a prescribed factor is (asymptotically and neglecting roundoff) only $O(\log h^{-1})$, as contrasted with $O(h^{-1})$ for SOR using the optimum relaxation parameter ω_b or for CCSI.

For model problems with $HV = VH$ and a known 'spectral interval' $[\alpha, \beta]$, a precise determination of the *optimum* parameters for a given total number of iterations, R, has been made by Jordan.[34]

---

[32] For discussions of the commutative case, see Birkhoff-Varga-Young [62, part II], and R. E. Lynch, J. R. Rice, and D. H. Thomas, J. SIAM 13 (1965), 995—1006.

[33] See Douglas-Rachford [56, p. 437].

[34] See Wachspress [66, p. 185]; also E. L. Wachspress, J. SIAM 10 (1962), 339–350; ibid. 11 (1963), 994–1016; C. de Boor and J. R. Rice, ibid. 11 (1963), 159–169; ibid. 12 (1964), 892–896; R. B. Kellogg and J. Spanier, Math. Comp. 19 (1965), 448–452.

9. Stationary ADI. Unfortunately, as was first pointed out in Birkhoff-Varga [59], the permutability conditions (8.5) are rarely fulfilled. As was proved there, we have the following result (cf. also [V, p. 234]).

THEOREM 9. *For the standard 5-point approximation to the linear source problem on a uniform rectangular mesh, $HV = VH$ only for DE's of the special 'separable variable' form*

$$A(x)u_{xx} + C(y)u_{yy} + [D(x) + E(y)]u = f(x, y), \tag{9.1}$$

in a rectangle having sides parallel to the axes.

Hence like the FFT of Chapter 4, §5, elliptic ADI is not a *general* 'fast' direct method. Nevertheless, it is always competitive with SOR because even 'stationary' ADI methods using a single parameter ρ have the same order of convergence as optimized SOR (see Birkhoff-Varga-Young [62, Theorem 20.1]). Specifically,[35] the Peaceman-Rachford method (8.5)–(8.5′) is always *convergent* if one chooses $D = E = \rho I - \Sigma/2$, where ρ is a fixed positive number, making $D + \Sigma/2$ positive definite.

In detail, we define, as usual, the error vector as the difference $\mathbf{e}^{(m)} = \mathbf{u} - \mathbf{u}^{(m)}$ between the exact solution $\mathbf{u}$ of (8.2) and the approximate solution $\mathbf{u}^{(m)}$ after the m-th iteration. For simplicity, we set $D = E = \rho I$. A straightforward calculation shows that, for the Peaceman-Rachford method, the effect of a single iteration of (8.4)–(8.4′) is to multiply the error vector $\mathbf{e}^{(m)}$ by the error reduction matrix T_ρ defined by

$$T_\rho = (V + \Sigma + \rho I)^{-1}(H - \rho I)(H + \Sigma + \rho I)^{-1}(V - \rho I).$$

If one sets $D_m = \rho I - \Sigma/2 = E_m$ also for the generalized Douglas-Rachford method (8.6)–(8.6′), then its error reduction matrix is W_ρ:

$$W_\rho = (V_1 + \rho I)^{-1}(H_1 + \rho I)^{-1}(H_1 V_1 + \rho^2 I).$$

We next prove a lemma which expresses the algebraic content of a theorem of Wachspress-Habetler [60, Theorem 1].

LEMMA 1. *Let P and Q be positive definite real matrices, with S symmetric.*[36] *Then $Q = (P - S)(P + S)^{-1}$ is a contraction (norm-reducing)*[37] *in the norm $\|\mathbf{x}\| = (\mathbf{x}^T S^{-1}\mathbf{x})^{1/2}$.*

Proof. For any norm $\|\mathbf{x}\|$, the statement that Q is norm-reducing is equivalent to the statement that $\|(S - P)\mathbf{y}\|^2 < \|(S + P)\mathbf{y}\|^2$ for every nonzero vector $\mathbf{y} = (P + S)^{-1}\mathbf{x}$. In turn, this is equivalent for the norm $\|\mathbf{x}\| = (\mathbf{x}^T S^{-1}\mathbf{x})^{1/2}$ to the statement that

---

[35] Theorem 1 of Wachspress-Habetler [60].

[36] Note that P is *not* assumed to be symmetric, but only to be such that $\mathbf{x}^T(P + P^T)\mathbf{x} > 0$, for all real $\mathbf{x} \neq \mathbf{0}$.

[37] The phrase 'norm-reducing' here refers to the Euclidean norm only in special cases.

$$\mathbf{y}^T(P+S)^T S^{-1}(P+S)\mathbf{y} > \mathbf{y}(P-S)^T S^{-1}(P-S)\mathbf{y}$$

for all nonzero $\mathbf{y}$. Expanding the bilinear terms and cancelling, this is equivalent to the condition that $\mathbf{y}^T(P+P^T)\mathbf{y} > 0$ for all nonzero $\mathbf{y}$. But this is the hypothesis that P is positive definite.

COROLLARY. *In Lemma* 1 *the spectral radius* $\rho(Q)$ *is less than unity.*

This follows from Lemma 1 and the following general result on matrices:

$$\rho(M) \leqslant \max_{\|\mathbf{x}\| \neq 0} \frac{\|M\mathbf{x}\|}{\|\mathbf{x}\|}, \quad \text{for any norm} \quad \|\cdot\|.$$

Actually, $\rho(M)$ is the infimum of $\max\{\|M\mathbf{x}\|/\|\mathbf{x}\|\}$ taken over all Euclidean (inner product) norms.

THEOREM 10. *Any stationary ADI process* (8.5)–(8.5′) *with all* $D_m = D$ *and all* $E_m = E$ *is convergent, provided* $\Sigma + D + E$ *is symmetric and positive definite,* and $2H+\Sigma+D-E$ *and* $2V+\Sigma+E-D$ *are positive definite.*[38]

The hypotheses of Theorem 10 are fulfilled for $D=\rho I-\theta\Sigma$, $E=\rho' I-\theta'\Sigma$ for any ρ, $\rho' > 0$ and any θ, θ' with $0 \leqslant \theta, \theta' < 2$. Substituting into (8.5)–(8.5′), we obtain the following results.

COROLLARY. *If* ρ, $\rho' > 0$ *and* $0 \leqslant \theta \leqslant 2$, *then the stationary ADI method defined with* $\theta' = 2-\theta$ *by*

$$[H+(\theta/2)\Sigma+\rho I]\mathbf{u}^{(m+1/2)} = \mathbf{k} - [V+(\theta'/2)\Sigma-\rho I]\mathbf{u}^{(m)},$$

$$[V+(\theta/2)\Sigma+\rho I]\mathbf{u}^{(m+1)} = \mathbf{k} - [H+(\theta'/2)\Sigma-\rho I]\mathbf{u}^{(m+1/2)},$$

is convergent. In fact, it is norm-reducing for the norm defined by

$$\|\mathbf{x}\| = \mathbf{x}^T(\Sigma+D+E)^{-1}\mathbf{x} = \mathbf{x}^T\{(\rho+\rho')I + [(\theta+\theta')/2]\Sigma\}^{-1}\mathbf{x}.$$

Operation count. Because ADI requires solving two tridiagonal systems of equations per iteration, the fact that stationary ADI with optimum ρ converges after about as many iterations as SOR with $\omega=\omega_b$ does not mean that it is equally efficient. The main advantage of ADI over SOR is its potential for greatly reducing the number of cycles by proper choice of variable ρ_m. (Its main disadvantage is the seeming impossibility of providing an algorithm for doing this with *nonpermutable* H, V, Σ.)

In this connection, we recall a very interesting theorem of Willis Guilinger.[39] Utilizing the *smoothness* of solutions of elliptic problems (Chapter 2, §8) on *convex* domains, Guilinger showed that the stationary Peaceman-Rachford method (with a single, appropriately chosen parameter ρ depending on h) can be made to reduce the error by a given factor

[38] This result, in the case $D=E$, was first given by Wachspress-Habetler [60]. For the analogous result on W, see Birkhoff-Varga [59].

[39] W. H. Guilinger, J. Math. Anal. Appl. 11 (1964), 261–277.

in a number of steps which is *independent of the mesh length*, provided that ρ is appropriately chosen. For an extension of this to a sequence of parameters in the case of *permutable matrices*, see Lynch-Rice [68].

Unfortunately, it seems to be impossible to make rigorous extensions of the preceding theoretical results to most problems with variable coefficients or in nonrectangular regions (see Birkhoff-Varga [59] and Birkhoff-Varga-Young [62]). Whereas the theory of SOR applies rigorously to general source problems, the experimentally observed success of ADI is in general hard to explain, and even harder to predict. Very favorable results have recently been reported by Gelinas-Doss-Miller [81], however.[40]

Three-dimensional ADI. For ADI methods in three or more dimensions, see J. Douglas, Numer. Math. 4 (1962), 41–63, and J. Douglas, R. B. Kellogg, and R. S. Varga, Math. Comp. 17 (1963), 279–282. To predict their rate of convergence is *very* difficult. For example, as Price and Varga have shown (cf. [V, p. 245]), they can *diverge* even for fixed parameters.

10. Multigrid; mesh refinement. Today, the best subroutines for solving ordinary DE's include provisions for automatically changing the mesh length so as to keep the error below a specified tolerance. Similar provisions can be made for elliptic problem solving, and we will conclude this chapter by describing one of them.

The 'multigrid' method as developed by Achi Brandt and others[41] is designed to optimize the interplay between discretization and solution processes. When high accuracy is wanted, the ultimate discretization may require a very fine mesh, giving rise to a large condition number $\kappa_h = \kappa(A_h)$, resulting in very slow convergence for iterative processes. By using intelligently a *sequence* of grids ('levels') of increasing fineness, a reduction by orders of magnitude in the 'computational complexity' can often be achieved.

---

[40] See O. Widlund, Math. Comp. 20 (1966), 500–515. For the effectiveness of ADI methods on nonrectangular meshes, see J. Spanier, in Ralston-Wilf [67, pp. 215–245]; and R. B. Kellogg, Math. Comp. 18 (1964), 203–210.

[41] R. P. Fedorenko, Zh. vych. mat. 1 (1961), 922–927; English translation, USSR Comput. Math. and Math. Phys. 1 (1962), 1092–1096. See also N. S. Bakhvalov, Zh. vych. mat. 6 (1966), 861–883, English translation, USSR Comput. Math. and Math. Phys. 6 (1966), 101–135. See Brandt [77] for historical notes and specific details of the method. See also R. A. Nicolaides, J. Comp. Phys. 19 (1975), 418–431; A. Brandt in Rice [77, pp. 277–317]; in Parter [79, pp. 53–147]; in Schultz [81]; and R. E. Alcouffe, A. Brandt, J. E. Dendy, and J. W. Painter, SIAM J. Sci. Stat. Comp. 2 (1981), 430–454. ELLPACK contains software described by H. Foerster and K. Witsch in *Multigrid Methods*, W. Hackbusch and U. Trottenberg, eds., Lecture Notes in Mathematics, Vol. 960, Springer, 1982.

For simplicity, consider the method as applied to the Model Problem: solve $-\nabla^2 u = f(x,y)$ in the unit square S, for the boundary condition $u \equiv 0$ on ∂S. A sequence of square grids Ω_j with mesh lengths $h_j = 2^{-j}$ is ideal for this problem; either the 'standard' 5-point or $O(h^6)$ 9-point difference can be used.

Smoothing. The Gauss-Seidel method (Chapter 4, §8) is very efficient for *smoothing* the iteration error $\mathbf{E}^{(m)} = \mathbf{U} - \mathbf{U}^{(m)}$. With the natural ordering, at interior mesh-points this error satisfies

$$E_{j,k}^{(m+1)} = \frac{1}{4}(E_{j-1,k}^{(m+1)} + E_{j,k-1}^{(m+1)} + E_{j+1,k}^{(m)} + E_{j,k+1}^{(m)}).$$

Consider a Fourier component of the error: $a^{(m)} \exp(i\pi h[jp + kq])$. A single application of Gauss-Seidel reduces the amplitude by the factor

$$\sigma_{p,q} = \left|\frac{a^{(m+1)}}{a^{(m)}}\right| = \left|\frac{e^{i\pi ph} + e^{i\pi qh}}{4 - e^{-i\pi ph} - e^{-i\pi qh}}\right|.$$

For large wavenumber (short wavelength) components, say $|ph|, |qh| \geqslant 1/2$, we have $\sigma_{p,q} \leqslant 1/2$; consequently, after a few iterations, these error components become negligible compared with the smallest wavenumber components, for which $\sigma_{p,q} = 1 - O(h^2)$. In other words, the error is dominated by components with $|ph|, |qh| \leqslant 1/2$, that is, by components associated with a mesh having twice the spacing, namely a spacing of $2h$.

Therefore, the computation can be continued on the coarser mesh. A few iterations on the coarse mesh make the short wavelength error components negligible on the coarser mesh and one can proceed to an even coarser mesh, and so on. Finally, the mesh becomes so coarse that a direct method can be used to solve the difference equation.

This solution, on the coarsest mesh with spacing h_1, gives the long wavelength correction to the approximation already obtained on the finer mesh with spacing $h_2 = h_1/2$ and an interpolant to it is added to the previously obtained approximation. A few iterations eliminate the short wavelength error components introduced by the interpolation. Then interpolation is used to improve the accuracy of the solution on the next mesh with spacing $h_3 = h_2/2$, and so on.

Multigrid. In detail, the multigrid method uses a nest of meshes: $\Omega_1 \subset \Omega_2 \subset \cdots \subset \Omega_n$ with mesh spacing $h_j = h_{j-1}/2$. Let $R_j : F_j \longmapsto F_{j-1}$ denote the restriction of a function F_j on Ω_j to F_{j-1} on Ω_{j-1}. Let $I_j : G_j \longmapsto G_{j+1}$ denote interpolation, where G_{j+1} on Ω_{j+1} is the interpolant (say, biquadratic or bicubic; see Chapter 7) to G_j on Ω_j. Let U_j on Ω_j denote the solution of $-\nabla_j^2 U = f$, where $-\nabla_j^2$ is the discrete Laplacian on a square grid with mesh length h_j.

One first solves $-\nabla_1^2 U_1 = f$. The interpolant $I_1 U_1 = U_2^{(0)}$ is taken as the first approximation $U_2^{(0)}$ of the solution of

$$-\nabla_2^2 U_2 = f; \tag{10.1}$$

$U_2^{(m)}$ (with m small) is computed by Gauss-Seidel iteration and its residual, $r = f + \nabla_2^2 U_2^{(m)}$, is primarily made up of long wavelength components.

Next one restricts the residual and the approximation to Ω_1 and solves

$$-\nabla_1^2 u = R_1[f + \nabla_2^2 U_2^{(m)}] - \nabla_1^2 R_1[U_2^{(m)}], \tag{10.2}$$

noting that this is equivalent to

$$-\nabla_1^2(u - R_1[U_2^{(m)}]) = R_1[f + \nabla_2^2 U_2^{(m)}]. \tag{10.3}$$

Thus $v = u - R_1[U_2^{(m)}]$ gives the dominant long wavelength correction on Ω_1 to $U_2^{(m)}$. One now interpolates again and adds this correction to obtain $U_2^{(m)} + I_2[v]$ which is taken as a new approximation $U_2^{(0)}$ to the solution of (10.1). A few Gauss-Seidel iterations give a new estimate, which should now be an accurate solution of (10.1). It is used as the initial estimate $U_3^{(0)}$ of the solution of

$$-\nabla_3^2 U_3 = f. \tag{10.4}$$

After eliminating the short wavelength error components, one applies Gauss-Seidel to

$$-\nabla_2^2 u_2 = R_2[f + \nabla_3^2 U_3^{(m)}] - \nabla_2^2 R_2[U_3^{(m)}] \tag{10.5}$$

to get $u_2^{(m)}$. One then solves (10.2), replacing $U_2^{(m)}$ with $U_2^{(m)}$ on the right side.

The process is repeated: after iterating on a fine grid Ω_k, the residual is restricted to the coarser grid Ω_{k-1} and iterations are performed on it. This process is repeated on $\Omega_{k-2}, \cdots$, until one has computed iterations on Ω_1. Then better approximations are obtained on successively finer grids, by the processes of interpolation, correction, and iteration, each carried out when changing from Ω_j to Ω_{j+1}, for $j = 1, \cdots, k-1$. One then does these processes again to get an approximation on Ω_{k+1}.

Local mesh refinement. In many problems, greater accuracy can be achieved for a given amount of computation by using *local* mesh refinements near a few singularities or 'stress concentrations', than by the multigrid method described above — e.g., than by making *uniform* refinements of a uniform (typically square) mesh. Many schemes for making such local mesh refinements have been proposed. The most systematic study is by Babuska and Rheinboldt.[42] Their papers, which apply finite element methods (Chapter 7) to problems in structural mechanics, should be studied by anyone wishing to analyze schemes of mesh refinement in depth. As they point out, there is also a tradeoff between using a finer mesh and using higher-order approximations.

[42] See their article in Chap. 3 of *Computational Methods in Nonlinear Mechanics*, J. T. Oden ed., North-Holland, 1980, and the references given there. See also their articles in Birkhoff-Schoenstadt [84].

Chapter 6

Direct Variational Methods

1. Introduction. In Chapter 2, §7 and elsewhere, we have described some variational principles which characterize the solutions of various self-adjoint elliptic boundary value problems. These principles are often rooted in ideas from mathematical physics. In particular, every configuration of stable static equilibrium in classical (Lagrangian) mechanics minimizes a suitable potential energy function $\mathbf{J}$. In the theory of 'small oscillations', $\mathbf{J}$ is an energy integral whose integrand is quadratic in a 'displacement function' $u(\mathbf{x})$ and its derivatives of order up to k. The associated Euler-Lagrange DE equivalent to $\delta\mathbf{J} = 0$, where $\delta\mathbf{J}$ is the first variation of $\mathbf{J}$, is then a linear elliptic DE of order $2k$.

Such variational principles apply, for example, to the source problem

$$(1.1)\quad \mathbf{L}[u] = -\nabla\cdot[p(\mathbf{x})\nabla u] + q(\mathbf{x})u = f(\mathbf{x}), \qquad p > 0, \quad q \geqslant 0,$$

whose solution minimizes the integral

$$(1.2)\qquad \mathbf{J}[u] = \frac{1}{2}\int_{\Omega}[p(\mathbf{x})\nabla u\cdot\nabla u + q(\mathbf{x})u^2 - 2f(\mathbf{x})u(\mathbf{x})]\,dR,$$

for given boundary values. The first variation $\delta\mathbf{J}[u]$ of $\mathbf{J}$ is zero for u satisfying Dirichlet-type boundary conditions. Indeed, assuming smoothness on the closure $\bar{\Omega} = \Omega \cup \Gamma$ of Ω, one can prove by elementary arguments the following result.[1]

THEOREM 1. *A function $u \in C^2(\Omega) \cap C(\bar{\Omega})$ satisfies (1.1) in Ω if and only if it strictly minimizes the $\mathbf{J}[v]$ of (1.2), in the class of all $v \in C^2(\Omega) \cap C(\bar{\Omega})$ assuming the same boundary values.*

Sketch of proof. Consider the bilinear functional

$$(1.3)\quad \mathbf{P}\langle u,v\rangle = \frac{1}{2}\int_{\Omega}[p(\mathbf{x})\nabla u\cdot\nabla v + q(\mathbf{x})uv]\,dR, \qquad p > 0, \quad q \geqslant 0.$$

It is easy to verify that $\mathbf{P}$ is an *inner product* (Chapter 2, §5). Writing $v = u + w$, whence $w \equiv 0$ on Γ, we have by the divergence theorem, since $\nabla\cdot(pw\nabla u) = p\nabla u\cdot\nabla w + w\nabla\cdot(p\nabla u)$, that

[1] See also [KK, Chap. IV]. See Gilbarg-Trudinger [77, §10.5] for the case of quasi-linear operators.

$$\mathbf{J}[v]-\mathbf{J}[u]=\frac{1}{2}\int_\Omega [2p\nabla u\cdot\nabla w+p\nabla w\cdot\nabla w+2quw+qw^2-2fw]\,dR$$
$$=\mathbf{P}\langle w,w\rangle+2\int_\Omega w[-\nabla\cdot(p\nabla u)+qu-f]\,dR.$$

The last integral vanishes (i.e., $\delta\mathbf{J}=0$) if $\mathbf{L}[u]=-\nabla\cdot(p\nabla u)+qu=f$; moreover, $\mathbf{P}\langle w,w\rangle>0$ for $v\neq u$, since the integrand in (1.3) is positive in some region unless $w=v-w\equiv const.=0$.

Example 1. The inner product associated with the *ordinary* differential operator d^4/dx^4 is

$$\mathbf{E}\langle u,v\rangle=\int_a^b u''(x)v''(x)\,dx.$$

For any piecewise smooth $u\in C^2[a,b]$ and $v\in C^1[a,b]$ we can integrate twice by parts to get

$$\mathbf{E}\langle u,v\rangle=\sum_{i=1}^n [u''v'-u'''v]\Big|_{x_{i-1}}^{x_i}+\sum_{i=1}^n\int_{x_{i-1}}^{x_i} u^{iv}v\,dx,$$

where x_i, $i=1,\cdots,n-1$, are the points of discontinuity of u''' and/or v''. The values of $u''v'$ cancel at these x_i, while v is continuous where u''' has jumps of $\Delta u_i'''=u'''(x_i^+)-u'''(x_i^-)$. Therefore

$$\mathbf{E}\langle u,v\rangle=\sum_{i=1}^{n-1} v_i\Delta u_i'''+\sum_{i=1}^{n-1}\int_{x_{i-1}}^{x_i} u^{iv}(x)v(x)\,dx$$
$$+u''(b)v'(b)-u''(a)v'(a)+v(a)u'''(a^+)-v(b)u'''(b^-). \tag{1.4}$$

The identity (1.4) makes it easy to prove the following.

LEMMA. *If* $u\in C^2[a,b]$ *is a cubic spline function with joints at the* x_i, *then*

$$\mathbf{E}\langle u,v\rangle=0 \tag{1.5}$$

for all $v\in C^1[a,b]$ *satisfying* $v_i=0$ *for all* x_i *and* $v'(a)=v'(b)=0$.

Proof. All terms in (1.4) vanish: the last because $u(x)$ is piecewise cubic whence $u^{iv}(x)\equiv 0$, and the other terms because of the continuity conditions on the u_i''' and v_i, and because $v'(a)=v'(b)=0$.

As a corollary, we have

THEOREM 2. *Among all piecewise smooth* $u\in C^1[a,b]$ *having given* $u'(a)$, $u'(b)$, *and* $u_i=u(x_i)$, $i=0,\cdots,n$, *the cubic spline interpolant minimizes* $\mathbf{E}\langle u,u\rangle$.

Proof. By (1.5), for all $u+v=w$ satisfying the interpolation conditions, the v_i, $v'(a)$, and $v'(b)$ all vanish. Hence, by the lemma,

$$\mathbf{E}\langle u+v,u+v\rangle=\mathbf{E}\langle u,u\rangle+\mathbf{E}\langle v,v\rangle.$$

Moreover, $\mathbf{E}\langle v,v\rangle=0$ implies that $\int[v''(x)]^2\,dx=0$, and hence that $v=\alpha x+\beta$ is *linear*. Since $v(a)=v(b)=0$, this however implies $v(x)\equiv 0$.

A more sophisticated variational principle refers to a simply supported plate with Poisson ratio ν and load density $\rho(x,y)$, resting on the plane $z = 0$. Here the equilibrium condition is $\delta \mathbf{J} = 0$ for (Courant-Hilbert [53, p. 250])

$$(1.6) \quad \mathbf{J}[u] = \frac{1}{2}\int_\Omega [(\nabla^2 u)^2 - 2(1-\nu)(u_{xx}u_{yy} - u_{xy}^2) - 2\rho u]\, dx\, dy.$$

As was stated in Chapter 2, §7, the Euler-Lagrange DE for the minimization of (1.6) is $\nabla^4 u = 0$, regardless of ν; the integral over Ω of $u_{xx}u_{yy} - u_{xy}^2$ is absorbed into a 'boundary term' (boundary stress); see §8.

Variational principles apply equally well to *nonlinear* elliptic problems and eigenproblems. For example, the surfaces of constant mean curvature are those whose area, subject to the constraint of bounding a given total volume (e.g., of a liquid drop), has first variation zero (cf. Chapter 1, §9). Finally, the eigenfunctions of various self-adjoint linear elliptic DE's are the stationary points of suitable Rayleigh quotients $\mathbf{R}[u]$, and are the functions such that $\delta \mathbf{R} = 0$ (see Chapter 2, §7, and §7 below).

The beauty and simplicity of such variational principles have inspired many scientists. They were invoked by Dirichlet, Riemann, Kelvin, and Hilbert for proving existence theorems. They were utilized by Rayleigh, Ritz, and other later scientists to compute approximate numerical solutions of various elliptic problems that had defied exact analysis. These methods are *direct* variational methods in the sense of Courant.[2] Sobolev has given this phrase a very special computational meaning; we quote Mikhlin [64, Preface]: "Direct methods are those methods . . . which reduce . . . problems to the solution of a finite number of algebraic equations".

In this chapter we shall describe some general methods which have been used to reduce elliptic problems to 'direct' variational form, and some of the computational schemes that have been designed to solve the resulting algebraic equations.

We will begin (in §§2–5) with the Rayleigh-Ritz and Galerkin methods for treating *linear* elliptic problems; i.e., for minimizing approximating *quadratic* functionals $\mathbf{J}[u]$. In particular, we will explain how these relate to the finite element methods (FEM) that have become so popular in the past two decades, and which we will discuss systematically in Chapter 7.

Partly for the sake of contrast, we next take up the *non*variational *collocation* method. This has been used successfully on a variety of small problems, and avoids the need for performing (perhaps messy) approximate numerical integration, such as is often required in applying direct variational methods.

---

[2] R. Courant, Jahresb. der Deutsche Math. Ver. 34 (1925), pp. 90–117.

In §7 we will study the *Rayleigh quotient* of a self-adjoint eigenproblem $\mathbf{L}[u] = \lambda \rho u$. This is a remarkable functional $\mathbf{R}[\phi] = \mathbf{V}[\phi]/\mathbf{T}[\phi]$, whose critical points (where $\delta \mathbf{R} = 0$) correspond to the *eigenfunctions* of $\mathbf{L}[u] = \lambda \rho u$.

We will then discuss, in §8, various 'domain constants' (like volume or surface area) associated with a given domain Ω. These include its electrostatic 'capacity', the least eigenvalue $\lambda_0(\Omega) = k_0^2$ of the Helmholtz equation on Ω, and so on. We will also consider the variation with Ω of $\lambda_0(\Omega)$ and other domain constants.

The last part of the chapter deals with *nonlinear* problems. We first treat their approximate solution by (quasi-) *Newton* methods which can be used to solve the resulting algebraic equations. Like the Galerkin method, these variational methods are not restricted to minimization problems.

In §9 we will exhibit typical variational principles associated with such nonlinear elliptic problems, going on in §§10–12 to explain various *minimization* (alias 'optimization') algorithms for actually computing solutions. Such algorithms minimize the functionals involved on appropriate approximating subspaces. After discussing Newton's method for solving *nonlinear* elliptic problems in §10, we turn our attention in §§11 and 12 to methods of 'unconstrained optimization', including gradient methods and Davidon's method. The chapter concludes with a brief discussion of fixpoint methods.

2. Rayleigh-Ritz method. The Rayleigh-Ritz method for finding approximate solutions of variational problems can be described very simply. If $\mathbf{J}[u]$ is the functional to be minimized (this is in practice an integral over Ω, with perhaps a 'boundary term' consisting of an integral over Γ), one first constructs an *approximating subspace* of functions $u(\mathbf{x},\boldsymbol{\gamma})$, $\boldsymbol{\gamma} = (\gamma_1, \cdots, \gamma_N)$, depending on N parameters γ_j. One next considers

$$\mathbf{J}[u(\mathbf{x},\boldsymbol{\gamma})] \equiv F(\boldsymbol{\gamma}) = F(\gamma_1, \cdots, \gamma_N), \tag{2.1}$$

and tries to minimize F in $\boldsymbol{\gamma}$-space.

Given any (positive definite) *quadratic* functional like those in the examples of the last section, and a linear approximating subspace with basis $\phi_1(\mathbf{x}), \cdots, \phi_N(\mathbf{x})$, we obviously have

$$\mathbf{J}[\gamma_1\phi_1(\mathbf{x}) + \cdots + \gamma_N\phi_N(\mathbf{x})] = \frac{1}{2}\sum_{i,j=1}^{N} a_{i,j}\gamma_i\gamma_j - \sum_{i=1}^{N} b_i\gamma_i + c, \tag{2.2}$$

where the $a_{i,j}$ and b_i are the second and first partial derivatives of F. To minimize this algebraic expression is straightforward; the condition is

$$\frac{\partial F}{\partial \gamma_i}(\boldsymbol{\gamma}) = \sum_{j=1}^{N} a_{i,j}\gamma_j - b_i = 0, \qquad j = 1, \cdots, N. \tag{2.3}$$

The solution of this linear system is unique if the symmetric matrix $A = ||a_{i,j}||$ is positive definite (as it is in typical physical problems). This problem is of the form $A\boldsymbol{\gamma} = \mathbf{b}$ discussed in Chapters 4 and 5.

Conversely, given any vector equation $A\mathbf{x} = \mathbf{b}$ with symmetric positive definite A, there is a corresponding variational problem with unique solution: find the $\mathbf{x}$ which minimizes $\mathbf{x}^T A\mathbf{x} - 2\mathbf{x}^T\mathbf{b}$ (see Chapter 4, §9).

In (2.2), $c = F(\mathbf{0})$ and

$$-2b_i = F_i = \frac{\partial F}{\partial \gamma_i}(\mathbf{0}), \qquad 2a_{i,j} = F_{i,j} = \frac{\partial^2 F}{\partial \gamma_i \partial \gamma_j}(\mathbf{0}),$$

so that $A = ||a_{i,j}||$ is a positive definite symmetric matrix. For $\mathbf{J}$ as in (1.2), the b_i and $a_{i,j}$ are

$$\begin{aligned} -2b_i &= F_i = \int_\Omega f(\mathbf{x})\phi_i(\mathbf{x})\, dR, \\ 2a_{i,j} &= F_{i,j} = \int_\Omega [p(\mathbf{x})\nabla\phi_i(\mathbf{x}) \cdot \nabla\phi_j(\mathbf{x}) + q(\mathbf{x})\phi_i(\mathbf{x})\phi_j(\mathbf{x})]\, dR. \end{aligned} \tag{2.4}$$

As in (2.3), the minimum occurs where $A\boldsymbol{\gamma} = \mathbf{b}$.

The preceding formulas again lead to algebraic problems almost identical to those which we treated in Chapters 4 and 5.

Example 2. Let Ω be a simply connected domain in the complex z-plane which contains $z = 0$ in its interior and has a smooth boundary. Then, among all complex analytic functions with $f(0) = 0$ and $f'(0) = 1$, that one which maps Ω onto a circular *disk* is characterized by either of the following two variational properties:[3]

$$\delta \iint_\Omega |f'(z)|^2\, dx\, dy = 0 \tag{2.5}$$

or

$$\delta \int_\Gamma |f'(z)|^2\, ds = 0. \tag{2.6}$$

One can take for the approximating subspace the set of all polynomials

$$f(z) = z + c_2 z^2 + \cdots + c_n z^n, \qquad c_k = a_k + ib_k.$$

This makes $N = 2n - 2$; the Rayleigh-Ritz method consists in minimizing the quadratic functional (2.5) (or (2.6)) on this subspace.

Sparsity. It is naturally easier to solve $A\mathbf{x} = \mathbf{b}$ if one uses coordinates with respect to which the Hessian $2A = ||\partial^2 f(\mathbf{0})/\partial\gamma_i\, \partial\gamma_j|| = ||F_{i,j}||$ is a sparse or at least well-conditioned matrix. Such coordinates always *exist*; in fact, when A is positive definite, there always exists a *coordinate system* making the Hessian the identity matrix. But for large N, to *compute* such

[3] In [KK, p. 366], it is shown that the function $f(z) = z + a_2 z^2 + \cdots$ which maps a given simply connected domain Ω onto a circle minimizes the integral of $|f'(z)|^2$ over Ω; i.e., minimizes the Dirichlet integral of $\mathrm{Re}\{f'\}$ or $\mathrm{Im}\{f'\}$.

a basis (i.e., to 'diagonalize' A) requires much more work than to solve (2.3).

Fortunately, one gets sparse matrices automatically, not only from difference approximations (where the 'sparsity' is great when the stencil is small), but also from finite element methods if one uses 'patch bases' of piecewise polynomial functions of the kind to be described in §3 and §4 and in Chapter 7. Finite element matrices are, however, usually less sparse than matrices arising from difference approximations.

Using patch bases or other local interpolation schemes, with $N < 1000$ unknowns, it is typically most efficient to solve (2.3) for the coefficients γ_i of the approximation by *band* elimination (Chapter 3, §5), or by one of the other direct sparse matrix methods discussed in Chapter 4, §§2–5.

3. Approximating subspaces. A systematic survey of 'finite element' schemes for approximating functions will be given in the next chapter. We will discuss here only a few very simple such schemes, whose invention predates the name 'finite element method'.

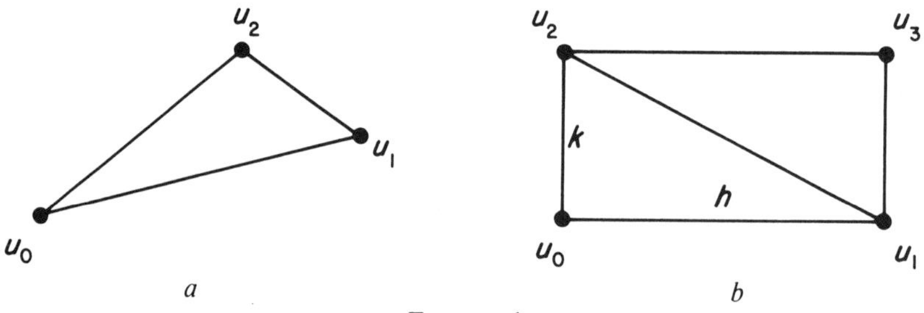

FIGURE 1

Example 3. 'Triangular elements', which interpolate linearly to corner values in each mesh triangle of a triangulation, provide one very versatile FEM for approximating solutions of second-order elliptic problems in the plane. If $u_j = u(x_j, y_j)$, $j = 0, 1, 2$, the triangle of Fig. 1a has area A, while the values of the coefficients of the linear interpolant $U = ax + by + c$ to the corner values u_j are given by $a = \alpha/2A$ and $b = \beta/2A$, where A, α, and β are given by the determinants

$$A = \frac{1}{2}\begin{vmatrix} x_0 & y_0 & 1 \\ x_1 & y_1 & 1 \\ x_2 & y_2 & 1 \end{vmatrix}, \qquad \alpha = \begin{vmatrix} u_0 & y_0 & 1 \\ u_1 & y_1 & 1 \\ u_2 & y_2 & 1 \end{vmatrix}, \qquad \beta = \begin{vmatrix} x_0 & u_0 & 1 \\ x_1 & u_1 & 1 \\ x_2 & u_2 & 1 \end{vmatrix}.$$

From these formulas, the Dirichlet integral can easily be calculated as

$$\mathrm{D}\langle U, U\rangle = \frac{1}{2}\iint |\nabla U|^2\,dx\,dy = \frac{A}{2}(a^2 + b^2) = \frac{1}{8A}(\alpha^2 + \beta^2).$$

This formula simplifies somewhat in the case of a *right* triangle with corner values $u_0 = u(0,0)$, $u_1 = u(h,0)$, and $u_2 = u(0,k)$, as in Fig. 1b. In this case, evidently $c = u_0$, $a = (u_1 - u_0)/h$, $b = (u_2 - u_0)/k$; the Dirichlet integral is therefore

$$\text{(3.1)} \qquad \frac{hk}{4}(a^2+b^2) = \frac{k}{4h}(u_0^2 - 2u_0u_1 + u_1^2) + \frac{h}{4k}(u_2^2 - 2u_0u_2 + u_0^2).$$

Rectangles. Finite element approximations are, in general, *not* equivalent to difference approximations such as the standard 5-point formula. A notable exception occurs with the Laplace equation. When a rectangle is bisected by a diagonal, as in Fig. 1b, the Dirichlet integral over the entire rectangle is, by (3.1),

$$\frac{h^2+k^2}{4hk}\sum u_j^2 - \frac{k}{2h}(u_0u_1 + u_2u_3) - \frac{h}{2k}(u_2u_0 + u_1u_3).$$

This is independent of the choice of diagonal, since u_0u_1 and u_2u_3 just refer to the horizontal, while u_2u_0 and u_1u_3 refer to the vertical sides.

Opposite corners of the rectangle are uncoupled (no terms in u_0u_3 or u_1u_2) in the formula displayed above; this proves part of the following elementary result.

THEOREM 3 (Courant). *Let a polygon* P *be subdivided by a rectangular mesh, and then triangulated by bisecting each rectangle along one diagonal. Let* Σ *be the approximating subspace of all continuous piecewise linear polynomials associated with this subdivision. The* $U \in \Sigma$ *which takes on given values at mesh points on the boundary of* P *and minimizes the Dirichlet integral on* P *has nodal values which satisfy the standard* 5*-point approximation to* $\nabla^2 u = 0$.

The proof is completed by comparing coefficients of $\partial \mathbf{D}\langle U, U\rangle/\partial U_0$ (in a quadrilateral containing (x_0, y_0) in its interior). Unfortunately, the analogues of the preceding statement do not hold for either the Poisson or the Helmholtz equation (Birkhoff-Gulati [74, p. 722]).

Pyramidal functions. For any triangulation, the 'pyramidal' (or 'tent') functions assuming the value 1 at one node and 0 at all others form convenient *basis* for the space of continuous, piecewise linear functions formed by piecing together triangular elements agreeing at nodes (hence along edges). The special case of a *square* mesh, subdivided into triangles by drawing diagonals through each mesh-point (see Fig. 2), was used extensively by Synge [57].

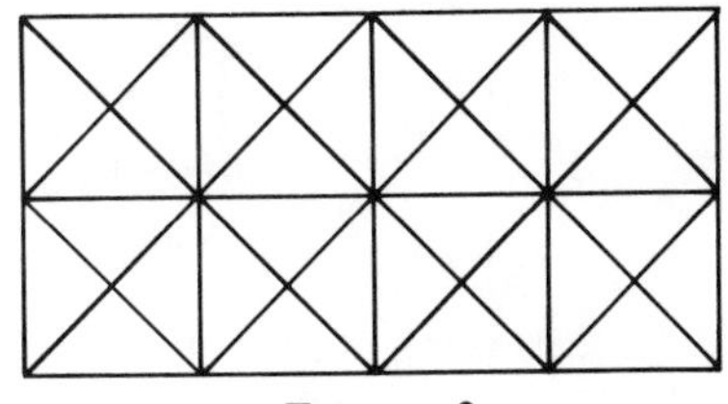

FIGURE 2

Remark. Although the Courant triangulation of Fig. 1 and the triangulation used by Synge [57] of Fig. 2 give *checkerboards* of 'red' and 'black' triangles, this does not imply Property Y in general.

Example 4. Given values $u_{i,j}$ of $u(x,y)$ at the grid points (or 'nodes') (x_i, y_j) of a rectangular mesh, we can interpolate a *piecewise bilinear*[4] function $U(x,y)$ to the nodal values $u_{i,j}$ of u; U is automatically continuous. This function is obtained easily in a 'canonical' mesh rectangle centered at the origin with vertices $(\pm h, \pm k)$ (see Fig. 3). There

$$U(x,y) = a_{00} + a_{10}x + a_{01}y + a_{11}xy, \tag{3.2}$$

with

$$4a_{00} = U_1 + U_2 + U_3 + U_4, \qquad 4ha_{10} = U_1 - U_2 - U_3 + U_4,$$
$$4ka_{01} = U_1 + U_2 - U_3 - U_4, \qquad 4hka_{11} = U_1 - U_2 + U_3 - U_4.$$

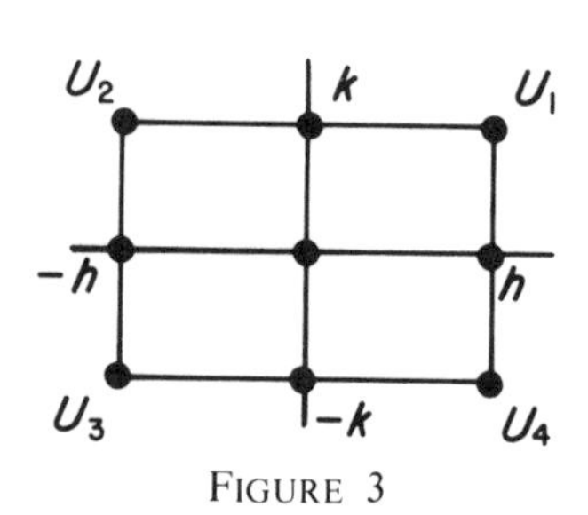

FIGURE 3

Other rectangles can be reduced to this case by translation. Along any mesh segment, say $y = k$, $U(x,k)$ is the linear interpolant to the two values of u at its ends; hence $U(x,y)$ is continuous across grid lines.

Since the mesh-rectangle has area $4hk$ and the gradient of U is

$$\nabla U = \text{grad } U = (a_{10} + a_{11}y, a_{01} + a_{1,1}x),$$

the resulting contribution of this cell R_m to the Dirichlet integral $\mathbf{D}_h\langle U, U\rangle = (1/2)\int \nabla U \cdot \nabla U \, dR$ over the entire domain Ω of the approximating function U is

$$\mathbf{D}_m = 2hk[a_{10}^2 + a_{01}^2 + \frac{1}{3}a_{11}^2(h^2 + k^2)]. \tag{3.3}$$

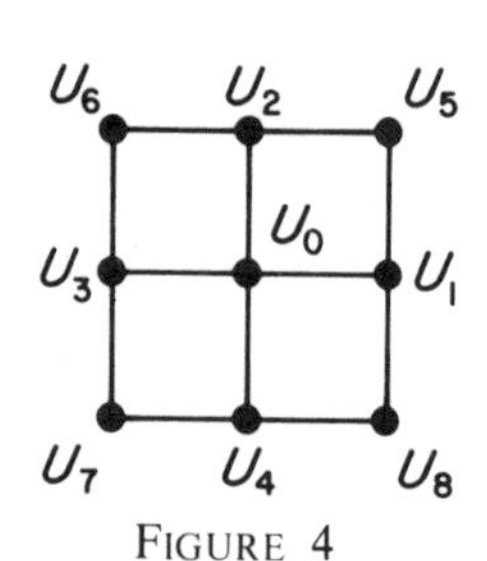

FIGURE 4

Square meshes. On a square mesh with $h = k$, the formulas become simpler. Thus the sum $\mathbf{D}_h\langle U, U\rangle$ of terms like (3.3) is minimized when the nodal values of U satisfy the following 9-point discretization of the Laplace equation (cf. Fig. 4):

$$8U_0 = U_1 + U_2 + U_3 + U_4 + U_5 + U_6 + U_7 + U_8.$$

Example 5. Consider the linear source problem in a *rectangular polygon* Ω consisting of squares of side h, with the boundary condition $u = g(\mathbf{y})$ on $\partial\Omega$. The Rayleigh-Ritz approximation U minimizes

$$\mathbf{J}[U] = \int\int \{p(\mathbf{x})[U_x^2 + U_y^2] + q(\mathbf{x})U^2 - 2f(\mathbf{x})U\} \, dx \, dy,$$

subject to the stated boundary conditions.

If $g(\mathbf{y})$ is piecewise linear, then we can take U in the space of continuous bilinear polynomials of Example 4. We can express the

---

[4] Piecewise bilinear interpolation was perhaps first used in Pólya [54]; see §7. See [FW, §24.2] for a summary of Pólya's paper.

approximation in terms of its nodal values $U_{i,j}$ as

$$U(x,y) = \sum U_{i,j}\phi_i(x)\phi_j(y)$$

where the 'tent function' basis elements are given by $\phi_k(z) = \phi(z-kh)$,

$$\phi(z) = 1 - |z|/h, \qquad z \in [-h,h].$$

Setting the derivative of $\mathbf{J}[U]$ with respect to $U_{i,j}$ equal to zero, one gets

$$\begin{aligned}\iint \{p(\mathbf{x})[U_x\phi_i'(x)\phi_j(y) + U_y\phi_i(x)\phi_j'(y)] \\ + [q(\mathbf{x})U - f(x,y)]\phi_i(x)\phi_j(y)\}\,dx\,dy = 0.\end{aligned} \tag{3.4}$$

Because the support[5] of $\phi_i(x)\phi_j(y)$ is the four mesh squares with vertex (ih,jh), (3.4) gives a linear relation among $U_{i,j}$ and its eight nearest neighbor nodal values.

For the modified Helmholtz equation, $-\nabla^2 u + k^2 u = f(\mathbf{x})$, with k^2 constant and 'source term' $f(\mathbf{x})$, this gives

$$\left(\frac{1}{3}\begin{bmatrix}-1 & -1 & -1\\ -1 & 8 & -1\\ -1 & -1 & -1\end{bmatrix} + \frac{k^2h^2}{36}\begin{bmatrix}1 & 4 & 1\\ 4 & 16 & 4\\ 1 & 4 & 1\end{bmatrix}\right) U_{i,j} = \int f(\mathbf{x})\phi_i(x)\phi_j(y)\,dx\,dy. \tag{3.5}$$

Quadrature. In addition to algebraic problems, FEM also often involve bivariate or multivariate numerical *quadrature*. Thus, the integral in (3.5) (and those in (3.4) when p and q are not constant) are approximated with quadrature formulas. For example, one can use the 4-point Gauss quadrature formula on each of the four mesh squares, e.g., on $[0,h]\times[0,h]$:

$$\int_0^h\int_0^h F(x,y)\,dx\,dy \approx \frac{h^2}{4}\sum F(h/2\pm\delta, h/2\pm\delta), \qquad \delta = h/2\sqrt{3}.$$

The points $h/2\pm\delta$ are zeros of the translated quadratic Legendre polynomial $P_2(-1+2x/h)$. The quadrature formula is exact for $F(x,y)$ a bicubic polynomial. Since $\phi_i(x)\phi_j(y)$ is bilinear, one is in effect approximating $f(\mathbf{x})$ in (3.5) by a piecewise biquadratic polynomial.

Alternatively, one can approximate $f(\mathbf{x})$ with its continuous piecewise bilinear interpolant at mesh-points. In this case the right side of (3.5) is replaced with

$$\frac{h^2}{36}\begin{bmatrix}1 & 4 & 1\\ 4 & 16 & 4\\ 1 & 4 & 1\end{bmatrix} f_{i,j}.$$

Error analysis. We now show that, if $u \in C^2$, then the bilinear interpolation error in u is $O(h^2)$, while that in $\nabla u = \operatorname{grad} u$ is $O(h)$.

[5] The 'support' of a function $f(\mathbf{x})$ is the set on which $f(\mathbf{x})$ does *not* vanish (i.e., where $f(\mathbf{x}) \neq 0$).

THEOREM 4. *Consider the error* $e(x,y) = U - u$ *of bilinear interpolation to the corner values* u_0, u_1, u_2, u_3 *in the rectangle* $\mathrm{R} = [0,h] \times [0,k]$. *If* $|u_{xx}| \leqslant M$ *and* $|u_{yy}| \leqslant M'$, *then*

$$|e(x,y)| \leqslant [Mx(h-x) + M'y(k-y)]/2,$$

for $(x,y) \in \mathrm{R}$, *and so* $|e(x,y)| \leqslant 8(Mh^2 + M'k^2)$.

Sketch of proof. It is easy to show that, for given corner values, $e(x,y)$ is a *strictly decreasing* function of u_{xx} and u_{yy}, and is determined by them — given two endpoint values on parallels through (x,y) to either the x-axis *or* the y-axis. Hence, the function

$$f(x,y) = \frac{1}{2}[Mx(h-x) + M'y(k-y)],$$

for given constant $f_{xx} = -M$ and $f_{yy} = -M'$, *maximizes* $e(x,y)$, while $-f(x,y)$ *minimizes* it.

Similarly, if $k \geqslant h$, the maximum error in ∇u is at most $17(M+M')h$.

Network analogies. We next consider variational principles that are based on *network* analogies. Indeed, the standard 5-point ΔE which we took as our prime example in Chapters 3–5 minimizes a quadratic 'energy dissipation' functional in the network analogy.[6] Other variational principles are associated with surface (and volume) integrals, such as arise naturally in the *finite element method*. In their most natural form, the latter yield 9-point (27-point in 3-dimensions) approximations to the Dirichlet problem, hence a different definition of 'discrete harmonic function' (see Chapter 3, §1).

Example 6. For Dirichlet-type boundary conditions, the 'discrete harmonic functions' $u_{i,j}$ of Duffin (Chapter 3, §1) minimize the sum of the squares of the first differences $\Delta_L u$ on a square mesh. Here $L = \overline{kk'}$ stands for an edge (or 'link') of the graph of the relevant square mesh. There are two types of edges: horizontal edges and vertical edges. A horizontal link joins a pair of nodes at, say $k = (ih, jh)$ and $k' = (ih+h, jh)$, and then, in the notation of Chapter 3, §4, $\Delta_L u = \Delta_x u_{i,j} = u_{i+1,j} - u_{i,j}$; when L is a vertical link, $\Delta_L u = \Delta_y u_{i,j} = u_{i,j+1} - u_{i,j}$.

Proof. In the notation of Fig. 5, the only terms in $\Sigma_L\, (\Delta_L u)^2$ involving $U_0 = u_{i,j}$ are

$$\sum_{k=1}^{4} (U_k - U_0)^2 = \sum_{k=1}^{4} U_k^2 - 2\sum_{k=1}^{4} U_k U_0 + 4U_0^2. \tag{3.6}$$

Calling this sum $\mathbf{J}$, we evidently have

$$\frac{1}{2}\frac{\partial \mathbf{J}}{\partial U_0} = 4U_0 - \sum_{k=1}^{4} U_k. \tag{3.7}$$

[6] J. C. Maxwell, *Electricity and Magnetism*, 3d ed., Oxford, 1904, pp. 403–408; Southwell

The usual 5-point approximation for the Poisson equation is obtained similarly: one subtracts $2h^2 U_0 f_0$ from both sides of (3.6); the condition $\delta\mathbf{J} = 0$ yields (3.7) with an extra $-h^2 f_0$ on the right side. Similar formulas (involving $2n+1$ points) clearly hold in $n > 2$ space dimensions.

Moreover, the highly accurate 9-point approximation $\mathbf{N}_h$ for the Laplace equation ((10.2) of Chapter 3) is equivalent to $\delta\mathbf{J}[U] = 0$ for

$$\mathbf{J}[U] = \sum [4(U_{i+1,j} - U_{i,j})^2 + 4(U_{i,j+1} - U_{i,j})^2 + (U_{i+1,j+1} - U_{i,j})^2 + (U_{i-1,j+1} - U_{i,j})^2].$$

That is, it corresponds to minimizing Maxwell's dissipation function

$$M[V] = \sum I_{kk'} \Delta_L V,$$

$I_{kk'} = I(\Delta_L V) =$ current, $V_k =$ voltage, in an electrical network whose diagonal wires have four times as much resistance as its horizontal and vertical wires (see Fig. 2).

Likewise, the usual 5-point discretization of *variable* coefficient elliptic DE's has a similar 'direct' variational formulation.

Example 7. The standard 5-point discretization of the linear source problem $-(pu_x)_x - (pu_y)_y + qu = f$ on a square network minimizes $\mathbf{J}[U] = \mathbf{P}_h\langle U, U\rangle - h^2 \Sigma\, U_{i,j} f_{i,j}$, with $\mathbf{P}_h$ as in Eq. (4.8) of Chapter 3, which we write as

$$\mathbf{J}[U] = \frac{1}{2} \sum p_L (\Delta_L U)^2 + \frac{h^2}{2} \sum [q_k (U_k)^2\, 2U_k f_k].$$

Here $\Delta_L U$ is as in Example 6 and the summation over L is over all links; p_L denotes the value of $p(x,y)$ at the midpoint of the *link* $L = \overline{kk'}$.

To prove this fact, we again consider the sum of all terms in $\mathbf{J}$ involving a given U_k. We then have:

$$\frac{\partial \mathbf{J}}{\partial U_k} = \sum p_{L_k} \Delta_{L_k} U + h^2 (q_k U_k - f_k),$$

where the summation is over the set of all links incident on the node k. Note that for $k = (i,j)$, $\Delta_{L_k} U = \pm \Delta_L U$, with sign the same as U_k in the difference $\Delta_L U$. With $k = 0 = (ih, jh)$, and again using the notation in Fig. 5, we get

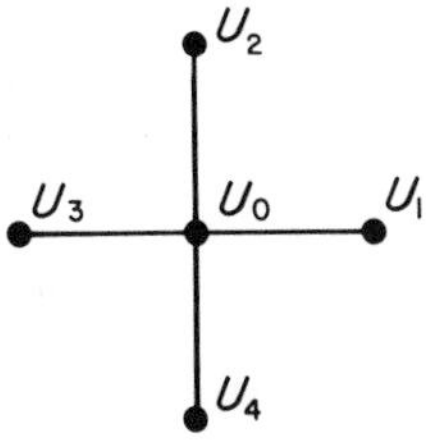

FIGURE 5

$$\frac{\partial \mathbf{J}}{\partial U_0} = p_{i-1/2,j}(U_0 - U_3) + p_{1+1/2,j}(U_0 - U_1) + p_{i,j-1/2}(U_0 - U_4) + p_{i,j+1/2}(U_0 - U_2) + h^2 q_{i,j} U_0 - h^2 f_{i,j}.$$

[40, Chap. VI].

4. Proving convergence. From the theoretical standpoint of proving *convergence* to an exact solution, neglecting roundoff, one wants to have a *nested sequence* of approximating subspaces S_h. To prove convergence, $\cup S_h$ must be *dense* in some function space which is known (often from deep function-theoretic considerations) to contain a unique solution. Thus, in Example 2 in §2, the parameter is related to the degree of a polynomial. Alternatively, such subspace sequences may be associated with sequences of repeated *mesh refinements*. Here $h = h(r)$ can be some mesh length that tends to 0 as $r \uparrow \infty$. Thus one might let $h = 2^{-r}$, for example, with mesh bisections like those used in the 'multigrid' scheme discussed in Chapter 5, §10. Of course, a very few (4 or less) successive mesh bisections are practical in most two-dimensional problems. (In $\mathbb{R}^3$, three mesh bisections would multiply the number of unknowns by 512 and the computational cost of solving $A\mathbf{u} = \mathbf{b}$ by an even greater factor.)

Variational methods can be applied most rigorously when the functional $\mathbf{J}[u]$ to be minimized has a *unique minimum*, and is continuous in some function space topology in which $\cup S_h$ is dense. In this case, it is easy to show that the 'minimizing sequence' of functions U_h which minimize $\mathbf{J}$ on S_h must approach the exact solution u as $r \to \infty$, and that $\mathbf{J}[U_h] \downarrow \mathbf{J}[u]$, the true minimum.

We will give other examples of such subspaces and error bounds in Chapter 7; continuous or continuously differentiable piecewise polynomial functions are often effective in practice.

Error bounds. Satisfactory error bounds can be derived whenever the exact solution of $\mathbf{L}[u] = f$ minimizes some quadratic functional $\mathbf{J}[u] = \mathbf{P}\langle u,u\rangle - 2\langle f,u\rangle$, in terms of $\|u\| = \mathbf{P}\langle u,u\rangle^{1/2}$, the 'energy' norm. Indeed, for all sufficiently smooth functions V taking on the same boundary values as u, we then have $2\mathbf{P}\langle u,V\rangle - 2\langle f,V\rangle = 0$; hence, setting $W = V - u$, we obtain

$$\begin{aligned} \mathbf{J}[V] &= \mathbf{P}\langle u,u\rangle + 2\mathbf{P}\langle u,W\rangle + \mathbf{P}\langle W,W\rangle - 2\langle f,u\rangle - 2\langle f,W\rangle \\ &= \mathbf{J}[u] + \mathbf{P}\langle W,W\rangle. \end{aligned} \tag{4.1}$$

For a mesh that is fine enough, we will therefore have $\|V - u\| \leqslant \epsilon$ for *some* V in the approximating subspace S. For instance, if u is piecewise linear on the boundary, then V might be the piecewise linear interpolant to u on a triangulation of Ω into triangles of side $O(h)$. If U minimizes $\mathbf{J}[U]$ in S, then evidently

$$\mathbf{J}[U] \leqslant \mathbf{J}[V] = \mathbf{J}[u] + \mathbf{P}\langle W,W\rangle = \mathbf{J}[u] + \|V - u\|^2,$$

and so $\mathbf{J}[U] - \mathbf{J}[u] \leqslant \epsilon^2$. — Now consider the special case that $V = U$ in (4.1); we then have

$$\mathbf{J}[U] - \mathbf{J}[u] = \mathbf{P}\langle U-u, U-u\rangle = \|U-u\|^2 \leqslant \epsilon^2,$$

which gives a very nice error bound.

Norms. The powerful and flexible mathematical concept of a *norm* on a function space (see Chapter 2, §5, and Chapter 3, §6) provides a cornucopia of numerical measures of approximation errors. Each norm defines a distance $||f-g||$ in function space, and different norms often correspond to different kinds of convergence. Thus the formulas

$$||f_n-f||_\infty \to 0, \qquad ||f_n-f||_1 \to 0, \quad \text{and} \quad ||f_n-f||_2 \to 0,$$

are equivalent to uniform, mean, and mean square convergence, respectively. The search for best uniform, best mean, and best mean square approximations to a given function u are standard problems in approximation theory (see Chapter 7, §3).

Given a norm $||\cdot||$ and a set S of functions, a function $U \in S$ such that $||U-u|| < ||f-u||$ for all $f \neq U$ in S is called the *best approximation* to u in S. Thus, in the space of trigonometric polynomials of degree n:

$$f(x) = \frac{1}{2}a_0 + a_1 \cos x + b_1 \sin x + \cdots + a_n \cos nx + b_n \sin nx,$$

the best mean square approximation to $u(x)$ on $[0,2\pi]$ is that function $U(x)$ whose coefficients are given by

$$a_k = \frac{1}{\pi}\int_0^{2\pi} u(x) \cos kx\, dx, \qquad b_k = \frac{1}{\pi}\int_0^{2\pi} u(x) \sin kx\, dx.$$

For different mathematical and physical problems, very different norms may be appropriate. Thus, for physical problems with a quadratic energy function, such as those of a vibrating string or loaded spring, the square root of the *energy* (kinetic plus potential) is often the most satisfactory norm. Mikhlin [64, Chap. II] calls such norms 'energy norms'. In many stochastic and population problems, $||p-q||_1$ is the appropriate norm.[7] It is very important to choose the 'right' norm or norms (or semi-norms) in analyzing approximations to physical problems.

For Rayleigh-Ritz methods, quadratic norms are the most convenient. This is because they lead to inner product spaces in which one can find a (the) best approximation by simple *orthogonal projection*. Moreover, the orthogonal projection of a given function u onto the subspace spanned by a given basis $\phi_1, \cdots, \phi_N$ is easy (in principle) to compute to within roundoff error. Since orthogonality amounts to asserting that $\langle u - \sum c_k \phi_k, \phi_j \rangle = 0$ for $j = 1, \cdots, N$, the c_k are determined by the linear algebraic equations having the 'Gram matrix', with entries $\langle \phi_k, \phi_j \rangle$, for coefficient-matrix:

$$\sum c_k \langle \phi_k, \phi_j \rangle = \langle u, \phi_j \rangle \qquad \text{for } J = 1, \cdots, N.$$

[7] Spaces of 'populations' are often best treated as *semi-ordered* linear spaces acted on by *positive* or nonnegative linear operators. Direct connections between order relations and numerical bounds are derived in L. Collatz, *Numerical Analysis and Functional Analysis*, Academic Press, 1966.

$$\sum c_k \langle \phi_k, \phi_j \rangle = \langle u, \phi_j \rangle \qquad \text{for } J = 1, \cdots, N.$$

If the ϕ_i are orthogonal, then $c_j = \langle u, \phi_j \rangle / \langle \phi_j, \phi_j \rangle$.

Sobolev spaces. Suitable function spaces for many applications are the Sobolev spaces[8] $H^s(\Omega)$ of all functions having (in an appropriate technical sense) Lebesgue square-integrable partial derivatives through order s on some domain Ω. The *norm* $\|\cdot\|$ in $H^s(\Omega)$ is defined by

$$\|u\|^2 = \int_\Omega \sum_{|\mathbf{i}|=0}^{s} |D_{\mathbf{i}}(u)|^2 \, dR. \tag{4.2}$$

In (4.2), we have adopted the convenient (and now standard) multi-index notations $\mathbf{i} = (i_1, \cdots, i_p)$, $|\mathbf{i}| = i_1 + \cdots + i_p$, and

$$D_{\mathbf{i}}(u) = \partial^{|\mathbf{i}|} u / \partial x_1^{i_1} \cdots \partial x_p^{i_p}.$$

All such 'Sobolev spaces' are Hilbert spaces. In any Sobolev space $H^s(\Omega)$ $(s \geqslant 0)$, the *subspace* $H_0^s(\Omega)$ of functions which vanish identically on the boundary $\Gamma = \partial\Omega$ of Ω plays a special role; it is *closed* if $s \geqslant 1$. An example is given by the Dirichlet norm, the square root of $\mathbf{D}\langle u, u \rangle$.

It is an important consequence of the Weierstrass approximation theorem that if Ω is compact, *polynomial* functions are *dense* in $H^s(\overline{\Omega})$.[9] *A fortiori*, so are piecewise polynomial functions. To construct dense subspaces of $H_0^s(\Omega)$, where Ω is the 'open' interior of $\overline{\Omega}$, is technically much more difficult.

Choice of basis. The monomials $1, x, x^2, \cdots, x^n$ form a very ill-conditioned basis for polynomials of degree n. For example, in the Hilbert space $L_2[0,1]$ (i.e., the space of square integrable functions with the norm $\|f\|_2$), for $\phi_k(x) = x^{k-1}$, the Gram matrix of inner products $\langle \phi_j, \phi_k \rangle = 1/(j+k-1)$ is the notorious 'Hilbert matrix'. Its condition number is about 10^8 for $n = 6$ and about 10^{12} for $n = 9$.[10] In such cases, it is often desirable to orthogonalize the ϕ_k; this diagonalizes the matrix. In the Hilbert space $L_2[-1,1]$, orthogonalization leads from the monomials x^k to the Legendre polynomials.[11]

Patch bases. For the piecewise linear and piecewise bilinear interpolants to nodal values discussed in §§2 and 3, the 'tent functions' having one nodal value $u(\mathbf{x}_j)$ equal to one and all the rest zero form an even simpler, also well-conditioned basis. Note that these are all 'patch

---

[8] See Adams [75].

[9] Cf. Chapter 7, §2.

[10] See Forsythe-Moler [67, Chap. 19]. Because of this, roundoff error will effectively destroy accuracy on computers which have floating point numbers with fewer than (about) $t = 7$ and $t = 13$ (decimal) digits, respectively; in such cases multiple precision is necessary.

[11] Orthogonalization on $[-1,1]$ with respect to the weight function $(1-x^2)^{-1/2}$ leads to the Chebyshev polynomials.

functions', whose *support* consists of only a few mesh cells; hence the coefficient matrix of $\langle \phi_j, \phi_k \rangle$ is very sparse.

Similar bases of patch functions can be constructed for the 'finite elements' and for the bicubic Hermite interpolants to be discussed in §6 and in Chapter 7. For the splines to be discussed in Chapter 7, §7, bases of B-spline 'patch functions' also give reasonably well-conditioned matrices.

5. Galerkin methods. It is well-known that, for *linear* elliptic problems (minimizing *quadratic* functionals), the Rayleigh-Ritz method of §2 is just a special case of the more general Galerkin method, in somewhat the same sense that the theory of orthogonal expansion is just a special case of the theory of biorthogonal expansion. This is explained clearly in [KK], in Collatz [60, pp. 31–32], in Mikhlin [71], in Strang-Fix [73, p. 116 ff.], and elsewhere. There it is observed that the Rayleigh-Ritz method is only applicable to *self-adjoint* elliptic problems.

The Galerkin method is more flexible than the Rayleigh-Ritz method, which it clarifies and generalizes. It presupposes *two* 'approximating subspaces' having the same (finite) number of dimensions: a space Φ of 'trial functions' and a space Ψ of 'test functions'. When $\Phi = \Psi$ and $\mathbf{L}$ is self-adjoint, the matrices A (constructed by good difference approximations, as in Chapter 3, or by the FEM of Chapter 7) should be symmetric, thus reducing the Galerkin method to the Rayleigh-Ritz method.

In general, suppose we are given an inhomogeneous linear elliptic problem, which we write symbolically as $\mathbf{L}[u] = f$. Picking convenient bases (e.g., of 'patch functions'; see §4), $\phi_1, \cdots, \phi_N$ for Φ and $\psi_1, \cdots, \psi_N$ for Ψ, the Galerkin method replaces $\mathbf{L}[u] = f$ by $\langle \mathbf{L}[U], \psi \rangle = \langle f, \psi \rangle$ for some bilinear functional and all $\psi \in \Psi$. This is equivalent to writing $U = \Sigma\, c_j \phi_j$ and $\mathbf{L}[U] = \Sigma\, c_j \mathbf{L}[\phi_j]$, and then solving

$$(5.1) \qquad \sum_{j=1}^{N} c_j \langle \mathbf{L}[\phi_j], \psi_k \rangle = \langle f, \psi_k \rangle, \quad k = 1, \cdots, N,$$

for the c_j (in §2 we used γ_j for these coefficients). Hence we have $A\mathbf{c} = \mathbf{b}$, where

$$(5.2) \qquad A = \|a_{j,k}\|, \qquad a_{j,k} = \langle \mathbf{L}[\phi_j], \psi_k \rangle, \qquad b_k = \langle f, \psi_k \rangle,$$

and $\mathbf{c}$ is the vector of coefficients of the approximating function $U = \Sigma\, c_j \phi_j$. In contrast to the Rayleigh-Ritz method, here L need *not* be self-adjoint and the matrix need *not* be symmetric.

Restated in slightly different terms, if $\mathbf{L}[u] = f$ is equivalent to $\delta \mathbf{J} = 0$ in (2.2), then one can apply (2.3) even if $\mathbf{J}$ is not minimized. Therefore, one can apply Gauss elimination or other direct methods to obtain approximate solutions of the given DE, essentially by looking for critical points

linear subspaces). All that is needed is that the second variation $\delta^2\mathbf{J}$ have eigenvalues bounded away from zero.[12]

Geometric interpretation. For any set of coefficients c_j, the *residual* r of an approximate solution $U = \sum c_j \phi_j$ is by definition the difference $r = f - \mathbf{L}[U]$. When one chooses the coefficients as in (5.1), so that $\langle \mathbf{L}[U], \psi_k \rangle = \langle f, \psi_k \rangle$ for $k = 1, \cdots, m$, then necessarily $\langle r, \psi_k \rangle = 0$ for all k. Thus the residual is *orthogonal* to the space Ψ of test functions.

Least squares.[13] As a special case, if one takes as a basis of 'test' functions the $\psi_j = L[\phi_j]$, where the ϕ_j are a basis of the space of 'trial' functions, then one has the method of 'least squares': The coefficients of the approximation $U = \sum c_j \phi_j$ are obtained by solving

$$\sum_j c_j \langle \mathbf{L}[\phi_j], \mathbf{L}[\phi_k] \rangle = \langle f, \mathbf{L}[\phi_k] \rangle, \qquad k = 1, \cdots, N. \tag{5.3}$$

The matrix is symmetric. The *residual* is $r = f - \mathbf{L}[U]$ and since it is orthogonal to the space of test functions, it has minimal norm:

$$\langle r, r \rangle \leqslant \langle f - \mathbf{L}[V], f - \mathbf{L}[V] \rangle, \tag{5.4}$$

for any 'trial' function $V = \sum d_j \phi_j$.

Unfortunately, it is in general very hard to give *a priori* bounds for errors arising from the Galerkin method unless A is positive definite as in (5.3). This is because it is rarely possible to predict the least eigenvalue (or singular value) of A; hence it is also hard to predict the norm of $[\delta^2\mathbf{J}]^{-1}$. For a theoretical analysis of the errors, see I. Babuska in Aziz [72].[14]

Affine subspaces. There are many variants of the Rayleigh-Ritz and Galerkin methods. For example, since $g(\mathbf{y})$ is not usually exactly matched by trial function values on Γ, it is often convenient to use an *affine* subspace of approximating functions, all of which satisfy given Dirichlet boundary conditions. To do this, one first constructs on Ω a function $v(\mathbf{x})$ which satisfies $v(\mathbf{y}) = g(\mathbf{y})$ on Γ, i.e., which interpolates to the given boundary conditions (e.g., by blending; see Chapter 7, §11), and then selects from some standard *linear* space a basis $\{\phi_k\}$ of functions which vanish on Γ. One then sets

$$U(\mathbf{x}) = v(\mathbf{x}) + \sum_{k=1}^{N} a_k \phi_k(\mathbf{x}).$$

---

[12] The matrix $\delta^2\mathbf{J}$ is just the matrix A in the Galerkin approximation (5.1)–(5.2).

[13] Cf. Chapter 8, §3.

[14] Other references are M. Keldych, Izv. Akad. Nauk SSSR 6 (1942), 309–330; D. Gilbarg in Langer [60a]; and Ciarlet [78].

6. Collocation.[15] Although conceptually attractive for self-adjoint problems associated with a minimum principle, the Rayleigh-Ritz-Galerkin methods of §§2–5 requires the evaluation of integrals, often by numerical quadratures requiring 9 or more function evaluations per mesh rectangle. Thus, even if the trial functions are piecewise bilinear, the evaluation of

$$\frac{1}{2}\int\int p(x,y)[U_x^2+U_y^2]\,dx\,dy$$

over a mesh rectangle may not be possible in closed form. Moreover, the accurate evaluation of double integrals over irregular boundary 'cells' (finite elements) may be even harder.

An alternative method needing fewer function evaluations is the *collocation method*. In principle, this requires the DE, say

$$\mathbf{L}[u]=f(\mathbf{x}),\qquad \mathbf{x}\in\Omega, \tag{6.1}$$

and Dirichlet boundary conditions,

$$u=g(\mathbf{y}),\qquad \mathbf{y}\in\Gamma=\partial\Omega, \tag{6.2}$$

to be satisfied at each of a finite number of points called the 'collocation points'.

One selects from some appropriate *linear* space a basis $\{\phi_k\}$ of functions and then sets

$$U(\mathbf{x})=\sum_{k=1}^{N} a_k\phi_k(\mathbf{x}). \tag{6.3}$$

The collocation method does not use a minimization procedure. Instead, it determines the coefficients a_k by the condition that U must satisfy (6.1) at I specified interior collocation points $\mathbf{x}_1,\cdots,\mathbf{x}_I$ in Ω, and (6.2) at J boundary collocation points $\mathbf{y}_1,\cdots,\mathbf{y}_J$ on Γ. Taken together, these give $I+J=N$ linear algebraic equations:

$$\sum_{k=1}^{I}\alpha_k\mathbf{L}[\phi_k](\mathbf{x}_i)=f(\mathbf{x}_i),\qquad \mathbf{x}_i,\ i=1,\cdots,I, \tag{6.4}$$

$$\sum_{k=1}^{J}\alpha_k\phi_k(\mathbf{y}_j)=g(\mathbf{y}_j),\qquad \mathbf{y}_j,\ j=1,\cdots,J. \tag{6.4'}$$

In some special cases (6.4)–(6.4′) simplify. Thus for the Dirichlet problem, with $\mathbf{L}=-\nabla^2$ and $f\equiv 0$, the use of harmonic basis elements ϕ_k eliminates (6.4) and all the collocation points are on the boundary Γ. An

[15] Collocation is explained at greater length in Collatz [60, §§1.4.2 and 5.4.2]; R. M. Prenter and R. D. Russell, SIAM J. Numer. Anal. 13 (1976), 923–939; and E. N. Houstis, BIT 18 (1978), 301–320.

interesting special case is provided by the disk $|r| \leqslant a$. In this case, 1 and the harmonic polynomials $r^k \cos k\theta$ and $r^k \sin k\theta$, $k = 1, \cdots, n$, form such a basis (see Chapter 2, §2). One can then use the collocation points $\theta_j = 2\pi j/(2n+1)$. In this case, convergence to the solution is assured by classic theorems of Jackson [30, Chap. 4].

For general linear source problems of the form $-\nabla \cdot (p\nabla u) + qu = f$, one obtains the coefficients a_k more efficiently by using a patch basis for the set $\phi_1, \cdots, \phi_N$. For, the natural ordering of equations and unknowns then leads to a vector equation whose matrix is banded.

For second-order problems, the subspace of continuous piecewise bilinear functions, although adequate for solving problems by the Rayleigh-Ritz method, is not suitable for the collocation method because second derivatives of such functions are everywhere either 0 or ∞. To solve second-order problems in rectangular polygons by collocation, a recommended linear space is that of *piecewise bicubic, continuously differentiable* functions to be studied in Chapter 7, §5. For a square mesh with mesh lines $x_i = ih$, $y_j = jh$, one can use as basis elements $\psi_r(x-ih)\psi_s(y-jh)$, $r, s = 0, 1$, where

$$\psi_0(x) = \frac{1}{h^3}(h^3 - 3x^2h + 2|x|^3), \qquad \psi_1(x) = \frac{1}{h^2}x(|x|-h)^2. \tag{6.5}$$

Or, one could use instead the subspace of continuous piecewise biquadratic elements to be constructed in Chapter 7, §4.

One of us has collaborated in numerical experiments (see Houstis et al. [78]) for a variety of second-order elliptic problems using Gauss collocation points on rectangular meshes to discretize bicubic Hermite finite element approximations having $O(h^4)$ accuracy. These experiments indicate that with a moderate number ($N < 300$) of unknowns, collocation is often a cheaper way of attaining a given accuracy $||u - U|| < \epsilon$, with $\epsilon / ||u||$ in the range 10^{-3} to 10^{-6}, than the Rayleigh-Ritz method.[16]

Unfortunately, for these and other self-adjoint problems, collocation gives a matrix which is in general *not* symmetric. Therefore, for *large* N, the Rayleigh-Ritz method of §2 requires less work than the collocation method. This is true asymptotically even if many fewer function evaluations per mesh rectangle are required for collocation than for the quadrature formulas of the Rayleigh-Ritz method. The reason is, that the number of function evaluations is proportional to the number of unknowns N, whereas the work in solving the band matrix problem is proportional to N^2 and the solution of a symmetric linear system is

[16] Somewhat different results were reported by A. Weiser, S. C. Eisenstat, and M. H. Schultz, SIAM J. Numer. Anal. 17 (1980), 908–929; but see also E. N. Houstis, R. E. Lynch, and J. R. Rice, SIAM J. Numer. Anal. 21 (1984), 695–715.

obtained with half the work of solving a nonsymmetric one with the same band structure. By way of compensation, the collocation method can be applied to elliptic problems not having a minimum principle.

Like the Galerkin method, collocation is not related to self-adjoint problems. Indeed, the collocation method can be interpreted as a Galerkin or a 'weighted residual' method in which a basis of 'test' functions consists of delta functions $\psi_j = \delta(\mathbf{x} - \mathbf{x}_j)$, where the $\mathbf{x}_j$ are the interior and the boundary collocation points; see Strang-Fix [73, §2.3] and Zienkiewicz, [77, p. 50]).

Selection of collocation points. We now outline the method used in ELLPACK to place collocation points for bicubic Hermite approximating functions.[17]

The approximation is $U(x,y) = \sum V_{i,j}(x,y)$, where the sum is over all mesh-points and, with $\xi = x - x_i$, $\eta = y - y_j$ (see (6.5)),

$$V_{i,j} = U(x_i,y_j)\psi_0(\xi)\psi_0(\eta) + U_x(x_i,y_j)\psi_1(\xi)\psi_0(\eta) + U_y(x_i,y_j)\psi_0(\xi)\psi_1(\eta) + U_{xy}(x_i,y_j)\psi_1(\xi)\psi_1(\eta). \tag{6.6}$$

Thus, four unknowns, $U(x_i,y_j), \cdots, U_{xy}(x_i,y_j)$, are associated with each mesh-point.

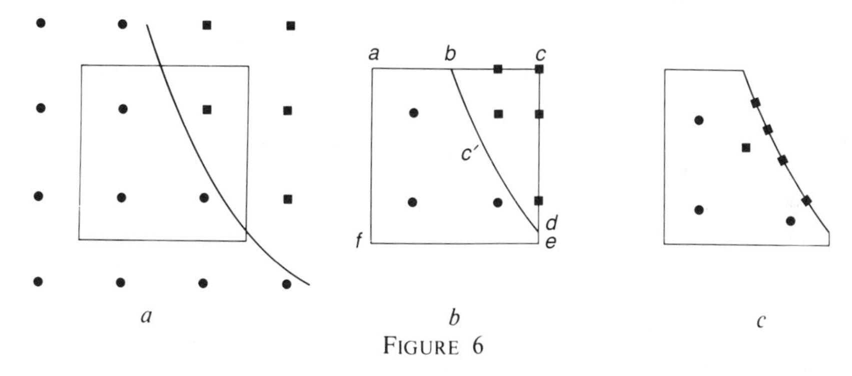

FIGURE 6

If a mesh square, say $[0,h] \times [0,h]$, is in the interior of the domain, Ω, then four collocation points are placed in it at the 'Gauss points': $(h/2 \pm \delta h, h/2 \pm \delta h)$, $\delta = 1/2\sqrt{3}$, where $\pm 2\delta$ are the zeros of the quadratic Legendre polynomial $P_2(z)$. Each of these points gives rise to one equation in the system (6.4).

Choice of boundary collocation points, $\mathbf{y}_j$, in (6.4′) depends on which edges of a mesh square are cut by the boundary.[18] Consider the case

[17] E. N. Houstis, W. F. Mitchell, and J. R. Rice, to appear in TOMS.

[18] There are five cases; we describe one; for the others see Houstis, Mitchell, and Rice, op. cit.

illustrated in Fig. 6. In Fig. 6a, the sixteen collocation points for the four mesh-points are shown for the case that the mesh square is in the interior; the small squares indicate collocation points which are in the exterior of Ω and these are moved to the boundary. The first step is to move the four points outside of the mesh square to its boundary as indicated in Fig. 6b. Next the mesh square is mapped into the 'pentagon' $abc'defa$ in the following way. The line segments bc and cd go onto the curved side $bc'd$; the parametrization for this is the same as in the user-supplied BOUNDARY segment in an ELLPACK program (see Chapter 9, §3). For example if the line segments bc and cd were of equal length and if the boundary were parametrized by arclength, then the point c would be mapped onto the middle of the boundary piece. The resulting distribution of interior and boundary collocation points for the mesh square is indicated in Fig. 6c.

Normalization. In (6.4)–(6.4′), it is desirable to 'normalize' by dividing each equation through by its largest coefficient. Otherwise, the coefficients of (6.4) are $O(h^{-2})$ larger than those of (6.4′), and a linear equations solver which pivots (but does not scale) will process first all the interior equations, leaving the boundary equations until last. This can result in unexpectedly large roundoff errors.[19]

7. Eigenproblems. The eigenfunctions of many classical vibration problems are those functions which define *critical* (or 'stationary') *points* of the Rayleigh quotient. This is defined as the ratio

$$\mathbf{R}[u] = \mathbf{V}[u]/\mathbf{T}[u] = (\text{potential energy})/(\text{kinetic energy}); \tag{7.1}$$

see Chapter 2, §6. The m-th eigenfunction ϕ_m is then characterized by the following *minimax property*: $\mathbf{R}[\phi_m]$ minimizes the value of $\mathbf{R}[u]$ on the subspace orthogonal to $\phi_1, \cdots, \phi_{m-1}$; $\mathbf{R}[\phi_m] = \lambda_m$ is the m-th eigenvalue; and the m-dimensional subspace spanned by $\phi_1, \cdots, \phi_m$ minimizes the maximum $\mathbf{R}[c_1\phi_1 + \cdots + c_m\phi_m]$, considered as a functional on the manifold of all m-dimensional subspaces. In particular, the 'lowest eigenfunction' ϕ_1 (i.e., the eigenfunction with least eigenvalue λ_1) minimizes $\mathbf{R}[u]$, and we have $\lambda_1 \leqslant \lambda_2 \leqslant \lambda_3 \leqslant \cdots$.

For example, consider the problem of determining the eigenfunctions of the Laplace operator; i.e., the solutions of the Helmholtz equation $\nabla^2 u + k^2 u = 0$, for the boundary condition $u \equiv 0$ on Γ. In this case, $\mathbf{T}[u] = \langle u,u \rangle = \int u^2\, dR$, while $\mathbf{V}[u] = \mathbf{D}\langle u,u \rangle$ is the Dirichlet integral and $\lambda = k^2$.

Rayleigh's principle. As Rayleigh recognized, to add inertia to any mechanical system increases $\mathbf{T}[u]$ and lowers all eigenvalues, while adding

[19] See W. R. Dyksen and J. R. Rice in Birkhoff-Schoenstadt [84, pp. 467–480].

stiffness increases them by increasing $\mathbf{V}[u]$.[20] Equivalently, the ϕ_m are the critical ('stationary') points of the potential ('strain') energy function on the unit sphere $\mathbf{S}: \{u \mid \mathbf{T}[u]=1\}$ in the Hilbert space defined by the norm $\mathbf{T}[u]^{1/2}$ associated with the kinetic (inertial) energy.

Discretization. The preceding variational properties can be 'arithmetized' in principle (see Chapter 3, §1) to any desired accuracy by constructing a nested sequence of finite-dimensional *approximating subspaces* $\Psi = \Psi_h$ with dense union.

Letting $\psi_1, \cdots, \psi_N$, $N = N(h)$, be a basis for Ψ, one can compute the critical points of $\mathbf{R}[u]$ *restricted to* Ψ as follows. Set $\boldsymbol{\alpha} = \sum \alpha_k \psi_k$ for a general $\boldsymbol{\alpha} \in \Psi$; then an easy algebraic computation gives

$$\mathbf{R}[\boldsymbol{\alpha}] = \mathbf{V}[\textstyle\sum \alpha_k \psi_k]/\mathbf{T}[\sum \alpha_k \psi_k] = \sum v_{jk}\alpha_j\alpha_k / \sum t_{jk}\alpha_j\alpha_k, \tag{7.2}$$

where $t_{jk} = \langle \psi_j, T\psi_k \rangle$ and $v_{jk} = \langle \psi_j, V\psi_k \rangle$ denote the entries of the matrices of inner products associated with the kinetic and potential energies, respectively. Then the 'approximate eigenvectors' $\boldsymbol{\gamma}_i \neq \mathbf{0}$ are the vectors $\boldsymbol{\gamma}_i = \sum c_{ij}\psi_j$, where the *discrete Rayleigh quotient*

$$\mathbf{R}[\boldsymbol{\gamma}_i] = \textstyle\sum v_{jk} c_{ij} c_{ik} / \sum t_{jk} c_{ij} c_{ik}$$

is stationary. Moreover, the approximate *eigenvalues* $\lambda_i = \lambda_{h,i}$ are the values $\mathbf{R}[\boldsymbol{\gamma}_i]$ of the Rayleigh quotient at the $\boldsymbol{\gamma}_i$.

The algebraic counterpart of Rayleigh's principle of mechanics is Weyl's monotonicity theorem: adding a positive definite symmetric matrix to $\mathbf{V}[u]$ increases all eigenvalues; adding such a matrix to $\mathbf{T}[u]$ decreases them.

We derive here also the following result.

THEOREM 5. *If T and V are positive definite symmetric matrices, then any two generalized eigenvectors $\boldsymbol{\alpha}$ and $\boldsymbol{\beta}$ satisfying*

$$V\boldsymbol{\alpha} = \lambda T\boldsymbol{\alpha}, \qquad V\boldsymbol{\beta} = \mu T\boldsymbol{\beta}, \quad \lambda \neq \mu, \tag{7.3}$$

are orthogonal with respect to the inner products $\langle \boldsymbol{\alpha}, V\boldsymbol{\beta} \rangle = \boldsymbol{\alpha}^T V\boldsymbol{\beta}$ and $\langle \boldsymbol{\alpha}, T\boldsymbol{\beta} \rangle$.

Proof. One easily derives

$$\begin{aligned} \mu\langle \boldsymbol{\alpha}, T\boldsymbol{\beta} \rangle = \langle \boldsymbol{\alpha}, \mu T\boldsymbol{\beta} \rangle &= \langle \boldsymbol{\alpha}, V\boldsymbol{\beta} \rangle = \langle V\boldsymbol{\alpha}, \boldsymbol{\beta} \rangle \\ &= \langle \lambda T\boldsymbol{\alpha}, \boldsymbol{\beta} \rangle = \lambda \langle T\boldsymbol{\alpha}, \boldsymbol{\beta} \rangle = \lambda \langle \boldsymbol{\alpha}, T\boldsymbol{\beta} \rangle. \end{aligned}$$

Since $\mu \neq \lambda$, this implies $\langle \boldsymbol{\alpha}, T\boldsymbol{\beta} \rangle = 0$. The proof that $\langle \boldsymbol{\alpha}, V\boldsymbol{\beta} \rangle = 0$ is similar.

[20] Rayleigh, Proc. London Math. Soc. 4 (1873), 357–358; H. Poincaré, Amer. J. Math. 12 (1890), 211–294, §2; Temple and Bickley [33]. See L. Collatz, *Eigenwertaufgaben mit technischen Anwendungen*, 2. Auflage, Akademische Verlagsgesellschaft Geest und Portig, 1963 for a comprehensive exposition of that state of knowledge at the time.

Computational problem. To compute the $\boldsymbol{\gamma}_j$ and λ_j defined above, in any space $\Psi = \Psi_h$, is a generalized *matrix eigenvalue problem*: $V\boldsymbol{\gamma} = \lambda T\boldsymbol{\gamma}$. It can be solved automatically for $N < 200$ by techniques discussed in Wilkinson [65] and implemented in EISPACK. The computed λ_j are *upper bounds* to the true eigenvalues (if roundoff errors are negligible).

Example 8. We now present Pólya's treatment of the Helmholtz equation equation in a square by piecewise bilinear polynomials on a uniform mesh of side h. For the subspace of such functions, the 'tent functions'

$$\psi_k(x,y) = \psi_{i,j}(x,y) = \chi(x-ih)\chi(y-jh),$$

where $\chi(t)$ is $1-|t|/h$ if $|t| \leqslant h$ and zero otherwise, form a convenient basis. This is because

$$U(x,y) = \sum U_{i,j}\psi_{i,j}(x,y)$$

satisfies $U(x_i,y_j) = U_{i,j}$ for all mesh-points (x_i,y_j). For the Helmholtz equations $\nabla^2 u + \lambda u = 0$ with $u \equiv 0$ on Γ, this leads to the generalized eigenvalue problem $A\mathbf{U} = \mu B\mathbf{U}$.

The two sides of the equation for this eigenproblem can be derived by properly interpreting the stencils of (3.5),

$$\frac{1}{3h^2}\begin{bmatrix} -1 & -1 & -1 \\ -1 & 8 & -1 \\ -1 & -1 & -1 \end{bmatrix} U_{i,j} = \frac{\mu}{36}\begin{bmatrix} 1 & 4 & 1 \\ 4 & 16 & 4 \\ 1 & 4 & 1 \end{bmatrix} U_{i,j}, \tag{7.4}$$

first obtained by Pólya [54]. For Ω the square $[0,\pi]\times[0,\pi]$,

$$\begin{aligned} \mu &= \frac{6}{h^2}\,\frac{4-\cos(mh)-\cos(nh)-2\cos(mh)\cos(mh)}{4+\cos(mh)+\cos(nh)+\cos(mh)\cos(mh)} \\ &= (m^2+n^2)+(m^4+n^4)h^2/12+O(h^4) = \lambda + O(h^2), \end{aligned}$$

and $U_{i,j} = \sin(imh)\sin(jnh)$ which interpolates at the mesh-points to the eigenfunction $\sin(mx)\sin(ny)$ of $-\nabla^2$. Each eigenvalue μ of the discrete system exceeds the corresponding eigenvalue (m^2+n^2) of $-\nabla^2$, because of Poincaré's minimax principle for the Rayleigh quotient $\mathbf{R}[U] = \mathbf{D}\langle U,U\rangle/\langle U,U\rangle$.

Similarly, use of continuous linear polynomials on equilateral triangles gives

$$\frac{2}{3h^2}\left[6U_{i,j} - \sum_{k=1}^{6} U_k\right] = \frac{\sigma_h}{12}\left[6U_{i,j} + \sum_{k=1}^{6} U_k\right], \tag{7.5}$$

where the summation is over the six mesh-point nearest (x_i,y_j).

Example 9. Consider the eigenvalue problem $-\nabla^2 u = \lambda u$ with $u \equiv 0$ on the boundary $\partial\mathrm{H}$ of a regular hexagon H inscribed in the unit circle. Let H_h denote the points of an equilateral triangular mesh with spacing h. Then the solutions of

$$(7.6) \qquad \frac{2}{3h^2}\Big[6U_{i,j} - \sum_{k=1}^{6} U_k\Big] = \mu_h U_{i,j}, \qquad (x_i, y_j) \in \mathrm{H}_h,$$

where the summation is over the six mesh-point nearest (x_i, y_j), with $U_{i,j} = 0$ on $\partial \mathrm{H}$, give approximations to the eigenfunctions u and eigenvalues λ. The eigenvalue in (7.5) is given by $\sigma_h = \mu_h/(1 - h^2\mu_h/8)$ and the eigenvalue λ_h in Example 4 of Chapter 3, §3, is given by

$$\lambda_h = \frac{8}{h^2}\big[-1 + \sqrt{1 + h^2\mu_h/4}\big].$$

The smallest eigenvalue of (7.6) was obtained with EISPACK; by symmetry, the number N of unknown nodal values was only about 1/12-th the number of mesh-points in H_h. The results are listed in Table 1. Aitken δ^2 extrapolation (Conte-deBoor [80, p. 98]) of these values for $1/h =$ 4, 8, and 16, yields 7.155094, 7.155348, and 7.154878, from the data for μ_h, λ_h, and σ_h, respectively. Assuming that the estimates behave like $\lambda + Kh^\nu$, we find that $\nu \approx 2.05$, 2.91, and 1.93, respectively.

TABLE 1. Smallest eigenvalues of H_h

$1/h$	N	μ_h	λ_h	σ_h
2	2	6.277997	7.055898	7.810274
4	6	6.942049	7.141258	7.340140
8	20	7.103500	7.153473	7.203441
12	42	7.132556	7.154774	7.176992
16	72	7.142600	7.155099	7.167597
20	110	7.147215	7.155215	7.163214

Lower bounds. The Rayleigh-Ritz method gives *upper* bounds to eigenvalues. Lower bounds can also be computed by methods originated by Weinstein and developed into computational techniques by Bazeley.[21]

Error bounds. As in Birkhoff-de Boor-Swartz-Wendroff [66], one can obtain error bounds for the computed eigenvalues $\lambda_m = \lambda_{m,h}$ by first establishing approximation theorems for the denseness of the *approximating* subspace, with respect to the norms. These bounds are typically of the form

$$(7.7) \qquad \|\Phi_m - \phi_m\| \leq K \|\Phi_m\| N^p,$$

where N is the dimension of the approximating subspace, and so refers to the numerator *and* to the denominator of the Rayleigh quotient. By combining these inequalities, one can obtain bounds on the errors of the $\lambda_m = \mathbf{R}[\phi_m]$; these typically involve Sobolev norms. A favorable point is

[21] See Gould [64] and G. J. Fix, Proc. Nat. Acad. Sci. U.S.A. 61 (1968), 1219–1223.

the fact that the error in λ_m is proportional to the *square* of the error in the ϕ_m.

A rather complete review of approximate solutions of elliptic eigenvalue problems by variational methods has been given by G. Fix and G. Birkhoff in Birkhoff-Varga [69, pp. 111–151], together with an extensive bibliography.

8. Domain constants. In physics and engineering problems, one is often less interested in the detailed behavior of the solution as a function than in some numerical constant associated with its domain, such as the electrostatic capacity, a hydrodynamical added mass coefficient, the lowest eigenvalue, or the fraction $f(\mathbf{k})$ of the wave energy scattered as a function of wave vector $\mathbf{k}$. These constants are called *domain constants*; many of them can also be characterized by variational principles. For example, the electrostatic capacity of a domain Ω is the reciprocal of the minimum self-energy

$$\int\int \frac{1}{|\mathbf{x}-\boldsymbol{\xi}|}\, de(\mathbf{x})\, de(\boldsymbol{\xi}) \qquad \text{for } \int de(\mathbf{x}) = 1,$$

of a unit total charge distributed over $\bar{\Omega}$ (Gauss principle). The charge in equilibrium is all on the surface Γ of Ω.[22]

Likewise, the *torsional rigidity* of Ω (a domain constant of elasticity theory; see Pólya-Szegö [51, pp. 88–89]) is $\int\int u\,dx\,dy$ for the solution of $-\nabla^2 u = 2$ with $u \equiv 0$ on Γ; cf. Example 1 of Chapter 1, §3. If $\phi(x,y)$ is the solution of this problem, then $z = \phi(x,y)$ represents a homogeneous elastic *membrane* (or soap bubble) under uniform tension equalizing a small constant difference between the air pressures on its two sides. (This is the celebrated *membrane analogy* of elasticity theory.)

In the membrane interpretation, $\mathbf{D}\langle\phi,\phi\rangle$ represents the linearized increment δS in the membrane area, whose energy is $\tau\,\delta S$ if τ is the membrane tension. Hence $u = \phi(x,y)$ maximizes the ratio $[\int\int u\,dx\,dy]^2/\mathbf{D}\langle u,u\rangle$. It is also proportional to the volumetric discharge rate (per unit time) of parallel ('generalized Poiseuille') viscous flow under a given pressure gradient, for laminar viscous flow through a straight cylindrical pipe of cross-section Ω.

For some variational principles (e.g., those for the lowest eigenvalue, the capacity, and the torsional rigidity), one can prove rigorously that the constant varies *monotonically* with the domain. It is this monotonicity that was utilized by P. Garabedian [56] in his analysis of the spectral radius of the SOR iteration matrix, as applied to the Dirichlet problem (cf. Chapter 4, §11 and §15).

---

[22] [K, p. 176]; Pólya-Szegö [51]; Bergman-Schiffer [53, p. 185, and pp. 360–371].

Again, straightforward similarity arguments often show that a domain constant $C = C(\Omega)$ satisfies $C = Kd^p$ for geometrically similar domains, where d denotes the domain diameter and K is a constant depending only on the *shape* of the domain. Thus, for domains of the same shape, the capacity is proportional to d $(p = 1)$, and the torsional rigidity to d^4.

For *convex* domains it follows that if similarity and monotonicity principles both hold, then the constant varies *continuously* with the domain. Indeed, it may vary *differentiably*: we may have a variational formula

$$\delta C = \int_\Gamma K(\mathbf{y})\, \delta n(\mathbf{y}),$$

where $K(\mathbf{y})$ is a kernel depending only on Ω and $\delta n(\mathbf{y})$ is the differential of normal displacement.

Using such principles, experts can often estimate such domain constants to within a few percent by simple rules of thumb — e.g., by assuming that they depend only on the area, or that mean flow velocity is a function only of the 'hydraulic diameter' ($4 \times$ area/radius). Thus for a unit square, the eigenfunction with lowest eigenvalue of the Laplacian is $\sin \pi x \sin \pi y$; its eigenvalue is $2\pi^2 \approx 19.7$. For the circle of equal area (radius $1/\sqrt{\pi}$), the lowest eigenfunction ϕ_1 is $J_0(r j_{0,1}/\sqrt{\pi})$, where $j_{0,1} = 2.4048\cdots$ is the first positive zero of $J_0(x)$; its eigenvalue is about 18.2. Both 19.7 and 18.2 are within $\pm 5\%$ of 19.0.

Although domain constants often vary continuously, or even monotonically with the domain, the following is a remarkable exception.

Babuska paradox. Let Δ be the unit disk, and let Δ_n be the regular n-sided polygon inscribed in Δ. Consider the following boundary value problem:

$$\nabla^4 u = 1 \text{ in } \Omega, \qquad u = \partial^2 u/\partial n^2 = 0 \text{ on } \Gamma. \tag{8.1}$$

If $\phi(\mathbf{x})$ is the solution of (8.1) for $\Omega = \Delta$, and $\phi_n(\mathbf{x})$ its solution for $\Omega = \Delta_n$, one would expect to have the $\phi_n(\mathbf{x})$ approach $\phi(\mathbf{x})$ as $n \to \infty$. The Babuska paradox consists in the fact that they do not.

A partial explanation of the Babuska paradox may be suggested by formula (1.6): the energy associated with the Poisson ratio can be transformed into a line integral[23] and presumably corresponds to a *boundary stress*. This could be appreciably affected by polygonal *corners*.[24]

---

[23] See H. L. Langhaar, J. Appl. Mech. 19 (1952), p. 228.

[24] See G. Birkhoff in Schoenberg [69, pp. 204–206] for a further explanation. See E. Reissner in Hull [70, pp. pp. 79–94], for an analysis of the boundary conditions and their rationale.

9. Nonlinear elliptic problems. In §§2–5, we described some 'direct' variational methods for solving *linear* elliptic boundary value problems. These methods were designed to locate the minima of positive definite *quadratic* functionals $\mathbf{J}[u]$, and their restrictions $F(\boldsymbol{\gamma})$ to finite-dimensional 'approximating subspaces'. For any such non-degenerate quadratic functional F, the variational condition $\delta F = 0$ characterizing its 'critical' (alias 'stationary' or 'equilibrium') points is equivalent, as in (2.1)–(2.2), to the system of simultaneous *linear* algebraic equations of the form

$$\frac{\partial F}{\partial x_i}(\mathbf{x}) = \sum_{j=1}^{N} a_{i,j}x_j - b_i = 0, \qquad i = 1, \cdots, N,$$

where $\mathbf{x} = (x_1, \cdots, x_N)$ and the coefficient-matrix $A = ||a_{i,j}||$ is the *Hessian* $||\partial^2 F/\partial x_i\, \partial x_j||$ of F. When F is quadratic, the matrix A and the vector $\mathbf{b}$ are *constant* (independent of $\mathbf{x}$).

Convex case. It is easy to show that a *quadratic* functional

$$F(\mathbf{x}) = \frac{1}{2}\sum a_{i,j}x_i x_j - \sum b_i x_i + c$$

has a unique minimum if and only if its Hessian $||a_{i,j}||$ is positive definite, and that this is also the condition for F to be *strictly convex*, in the sense that

$$F(\alpha\mathbf{x} + [1-\alpha]\mathbf{y}) < \alpha F(\mathbf{x}) + [1-\alpha]F(\mathbf{y}) \qquad \text{if } \mathbf{x} \neq \mathbf{y}, \tag{9.1}$$

for all $\alpha \in (0,1)$. For continuous F, (9.1) is equivalent to

$$F(\tfrac{1}{2}[\mathbf{x}+\mathbf{y}]) < \tfrac{1}{2}[F(\mathbf{x}) + F(\mathbf{y})] \qquad \text{if } \mathbf{x} \neq \mathbf{y}. \tag{9.1$'$}$$

The notion of strict convexity thus *generalizes* the concept of being 'positive definite' from quadratic to more general functionals.

For general $F \in C^2$, the facts are less simple. It is still true that F is *convex* if and only if its Hessian $||\partial^2 F/\partial x_i\, \partial x_j||$ is nonnegative definite, and that F is strictly convex if its Hessian is everywhere positive definite. However, $x_1^4 + x_2^4$ and $(x_1^2 + x_2^2)^2$ are strictly convex even though $\partial^2 F(\mathbf{0})/\partial x_i\, \partial x_j = 0$: the Hessian of F at $\mathbf{0}$ is the zero matrix.

In the rest of this chapter, we will consider the more difficult problem of locating maxima, minima, and other stationary points of *non*quadratic functionals $\mathbf{J}[u]$, such as are associated with *nonlinear* elliptic problems. It will become apparent that the case of strictly convex $\mathbf{J}$ is again the most amenable to mathematical analysis. This is because, as is easily shown, a strictly convex functional can have *at most one* critical point. Moreover, its value at any critical point is a strict global minimum.[25]

---

[25] For more information about the minima of convex functions (and convex programming), see R. T. Rockafellar, *Convex Analysis*, Princeton Univ. Press, 1970.

Such convex nonquadratic functionals include the *area* of a surface

$$\mathbf{J}[u]=\iint\sqrt{1+u_x^2+u_y^2}\,dx\,dy, \tag{9.2}$$

which is minimized by solutions of the Plateau problem (Chapter 1, §9). This functional is strictly convex, essentially because

$$\frac{d^2}{d\lambda^2}\sqrt{a^2+\lambda^2u^2}=\frac{a^2u^2}{(a^2+u^2)^{3/2}}>0.$$

In §10, we will explain how Newton's method can be applied to minimize $\mathbf{J}$ in (9.2) for a substantial class of Dirichlet boundary conditions.

Vector equations. As we saw in Chapters 3–5, the most effective way to minimize a quadratic functional (such as $\mathbf{D}_h\langle u,u\rangle$) is often to solve an associated vector equation $A\mathbf{x}=\mathbf{b}$, or $A\mathbf{x}-\mathbf{b}=\mathbf{0}$. More generally, the problem of finding *stationary points* of any functional $F(\mathbf{x})$ is equivalent to solving $\mathbf{G}(\mathbf{x})=\mathbf{0}$, where $\mathbf{G}=\nabla F$ is the *gradient field* of F. For example, the eigenvector problem of finding the stationary points of the Rayleigh quotient $V(\mathbf{x})/T(\mathbf{x})$ on the unit sphere $\langle\mathbf{x},\mathbf{x}\rangle=1$ can be reduced to this problem.

The following easily proved result explains which vector equations $\mathbf{G}(\mathbf{x})=\mathbf{0}$ arise from gradient fields in this way.

THEOREM 6. *A vector field $\mathbf{G}(\mathbf{x})$ on a simply connected domain is the gradient field $\mathbf{G}=\nabla F$ of a scalar potential $F(\mathbf{x})$, locally, if and only if its Jacobian matrix $\|\partial G_i/\partial x_j\|$ is symmetric.*

Proof. This is the condition for the line integral

$$\int_a^b \mathbf{G}(\mathbf{x})^T\,d\mathbf{x}=\int_a^b \mathbf{G}(\mathbf{x})\cdot d\mathbf{x}$$

to be independent of the path, provided that the domain of definition of $\mathbf{G}$ is simply connected.

In the same vein, we will next give an alternative characterization of the convexity of a function $F(\mathbf{x})$ in terms of its *gradient field* $\mathbf{v}(\mathbf{x})=\nabla F(\mathbf{x})$.

As we observed above, unless the Hessian matrix

$$H=\|\partial^2F/\partial x_i\,\partial x_j\|=\|\partial v_i/\partial x_j\|=\|\partial v_j/\partial x_i\|$$

is singular (has determinant zero), a stationary point is a (strict) *minimum* of F if and only if the Hessian is *positive definite*. At other points, positive definiteness of H is a condition for (strict) convexity, again if H has full rank. In terms of $\mathbf{v}(\mathbf{x})$ and $(d\mathbf{v})_i=\sum_j(\partial v_i/\partial x_j)\,dx_j$, the local positive definiteness condition becomes

$$d\mathbf{v}\cdot d\mathbf{x}=(Hd\mathbf{x})\cdot d\mathbf{x}=d\mathbf{x}\cdot Hd\mathbf{x}>0,\qquad \text{if } d\mathbf{x}\neq\mathbf{0}, \tag{9.3}$$

implying the *global* strict convexity of $\int\mathbf{v}\cdot d\mathbf{x}=F$.[26]

[26] We omit proofs of these well-known facts, for which see R. T. Rockafellar, op. cit.

Monotone mappings. Now let $\mathbf{G}: \mathbf{x} \to \mathbf{G}(\mathbf{x})$ be *any* continuous mapping of $\mathbb{R}^n$ into itself; the case $\mathbf{G} = \nabla F$ of differentiable gradient fields is quite special. Such a mapping $\mathbf{G}$ is called *monotone*[27] when it satisfies

$$(9.4) \qquad \langle \mathbf{b} - \mathbf{a}, \mathbf{G}(\mathbf{b}) - \mathbf{G}(\mathbf{a}) \rangle \geqslant 0, \qquad \text{for all } \mathbf{a}, \mathbf{b}.$$

In the case $\mathbf{G} = \nabla F = \mathbf{v}$ of a gradient mapping, the inner product (9.4) is the line integral along $\mathbf{x}(\lambda) = \lambda\mathbf{b} + (1-\lambda)\mathbf{a}$:

$$\langle \mathbf{b} - \mathbf{a}, \mathbf{G}(\mathbf{b}) - \mathbf{G}(\mathbf{a}) \rangle = \langle \mathbf{b} - \mathbf{a}, \int d\mathbf{G} \rangle = \int \langle \mathbf{b} - \mathbf{a}, d\mathbf{G} \rangle .$$

In the case that $\mathbf{v} = \mathbf{G}$ is continuously differentiable (i.e., $F \in C^2$), this is always nonnegative if and only if $H = \|\partial v_i / \partial x_j\|$ is everywhere *positive semi-definite*. (By Theorem 6, $\|\partial v_i / \partial x_j\|$ is symmetric precisely when $\mathbf{v}(\mathbf{x})$ *is* locally a gradient field!) This proves the following result.

THEOREM 7. *A continuously differentiable function* $F: \mathbb{R}^n \mapsto \mathbb{R}$ *is convex if and only if the gradient mapping* $\mathbf{G}: \mathbb{R}^n \mapsto \mathbb{R}^n$ *defined by its gradient field* $\mathbf{G}(\mathbf{x}) = \nabla F(\mathbf{x})$ *is monotone.*

COROLLARY. *To minimize a convex functional* $F(\mathbf{x})$, *it suffices to solve the vector equation* $\mathbf{G}(\mathbf{x}) = \mathbf{0}$, *where* $\mathbf{G}(\mathbf{x})$ *is the gradient field of* $F(\mathbf{x})$.

Nonlinear networks. A variant of the monotonicity condition (9.4) was applied to 'nonlinear networks' by G. Birkhoff and J. B. Diaz. There they showed that if all the 'conductivity' functions $c_k(\Delta u)$ of the links are *increasing* (or nondecreasing) functions, then

$$\sum [c_k(\Delta u_k) - c_k(\Delta v_k)](\Delta u_k - \Delta v_k) \geqslant 0.$$

They used this fact to establish the uniqueness and (under mild asymptotic conditions 'at infinity' for very large $|u|$) the existence of flows through any network, for given terminal voltages (or currents).[28]

10. Newton's method. The oldest algorithm for solving vector equations like $\mathbf{G}(\mathbf{x}) = \mathbf{0}$ is Newton's method. It proceeds iteratively as follows.[29] Given a trial vector $\mathbf{x}^{(0)}$, first evaluate $\mathbf{b}^{(0)} = \mathbf{G}(\mathbf{x}^{(0)})$. Then consider the first variation

[27] The significance of this condition was discovered independently by E. H. Zarantonello, Felix Browder, and G. Minty; see G. Minty, Proc. Roy. Soc. A 257 (1960), 194–212.

[28] G. Birkhoff and J. B. Diaz, Quart. Appl. Math. 13 (1956), 432–443. See also G. Birkhoff and R. B. Kellogg, Proc. Symp. Generalized Networks, MRI Symposium Series 16, Brooklyn Polytechnic Press, New York; 1966; T. A. Porsching, SIAM J. Numer. Anal. 6 (1969), 437–449; and the references given in Ortega-Rheinboldt [70]. The rate of dissipation of electrical energy into heat is a convex functional of the currents $\sum \Delta u_k c(\Delta u_k)$.

[29] A. M. Ostrowski, *Solutions of Equations and Systems of Equations*, 2nd ed., Academic Press, 1966; C. G. Broyden, Math. Comp. 19 (1965), 577–593; J. M. Ortega and W. C. Rheinboldt, SIAM J. Numer. Anal. 4 (1967), 171–190; J. Math. Anal. Appl. 32 (1970), pp. 274–307.

(10.1) $$\delta \mathbf{b} = \mathbf{G}(\mathbf{x}^{(0)} + \delta\mathbf{x}) - \mathbf{G}(\mathbf{x}^{(0)}) = H(\mathbf{x}^{(0)})\delta\mathbf{x},$$

where

(10.1′) $$H_{i,j}(\mathbf{x}) = \frac{\partial G_i}{\partial x_j}(\mathbf{x}) = \frac{\partial^2 F}{\partial x_i \partial x_j}(\mathbf{x})$$

is the *Hessian* of F (and the Jacobian of $\mathbf{G}(\mathbf{x})$). Next, solve the *linear* system $H(\mathbf{x}^{(0)})\delta\mathbf{x}^{(0)} = -\mathbf{b}(\mathbf{x}^{(0)}) = -\mathbf{G}(\mathbf{x}^{(0)})$, and set

(10.2) $$\mathbf{x}^{(1)} = \mathbf{x}^{(0)} + \delta\mathbf{x}^{(0)} = \mathbf{x}^{(0)} - H^{-1}(\mathbf{x}^{0})\mathbf{b}^{(0)}.$$

Now repeat this process by computing $H(\mathbf{x}^{(1)})$ and by solving recursively the linear systems

(10.3) $$H(\mathbf{x}^{(m)})[\mathbf{x}^{(m+1)} - \mathbf{x}^{(m)}] = -\mathbf{b}^{(m)} = -\mathbf{G}(\mathbf{x}^{(m)}).$$

When $\mathbf{G} \in C^2$, this converges extremely rapidly *if* one has a good initial approximation $\mathbf{x}^{(0)}$ and if $H(\mathbf{x}^{(m)})$ is nonsingular. Indeed, if $||\mathbf{e}^{(0)}||$ is sufficiently small, then the error $\mathbf{e}^{(m)} = \mathbf{x} - \mathbf{x}^{(m)}$ is essentially *squared* at each iteration. More precisely, it satisfies

(10.4) $$||\mathbf{e}^{(m+1)}|| \leqslant M\,||\mathbf{e}^{(m)}||^2, \qquad M \text{ finite}.$$

Newton-Kantorovich method. Newton's method has a natural extension to Banach spaces, first analyzed rigorously by Kantorovich.[30] Given a nonlinear elliptic DE, $\mathbf{G}[u] = 0$, formal expansion gives

(10.5) $$\mathbf{G}[u + \epsilon w] = \mathbf{G}[u] + \epsilon \mathbf{L}[u, w] + O(\epsilon^2),$$

where $\mathbf{L}[u, w] = \mathbf{L}_u[w]$ is linear in w. For a given m-th approximate solution $u^{(m)}$, the next approximation $u^{(m+1)} = u^{(m)} + w$ is obtained by solving

(10.6) $$\mathbf{L}[u^{(m)}, u^{(m+1)}] = \mathbf{L}[u^{(m)}, u^{(m)}] - \mathbf{G}[u^{(m)}].$$

Kantorovich proved rigorously that, in any finite-dimensional (Euclidean) *or* infinite-dimensional (Hilbert or Banach) space, the resulting 'Newton-Kantorovich' method converges extremely rapidly *if* one has a good initial approximation $u^{(0)}$.

Example 10. Consider the Plateau problem (Chapter 1, §9) of minimizing the surface area $F[u] = \int_\Omega \sqrt{1 + u_x^2 + u_y^2}\,dx\,dy$. The corresponding variational equation $\mathbf{G}[u] = \mathbf{0}$ is

(10.7) $$\mathbf{G}[u] = (1 + u_y{}^2)u_{xx} - 2u_x u_y u_{xy} + (1 + u_x{}^2)u_{yy} = 0 \qquad \text{on} \quad \Omega.$$

[30] L. V. Kantorovich, Uspehi Mat. Nauk 3 (1948), 98–105. For the underlying theory see L. V. Kantorovich and G. P. Akilov, *Functional Analysis in Normed Spaces*, Pergamon Press, 1964, Chap. 18; Ortega-Rheinboldt [70]; or Rheinboldt [74].

The differential operator in (10.5), *linear* in w, is then

$$\text{(10.8)}\quad \begin{aligned} \mathbf{L}[u,w] = {} & (1+u_y{}^2)\,w_{xx} - 2u_x u_y w_{xy} + (1+u_x{}^2)\,w_{yy} \\ & + 2(u_x u_{yy} - u_y u_{xy})\,w_x + 2(u_y u_{xx} - u_x u_{xy})\,w_y . \end{aligned}$$

One picks $u^{(0)}$ and then, with $\mathbf{L}_m[w] = \mathbf{L}[u^{(m)},w]$, one solves

$$\text{(10.9)}\quad \mathbf{L}_m[u^{(m+1)}] = \mathbf{L}_m[u^{(m)}] - \mathbf{G}[u^{(m)}], \qquad m = 0,1,2,\cdots.$$

In practice, (10.9) is replaced with a discrete approximation:

$$\text{(10.9}')\quad \mathbf{L}_{h,m}[U^{(m+1)}] = \mathbf{L}_{h,m}[U^{(m)}] - \mathbf{G}[U^{(m)}], \qquad m = 0,1,2,\cdots.$$

For instance, we used collocation with bicubic Hermite elements and took $U^{(0)} \equiv 0$. The computed $U^{(1)}$ is then an approximation to the solution of the Laplace equation with given Dirichlet boundary conditions. Results for the boundary conditions

$$\text{(10.10)}\qquad g(\mathbf{y}) = \begin{cases} 0 & \text{for } x=0 \text{ or } y=0, \\ y & \text{for } x=1, \\ x^2 & \text{for } y=1, \end{cases}$$

are displayed in Table 2 (see Chapter 9, §10, for an ELLPACK program).

TABLE 2. $U^{(m)}$ *at* (1/2,1/2) *for Example* 10.

h	1/2	1/4	1/6	1/8	1/10
Unknowns	28	84	172	292	444
$U^{(1)}$	.1986607	.1986647	.1986699	.1986709	.1986709
$U^{(2)}$	.1728257	.1717290	.1716738	.1716648	.1716623
$U^{(3)}$	.1708038	.1703172	.1702908	.1702863	.1702849
$U^{(4)}$	.1708097	.1703144	.1702885	.1702839	.1702826
$U^{(5)}$	.1708096	.1703144	.1702885	.1702841	.1702827

Many successful applications of the Newton-Kantorovich method to elliptic problems may be found in the literature.[31] However, not only do 'Newton' methods often diverge, but at each iteration a DE like (10.6) must be solved, so that the method is not very economical in general. Moreover, one must often use extensive trial and error experiments to get near enough to the solution so that Newton's method will converge at all — e.g., so that $\|\mathbf{e}^{(0)}\| < 1/M$ (see (10.4)).

Sparse matrices. For well-designed discretizations of elliptic problems, the Jacobian matrices $\|\partial G_i/\partial x_j\|$ are often sparse. The cost of solving systems of linear algebraic equations like (10.3) and (10.9′) at each

[31] Anselone and Moore, J. Math. Anal. Appl. 13 (1966), 476–501.

iteration can be greatly reduced by taking advantage of this sparsity. Methods for solving sparse systems in this context have been discussed by S. C. Eisenstat, M. H. Schultz, and A. H. Sherman, in Springer Lecture Notes in Math. No. 430, D. L. Colton and R. P. Gilbert, eds., 1974.[32]

Quasi-Newton methods. Use of an approximation to the Hessian can also reduce the cost of each step, and use of more than just the latest approximation, $\mathbf{x}^{(m)}$, to generate $\mathbf{x}^{(m+1)}$ can sometimes speed up convergence. In his excellent introduction to these *quasi-Newton* methods,[33] Dennis first notes the equivalence of the problem of minimizing a general smooth function[34] on an open convex set (program UCMIN) with that of solving a system of nonlinear equations (program NLEQ), and he singles out the nonlinear least squares problem (program NL2) as of special importance. He then comments: "The algorithms in most common usage for all these problems are in virtually all cases variations of Newton's methods for NLEQ . . . also called the Newton-Raphson method".

Several such 'quasi-Newton' methods for solving nonlinear elliptic problems have been proposed in the literature. These methods stem from a fundamental paper of Broyden [65], who constructed an algorithm for solving $\mathbf{G}(\mathbf{x}) = \mathbf{0}$ analogous to Davidon's method for minimizing $F(\mathbf{x})$.

11. Steepest descent. Newton's method is only one of several *gradient methods*, that try to minimize a functional $\mathbf{J}[u]$ by analyzing its *gradient* $\mathbf{G}[u]$, as defined for functionals by the variational formula

$$\mathbf{J}[u+\delta u] = \mathbf{J}[u] + \mathbf{G}[u]\delta u,$$

and in $\mathbb{R}^n$ by the formula

$$F(\mathbf{x}+\delta\mathbf{x}) = F(\mathbf{x}) + \mathbf{G}(\mathbf{x})\delta\mathbf{x}, \qquad F(\mathbf{x}) \in C^1. \tag{11.1}$$

Another famous gradient method is Cauchy's method of 'steepest descent' (1847).[35] This tries to locate the minimum of $F(\mathbf{x})$ by following the solution curve of the vector DE

$$\mathbf{x}'(t) = -\mathbf{G}(\mathbf{x}), \tag{11.2}$$

where $\mathbf{G}(\mathbf{x}) = -\nabla F(\mathbf{x})$. Its theoretical basis is the following result.

---

[32] See also D. J. Rose and R. E. Bank, SIAM J. Numer. Analy. 17 (1980), 806–822.

[33] J. E. Dennis, Proc. Symp. Appl. Math. XXII, Am. Math. Soc., 1978, 19–49. See also G. C. Broyden, J. E. Dennis, and J. J. Moré, J. Inst. Math. Appl. 12 (1973), 223–245, and Dennis-Moré [77]. For the state of the field in 1970, see R. E. Bellman, *Methods of Nonlinear Analysis*, 2 vols., Academic Press, 1970, 1973.

[34] Technically, Dennis assumes $f \in \mathrm{Lip}^2(\Omega)$, i.e., that all second derivatives of f satisfy a Lipschitz condition: $\|f''(\mathbf{x}) - f''(\mathbf{y})\| \leqslant L\|\mathbf{x}-\mathbf{y}\|$.

[35] See G. E. Forsythe, Numer. Math. 11 (1968), 57–76; A. M. Ostrowski, op. cit.; and J. W. Daniel, Numer. Math. 10 (1967), 123–131.

THEOREM 8. *Let* $\mathbf{x}(t)$ *be the solution of* (11.2) *for the initial value* $\mathbf{x}(0) = \mathbf{c}$, *let the domain where* $F(\mathbf{x}) \leqslant M$ *be bounded, and let* $F(\mathbf{x}) \in C^1$ *have only one 'critical point'* $\mathbf{a}$ *where* $\nabla F(\mathbf{a}) = \mathbf{0}$. *Then* $\lim_{t \uparrow +\infty} \mathbf{x}(t) = \mathbf{a}$, *and* F *assumes its minimum value at* $\mathbf{a}$ (*i.e.,* $F(\mathbf{a}) < F(\mathbf{x})$ *for all* $\mathbf{x} \neq \mathbf{a}$).

Note that only the *first* derivatives $\partial F/\partial x_i$ need be computed to apply Cauchy's gradient method, and not its Hessian matrix $\|\partial^2 F/\partial x_i \partial x_j\|$. Note also that, when F has many critical points or even many minima, all limit points of any solution curve $\mathbf{x}(t)$ of (11.2) are critical points; at best, Cauchy's method of steepest descent will only find one of them.

Linear case. The solution of a properly discretized linear self-adjoint elliptic problem minimizes an inhomogeneous, positive definite quadratic form

$$F(U) = \frac{1}{2} \sum a_{jk} U_j U_k - \sum b_k U_k .$$

This minimium occurs where $\nabla F = A\mathbf{U} - \mathbf{b} = \mathbf{0}$, and so the paths of steepest descent satisfy

$$dU_k/dt = b_j - \sum a_{jk} U_k . \tag{11.3}$$

If we integrate (11.3) by the Euler-Cauchy polygon method, with time step $\Delta t = \omega$, we get the *Richardson method* (Chapter 4, §8) for solving the linear system $A\mathbf{U} = \mathbf{b}$:

$$\mathbf{U}^{(m+1)} = \mathbf{U}^{(m)} + \omega(\mathbf{b} - A\mathbf{U}^{(m)}) = \mathbf{U}^{(m)} + \omega \nabla \mathbf{F}(\mathbf{U}^{(m)}). \tag{11.4}$$

The ω which gives the fastest convergence is determined by the extreme eigenvalues of A (Theorem 5, Chapter 4, §10). For the Model Problem, $-\nabla_h^2 U = f$, $\omega = h^2/2$ is optimal.

Although the Euler-Cauchy polygon method (11.4) for solving (11.3) is much less accurate than higher-order methods (e.g., fourth-order Runge-Kutta) for computing the *path* of steepest descent, it is more efficient than higher-order methods when only the point $\mathbf{a}$ where $F(U)$ is minimized is to be located.

As we have seen in Chapters 4 and 5, many algorithms are more efficient by orders of magnitude than Richardson's method for solving (discretized, self-adjoint) linear elliptic problems; hence the method of steepest descent is not to be recommended for solving linear problems. However, it has the great advantage of being applicable also to *nonlinear* elliptic problems, like the Plateau problem, whose solution minimizes an appropriate nonquadratic functional. Specifically, for a given initial or 'trial' value $\mathbf{x}(0) = \mathbf{x}^{(0)}$, the autonomous system of ordinary DE's $x_i'(t) = -\partial F/\partial x_i$ (i.e., of $\mathbf{x}'(t) = -\nabla F$) normally converges to at least a *local* minimum of $F(\mathbf{x})$.[36] Moreover, if F is strictly *convex*, this minimum is global.

For such nonlinear variational problems, finite difference *or* finite element (see Chapter 7) discretizations should minimize sums $F(\mathbf{U}) = \Sigma F_j(\mathbf{U})$ of functionals F_j each depending on only a few U_k and easily differentiated explicitly. In such cases, the gradient (and the Hessian) are given by easily computed vectors $\nabla F(\mathbf{U})$ (and sparse matrices $||\partial^2 F/\partial U_j \partial U_k||$), so that the Euler-Cauchy polygon method is easy to implement.

For *non*convex functions, however, this *path of steepest descent* does *not* generally converge to the global minimum. To locate the global, or absolute minima of general nonconvex functions may require elaborate search techniques. One reason for this becomes apparent if one considers examples like $F(x) = 2x^2 - \pi x - 5\exp[-(10^3 x)^2]$, or, in $\mathbb{R}^n$,

$$F(\mathbf{x}) = \sum_{k=1}^{n} x_k^2 - \exp\left[-10^6 \sum_{k=1}^{n} (x_k - k^{-1/k})^2\right]. \tag{11.5}$$

Most efficient ('fast') minimization algorithms will converge to the *local* minimum near $\mathbf{x} = \mathbf{0}$ totally missing the *global* minimum near $\mathbf{x} = (1, 1/2^{1/2}, \cdots, 1/n^{1/n})$.

Condition number problem. Even in the (strictly convex) case of positive definite quadratic forms $\Sigma a_{jk} U_j U_k - 2\Sigma b_j U_j$, the method of steepest descent becomes slow and inaccurate if the matrix $||a_{jk}||$ has a very large condition number. This limitation, which is suggested by experience with the Richardson method of Chapter 4, is illustrated very simply by the following example.

Example 11. Consider $u = x^2 + \epsilon y^2$, where $\epsilon << 1$; the condition number is $1/\epsilon$. An easy calculation gives $-\nabla u = -2(x, \epsilon y)$, whence the parametric curve $dx/dt = -x$, $dy/dt = -\epsilon y$, with solution $x = e^{-t}$, $y = e^{-\epsilon t}$ through (1,1) satisfies $y = x^{\epsilon}$. For $\epsilon < 10^{-4}$, $x = 10^{-8}$ gives $y = 10^{-0.0008} \approx 0.998$, at which point roundoff errors would have become very destructive with low precision (10^{-7}) in single precision if the coordinate axes had not happened to be eigenvectors.

Gradient direction method. The Euler-Cauchy polygon method for computing the path of steepest descent — i.e., for integrating the 'autonomous system' $dx_i/dt = \partial F/\partial x_i$ — suggests the following iterative gradient direction method: Locate successive minima $u^{(m+1)} = u(\mathbf{x}^{(m+1)})$ along each of a sequence of straight line paths $\mathbf{x}^{(m+1)} = \mathbf{x}^{(m)} - \lambda \nabla F$ in the (negative) gradient directions. This 'gradient direction' method can also converge very slowly, again like Richardson's method.

[36] See C. B. Morrey in E. F. Beckenbach, ed., *Modern Mathematics for the Engineer*, McGraw-Hill, 1956, pp. 389–427.

Nonlinear least squares. The determination of the minima of a sum of squares of *nonlinear functions* is referred to as the 'nonlinear least squares' problem; such problems often arise in data-fitting. A widely used and apparently often successful algorithm for solving nonlinear least squares problems is the so-called Levenberg-Marquardt algorithm. We have had no experience with it, and refer the reader to the literature for its description and properties.[37]

However, we do call attention to another kind of difficulty that can arise in solving such problems. Given any set of N functions $g_j(x_i, \cdots, x_N)$ $j = 1, \cdots, N$, of N variables, it is obvious that the *minima* of the sum

$$F(\mathbf{x}) = \sum_{k=1}^{N} [g_j(x_1, \cdots, x_N)]^2$$

occur precisely at the points where $g_1(\mathbf{x}) = g_2(\mathbf{x}) = \cdots = g_N(\mathbf{x}) = 0$. That is, they occur at the *intersections* $\cap S_j$ of the hypersurfaces $S_j : g_j(\mathbf{x}) = 0$, and are *isolated* if the Hessian matrix $||\partial^2 F / \partial x_j \partial x_k||$ is nonsingular there. To locate all of them may be difficult, if the system of polynomial equations is hard to solve.

By constructing complicated hypersurfaces S_j with many known points of intersection, one thus obtains good examples for testing the 'robustness' of optimization algorithms. A simple one-parameter family of examples, which can be troublesome for gradient algorithms when ϵ is sufficiently small, is the following.

Example 12. Try to compute the minimum of the function[38]

$$F(x,y) = \epsilon^2 x^2 + (y - x^2)^2, \qquad \epsilon << 1.$$

Its graph has a narrow, steep 'canyon' of values near 0 running along the parabola $y = x^2$.

Actually, it is easy to defeat any general purpose computer program for locating minima, by quite simple functions of one variable such as the nonconvex function $f(x) = x - 10 \exp [10^3(x-2)]^2$, whose graph has a very slender spine pointing downward in the middle of an otherwise smooth slope. It is even easier to defeat such programs by analogous multivariate functions such as (11.5).

---

[37] M. R. Osborne, J. Austral. Math. Soc. 19 (1976), 343–357; see also his paper in Lootsma [72, pp. 171–190]. The algorithm is due to K. Levenberg, Quar. Appl. Math. 2 (1944), 164–168, and D. W. Marquardt, SIAM J. Appl. Math. 11 (1963), 431–441.

[38] A similar but more elaborate example due to Y. Bard, SIAM J. Numer. Anal. 7 (1970), 157–186, has been the subject of considerable special study.

12. Davidon's method; fixpoint methods.[39] Far more flexible than gradient methods for locating 'unconstrained minima' is the following 'variable metric' method due to Davidon. This method attempts to minimize $F(\mathbf{x})$, assumed to be reasonably smooth (of class C^2), as follows. Let $\mathbf{G}(\mathbf{x}) = \nabla F(\mathbf{x})$ be the gradient of F. In the absence of special information, set $\mathbf{x}^{(0)} = \mathbf{0}$ and $H^{(0)} = I$, the identity matrix of appropriate size. Davidon's method then proceeds as follows:

$$\mathbf{x}^{(m+1)} = \mathbf{x}^{(m)} - \lambda^{(m)} H^{(m)} \mathbf{G}^{(m)},$$

where $\lambda^{(m)} > 0$ is chosen so as to *minimize* F in the 'virtual gradient' direction, $H^{(m)}\mathbf{G}^{(m)}$ (with respect to the latest 'variable metric'). Since $H^{(0)} = I$, the first step of Davidon's method is identical with that of the gradient direction method.

A new metric is then set up, based on the following *secant* (forward difference) approximations:

$$\boldsymbol{\delta}^{(m)} = \mathbf{x}^{(m+1)} - \mathbf{x}^{(m)}, \qquad \boldsymbol{\gamma}^{(m)} = \mathbf{G}^{(m+1)} - \mathbf{G}^{(m)},$$

and

$$H^{(m+1)} = H^{(m)} - \frac{H^{(m)}\boldsymbol{\gamma}^{(m)}\boldsymbol{\gamma}^{(m)T}H^{(m)}}{\boldsymbol{\gamma}^{(m)T}H^{(m)}\boldsymbol{\gamma}^{(m)}} + \frac{\boldsymbol{\delta}^{(m)}\boldsymbol{\delta}^{(m)T}}{\boldsymbol{\delta}^{(m)T}\boldsymbol{\gamma}^{(m)}}.$$

Like the conjugate gradient method, Davidon's method would give the exact minimum of a *quadratic* function of N variables in at most N steps if there were no roundoff. Unlike the conjugate gradient method (applied to ∇F), it will do this even after entering a quadratic region from a non-quadratic region. For *strictly convex* functions with everywhere positive definite Hessians, it has been proved[40] that "the numbers $\langle \boldsymbol{\delta}^{(m)}, \mathbf{G}^{(m)} \rangle$ are positive, all the matrices $H^{(m)}$ are positive definite, and the definition of $\lambda^{(m)}$ implies the equation $\langle \mathbf{G}^{(m+1)}, \boldsymbol{\delta}^{(m)} \rangle = 0$," whence

$$\langle \boldsymbol{\gamma}^{(m)}, \boldsymbol{\delta}^{(m)} \rangle = \langle \mathbf{G}^{(m+1)} - \mathbf{G}^{(m)}, \boldsymbol{\delta}^{(m)} \rangle = -\langle \mathbf{G}^{(m)}, \boldsymbol{\delta}^{(m)} \rangle .$$

For the function in Example 12 with $\epsilon^2 = 0.01$ and $x^{(0)} = y^{(0)} = 1$, Davidon's method reduces the initial error, $(\mathbf{x}^T\mathbf{x})^{1/2}$, by a factor of 10 after 9 steps; another 5 steps reduces it to 7×10^{-7}. In contrast, 100 steps of steepest descent reduces the initial error only by a factor of 4/5.

[39] We follow here Powell's exposition in Lootsma [72, pp. 1–18]. For earlier expositions, see R. Fletcher and M. J. D. Powell, Comput. J. 6 (1963), 163–168; W. Davidon, ibid., 10 (1968), 406–410. See also C. G. Broyden, Math. Comp. 21 (1967), 368–381; J. Greenstadt, ibid., pp. 360–367; A. A. Goldstein and J. F. Price, Numer. Math. 10 (1967), 184–189; Y. Bard, Math. Comp. 22 (1968), 665–666; Fletcher [69, pp. 1–20]; M. J. D. Powell in Hull [70, pp. 43–61]; Powell [70].

[40] See R. Fletcher and M. J. D. Powell, Computer J. 6 (1963), 163–168; J. D. Pearson, ibid., 12 (1969), 171–178; R. Fletcher, ibid. 13 (1970), 317–322.

MINPACK. A 'minimization package' of computer programs for automatically computing minima of given functions has been developed at the Argonne National Laboratory under the leadership of Jorge Moré, K. E. Hillstrom, and others.[41] MINPACK, like the earlier 'Harwell package', sets a high standard of expertise.

In addition to solving nonlinear systems and finding minima, it contains a version (proposed by Hebden and refined by Moré) of the Levenberg-Marquardt algorithm for the nonlinear least squares problem. MINPACK has performed well on test cases, but the 'stopping problem' is again serious.

Fixpoint methods. The variational problem of locating the minima and other stationary points of a function F is equivalent not only to solving the vector equation $\mathbf{G}(\mathbf{x}) = \mathbf{0}$ for $\mathbf{G} = \nabla F$, but also to solving a closely related fixpoint problem. This is because, for *any* vector field $\mathbf{G}(\mathbf{x})$, the vector equation $\mathbf{G}(\mathbf{x}) = \mathbf{0}$ holds at $\mathbf{x}$ if and only if $\mathbf{\Phi}(\mathbf{x}) = \mathbf{x} + \mathbf{G}(\mathbf{x}) = \mathbf{x}$. There follows

THEOREM 9. *The problem of finding all the solutions of the vector equation* $\mathbf{G}(\mathbf{x}) = \mathbf{0}$ *is equivalent to that of finding all the fixed points of the function* $\mathbf{\Phi}(\mathbf{x}) = \mathbf{G}(\mathbf{x}) + \mathbf{x}$.

On the other hand, $\mathbf{G}(\mathbf{x}) = \mathbf{0}$ is equivalent to $\lambda\mathbf{G}(\mathbf{x}) = \mathbf{0}$ for any $\lambda \neq 0$. This raises the question: Which is the optimal λ to use in the iterative scheme

$$\mathbf{x}^{(m+1)} = \mathbf{\Phi}(\mathbf{x}^{(m)}) = \mathbf{x}^{(m)} - \lambda\mathbf{G}(\mathbf{x}^{(m)})?$$

This question is analogous to that of finding the optimal ω to use in solving $A\mathbf{u} = \mathbf{b}$ in Chapter 4, §10, by iterating

$$\mathbf{u}^{(m+1)} = \mathbf{u}^{(m)} + \omega[\mathbf{b} - A\mathbf{u}^{(m)}] = \mathbf{\Phi}(\mathbf{u}^{(m)}).$$

When $F \in C^2$ is *convex*, this is usually best done by computing the Hessian $||\partial^2 F/\partial x_i\, \partial x_j||$, and using methods similar to those of Chapters 4 and 5.

Global Newton methods. However, for *nonconvex* functions such as arise in eigenvector problems, it may be better to use 'global fixpoint' methods similar to the method originally used by L. E. J. Brouwer to prove his fixpoint theorem. Thanks to the power of contemporary computers, it has recently become practical to use such global methods.

A pioneer paper along these lines was that of F. H. Branin, Jr., in IBM J. Res. Dev. (1972), 504–522; see also S. Smale, J. Math. Ec. 3 (1977), 107–120. More closely related to Brouwer's work are papers by C. Eaves

[41] See J. J. Moré, B. S. Garbow, and K. E. Hillstrom, TOMS 7 (1981) 17–41; K. E. Hillstrom, TOMS 3 (1977), 305–315; K. L. Hiebert, TOMS 8 (1982), 5–20; and J. J. Moré, D. C. Sorensen, B. S. Garbow, and K. E. Hillstrom, in Cowell [84].

and H. Scarf, Math. of Operations Research 1 (1976), 1–27, and by R. B. Kellogg et al. in SIAM J. Num. Anal. 13 (1976), 473–483.

For example, the form of Newton's scheme given in (10.3) can be written, with $\alpha = 1/\Delta t$, as

$$H(\mathbf{x}^{(m)})\frac{\mathbf{x}^{(m+1)} - \mathbf{x}^{(m)}}{\Delta t} + \alpha \mathbf{G}(\mathbf{x}^{(m)}) = \mathbf{0},$$

which is the Euler-Cauchy polygon method for solving

$$\frac{d}{dt}[e^{\alpha t}\mathbf{G}(\mathbf{x}(t))] = \mathbf{0}.$$

For $\mathbf{x}(0) = \mathbf{x}^{(0)}$, this gives

$$\mathbf{G}(\mathbf{x}(t)) = e^{-\alpha t}\mathbf{x}^{(0)}.$$

By taking $\alpha = \text{sgn det}\, H(\mathbf{x}(t))$, Branin allowed the path $\mathbf{x}(t)$ to pass through points where the Hessian $H(\mathbf{x}(t))$ is singular, and Smale considered similar choices. H. B. Keller in de Boor-Golub [79, pp. 73–94] treated generalizations as well as methods for avoiding small time steps near singularities of the Hessian.

Chapter 7

Finite Element Approximations

1. Introduction. In Chapter 3, we defined a number of difference approximations to elliptic DE's, analyzed their orders of accuracy, and indicated some difficulties in applying them to general (plane) domains, especially near the boundary. In Chapter 6, §3, we described some simple alternative *finite element* approximations for solving elliptic problems, which can be applied when the solution satisfies some *variational principle*.

This chapter will study these and other higher-order finite element approximations more systematically, with special emphasis on their discretization (local and global truncation) errors. It will compare these not only with the errors arising from difference approximations, but also with those associated with *spline* approximations and *collocation*. We do this because most experts today consider the basic decision to be whether to use finite element or difference methods in discretizing elliptic problems. Finite element methods are generally preferred by structural engineers, whereas difference methods with $O(h^2)$ accuracy (or less!) retain their popularity in numerical fluid dynamics.

We will begin (in §§2–3) by recalling a few classical theorems about the approximation of functions. We will be very selective, because our main concern is with accurate *piecewise* approximation of functions of two or three variables on *general domains*, whereas most of the classical literature is concerned with univariate and global approximations.

Smooth functions of two variables are usually approximated quite accurately (to within 1%, say) with only moderately many (a few hundred?) unknowns, by suitably designed *piecewise polynomial interpolation* schemes. We will explain in §4 schemes of piecewise polynomial interpolation having $O(h^3)$ accuracy in *triangles* and *rectangles*, respectively.

In §5 we describe Hermite and univariate spline interpolation schemes having higher orders of accuracy (and using polynomials of higher degree). These are extended in §7 to rectangles and box-shaped regions; the variational properties of bicubic splines are also briefly discussed.

In §6, we describe a standard 'Lagrangian' scheme for continuous piecewise polynomial interpolation of any degree (and corresponding order of accuracy) in a general simplicial 'complex' — i.e., in any polyhedron that has been decomposed into simplices.

In §8, these schemes are generalized to rectangular polygons and rectangular 'polytopes', subdivided into rectangles or box-shaped regions. Hermite interpolation and approximation are easy to use; splines are more troublesome in domains not products of intervals.

Fortunately, bilinear blending can be used to obtain bivariate interpolants to arbitrary 'compatible' boundary functions on rectangles, triangles, and quadrilaterals with one curved side. Likewise, multilinear and higher-order blending can be used to produce analogous finite elements in $\mathbb{R}^n$, as is shown in §9. We conclude this chapter with brief descriptions of isoparametric elements, domain approximation by polygons, the 'penalty function' method, and 'singular' elements.

Historical remarks. The study of piecewise polynomial approximations to multivariate functions was of purely academic interest until around 1960. However, by 1912 L. E. J. Brouwer had already utilized piecewise linear functions to prove theorems about continuous mappings. Courant suggested using them to minimize approximately the Dirichlet integral as a more or less equivalent alternative to difference methods in 1925. Likewise, Pólya used piecewise bilinear functions to calculate approximate eigenvalues in 1952; and in 1957 J. L. Synge introduced 'pyramidal' basis functions (Chapter 6, §3) in his notable book on the hypercircle method.

But it was not until 1960 that Clough coined the apt name *finite elements* for the mathematical *pieces* which could be fitted together to solve a wide variety of variational problems in structural mechanics. He and Argyris had used the 'finite element method' (FEM) earlier, but only after 1960 did this approach become widely used by mathematically minded structural engineers. During the 1960's, the FEM grew explosively in breadth and depth, and already by 1967 O. C. Zienkiewicz had published the first edition (written with Y. K. Cheung) of his now classic treatise on the subject (Zienkiewicz, [77]).

It was also around 1960 that multivariate spline functions were first invented and used systematically, to approximate the shapes of sheet metal surfaces used on automobiles.[1] Like some of the 'finite elements' developed during the 1960's, bicubic splines enable one to fit general smooth surfaces with high (fourth-order) accuracy, using a moderate number of 'free' parameters. Hence they, and even more clearly bicubic Hermite interpolants, can be regarded as 'finite elements' of a special kind.

---

[1] Their successful application for this purpose led the General Motors Research Laboratories to sponsor a 1964 Symposium on the Approximation of Functions; see Garabedian [65].

The instant success of the preceding methods was largely due, of course, to the availability by 1960 of computers which could solve quickly and relatively cheaply the many (typically 200–2000) equations involved, usually by one of the methods described in Chapters 4 and 5.

2. Univariate interpolation. Our main concern in this chapter will be with methods for approximating smooth functions of two or more variables, because it is such methods that are most helpful for the numerical solution of elliptic problems. Fortunately for research workers, the bulk of this literature is recent (post-1960) and relatively accessible. It separates into two main parts: the part concerned with conforming and non-conforming 'finite elements', simplicial or box-shaped, and that dealing with bivariate and multivariate spline and Hermite approximation.

To understand multivariate approximation in depth, one must know something about approximating functions of *one* (real or complex) variable. This is treated in an enormous classical literature.[2] Although little of it is directly applicable to partial DE's, one can hardly exaggerate the fascination of its problems and the ingenuity of the methods that have been invented to solve them.

The most widely used approximation formulas are based on polynomial interpolation, which is exact for function values (though not for derivatives) at mesh-points. It is elementary that, given $x_0 < \cdots < x_n$ and $f(x_i) = y_i$, $i = 0, 1, \cdots, n$, there is one and only one polynomial $p(x)$ of degree n such that $p(x_i) = y_i$ for all i. Moreover, if $f \in C^{n+1}[I]$, where I is the interval spanned by x and $[x_0, x_n]$, the *interpolation error* is

$$f(x) - p(x) = e(x) = \frac{1}{(n+1)!} \prod_{i=0}^{n} (x - x_i) f^{(n+1)}(\xi), \tag{2.1}$$

for some $\xi \in I$ (Cauchy remainder formula).

We now fix x, n, and θ_i with $-1 \leqslant \theta_0 < \theta_1 < \cdots < \theta_n \leqslant 1$, and take a 'stencil' of points $x_i = x + \theta_i h$, $i = 0, \cdots n$, regarding h as a parameter. By (2.1), for some $\xi \in I$,

$$|e(x)| \leqslant \frac{h^{n+1}\Theta}{(n+1)!} |f^{(n+1)}(\xi)|, \qquad \text{where } \Theta = \prod_{i=0}^{n} |\theta_i| < 1,$$

because each $|\theta_i| \leqslant 1$. This proves the following result.

THEOREM 1. *If $f \in C^{n+1}(x-h, x+h)$ for some $h > 0$, the interpolation error of* (2.1) *satisfies, for $M = \Theta |f^{(n+1)}(\xi)|_{\max}/(n+1)!$,*

$$|e(x)| \leqslant Mh^{n+1}, \tag{2.2}$$

regardless of the θ_i. In other words, $e(x) = O(h^{n+1})$.

[2] See Davis [63] and the references given there; Jackson [30] is classic. The books by Lorentz-Jetter-Riemenschneider [83], Meinardus [67], Timan [63], Varga [71], and Walsh [60] also contain much relevant information.

More generally, for any function $f \in C^{n+1}[a,b]$, polynomial interpolation of degree n gives a j-th derivative whose errors are $O(h^{n+1-j})$ for $j = 0, 1, \cdots, n$, as $h \downarrow 0$ with fixed n, $h = \max|x_j - x|$.

We get Hermite interpolation as limiting cases of coincident sets of points. For example, the *cubic Hermite* interpolant through $x_0 = x_1 = a$, $x_2 = x_3 = b$ is that cubic polynomial such that

$$p(a) = f(a), \quad p'(a) = f'(a), \quad p(b) = f(b), \quad p'(b) = f'(b). \tag{2.3}$$

More generally the *Hermite interpolant of degree* $2k-1$ to a function $f(x)$ on $[a,b]$ is defined as the polynomial $p(x)$ of degree $2k-1$ or less such that $p^{(m)}(a) = f^{(m)}(a)$ and $p^{(m)}(b) = f^{(m)}(b)$ for $m = 0, \cdots, k-1$ (Davis, [63, p. 28]). Specifically, we have the following result.[3]

THEOREM 2. *Let* p *be the Hermite interpolant of degree* $2k-1$ *to* $f(x) \in C^{2k}[0,h]$; *then*

$$|f^{(r)}(x) - p^{(r)}(x)| \leqslant M_{k,r} h^{2k-r} \qquad \text{for } x \in [0,h], \tag{2.4}$$

where

$$M_{k,r} \leqslant \frac{1}{4^{k-r}(2k-2r)!r!} ||f^{(2k)}||_\infty. \tag{2.4'}$$

We will discuss Hermite interpolation further in §5.

Trigonometric interpolation. Analogous, but deeper theorems hold for interpolation to *periodic* functions by 'trigonometric polynomials' at equal-spaced interpolation points (a uniform mesh).[4] Thus, for given n, there is one and only one interpolant

$$g_{2n+1}(\theta) = \frac{1}{2} a_0 + \sum_{k=1}^{n} [a_k \cos k\theta + b_k \sin k\theta] \tag{2.5}$$

to the $f(\theta_j)$ at the $2n+1$ equally spaced mesh-points $\theta_j = jh$, $j = 0, \cdots, 2n$, $h = 2\pi/(2n+1)$. Likewise, there is one and only one function

$$g_{2n}(\theta) = \frac{1}{2} a_0 + \sum_{k=1}^{n-1} [a_k \cos k\theta + b_k \sin k\theta] + a_n \cos n\theta, \tag{2.6}$$

which interpolates to the $f(\theta_j)$ at the $2n$ equally spaced points $\theta_j = j\pi/n$, $j = 0, \cdots, 2n-1$.

---

[3] See Birkhoff-Schultz-Varga [68]; also P. G. Ciarlet, M. H. Schultz, and R. S. Varga, Numer. Math. 9 (1967), 394–430.

[4] Jackson [30], Davis [63]. Since $z^k = \cos k\theta + i \sin k\theta$, if $z = \exp(i\theta)$, the results are theorems on polynomial interpolation in the complex domain.

What is most interesting, these schemes of interpolation to *analytic* periodic functions by trigonometric polynomials of degree n through $2n$ or $2n+1$ equidistant points converges exponentially, in the sense that $|f(\theta)-g_n(\theta)| \leqslant K\rho^n$ for some $\rho < 1$, $K < \infty$, and for all n. Likewise for periodic $f \in C^k$, with $f^{(k)} \in \mathrm{Lip}\,\alpha$, the error is $O(n^{-k-\alpha})$.[5]

Functions analytic on $[-1,1]$. Related results hold for real functions $g(x)$ analytic on a real interval, which we can translate and scale to be $[-1,1]$. Since $[-1,1]$ is compact, $g(x)$ can be extended ('continued analytically') to a function $g(z)$ analytic in some ellipse with foci at ± 1. Setting $z=(t+t^{-1})/2$, we obtain a function $f(t)$ analytic on an annulus $-\epsilon < \log|t| < \epsilon$, $\epsilon > 0$, in the complex t-plane. Hence we can expand $f(t)$ in a Laurent series convergent in this annulus. Setting $t = r\,e^{i\theta}$, we get a real *even* function $\phi(\theta) = f(e^{i\theta})$ on the unit circle. As in the preceding paragraph, trigonometric interpolation to $\phi(\theta)$ at equidistant points $\theta_k = 2\pi k/N$ converges exponentially. But this corresponds to polynomial interpolation to $g(x) = g(\cos\theta) = f(t)$ at the zeros of the Chebyshev polynomials $T_n(x)$, which therefore also converges exponentially (see Lanczos [56, p. 245] and Varga [71]).

Runge's counterexample. So far we have discussed only positive (i.e., favorable) results. We now give a famous counterexample due to Runge, which illustrates the pitfalls of interpolating with polynomials of high degree over large intervals. Consider the restriction to $[-5,5]$ of the analytic function $f(x) = 1/(1+x^2)$, which is bounded on the entire real axis. Let $p_{2n}(x)$ be the polynomial interpolant of degree $2n$ to $f(x)$ on the uniform mesh $x_j = 5j/n$, where $j = -n, -n+1, \cdots, n-1, n$. Then it is not true that $\lim_{n\to\infty} p_{2n}(x) = f(x)$; instead, as n increases, the values of $p_{2n}(x)$ oscillate more and more wildly! The divergence is due to the singularities at $\pm i$; see Davis [63, Chap. 4].

3. Best approximation. In any subspace S of a normed function space V, a *best approximation* to a given $f \in V$ is a function $\phi \in S$ which *minimizes* the 'distance' $||\phi - f||$. In practice, we will be mainly interested in the norms $||\phi-f||_2$, $||\phi-f||_\infty$, and their analogues (e.g., in Sobolev spaces; see Chapter 6, §4) which involve derivatives as well as function values.

Quadratic norms associated with inner products have by far the most advantageous theoretical properties. This is because it is relatively easy (in principle!) to obtain the best mean square approximation, in any finite-dimensional subspace S with any Euclidean (i.e., inner product) norm, by orthogonal projection P onto S. Moreover, this orthogonal projection P is *linear*: $P[c_1f_1+c_2f_2] = c_1P[f_1]+c_2P[f_2]$.

---

[5] For $0<\alpha<1$, if $|g(x)-g(y)| \leqslant K|x-y|^\alpha$ for all $x, y \in (a,b)$, then $g \in \mathrm{Lip}\,\alpha$.

For example, much as in Chapter 6, §3, if

$$f(\theta) = \frac{1}{2}a_0 + \sum_{k=1}^{\infty} [a_k \cos k\theta + b_k \sin k\theta], \tag{3.1}$$

then the best mean square approximation to a periodic function $f(\theta)$ by a linear combination of 1, $\cos\theta$, $\sin\theta$, $\cdots$, $\cos n\theta$, $\sin n\theta$ (i.e., a 'trigonometric polynomial' of degree n) is simply the usual truncated Fourier series

$$f_n(\theta) = \frac{1}{2}a_0 + \sum_{k=1}^{n} [a_k \cos k\theta + b_k \sin k\theta]. \tag{3.2}$$

The f_n always converge to f in the mean square norm,

$$||f||_2 = \left[\int_0^{2\pi} [f(\theta)]^2 \, d\theta\right]^{1/2},$$

if f is square-integrable (i.e., if $||f||_2 < \infty$). For periodic $f \in C^1$, $f_n(\theta) \to f(\theta)$, uniformly. If $f \in C^m$ is of period 2π, then integrating by parts m times we obtain

$$\begin{Bmatrix} a_k \\ b_k \end{Bmatrix} = \frac{1}{\pi}\int u(x)\begin{Bmatrix} \cos \\ \sin \end{Bmatrix} kx \, dx = \frac{1}{\pi k^m}\int u^{(m)}(x)\begin{Bmatrix} \cos \\ \sin \end{Bmatrix} kx \, dx \leqslant \frac{2U_m}{k^m},$$

where $U_m = \max |u^{(m)}|$.

As with trigonometric *interpolation* (see (2.5)–(2.6)), if f is an analytic periodic function, then $||f_n - f||_\infty$ tends to zero exponentially fast. However, there exist continuous periodic functions f whose best mean square approximations f_n (as defined by (3.2)) do not converge pointwise to f.

On the other hand, for any $f \in C$ and $\epsilon > 0$, there always *exists* some trigonometric polynomial of degree $n = n(f,\epsilon)$:

$$g_n(\theta) = \frac{1}{2}A_{n,0} + \sum_{k=1}^{n} [A_{n,k} \cos k\theta + B_{n,k} \sin k\theta] \tag{3.3}$$

such that $||g_n - f||_\infty < \epsilon$.

Modified sine series. The choice of a good family of approximating subspaces may be hightly problem-dependent. This fact is illustrated very clearly by the use of 'modifed sine series' to approximate functions $u(x) \in C^{2m}[0,\pi]$. If a function $v \in C^{2m}$ defined on this interval is continued by odd reflection, so that $v(-x) = -v(x) = v(2\pi - x)$, one obtains a function in $C^{2m}(-\infty,\infty)$ if and only if

$$v^{(2j)}(0) = v^{(2j)}(\pi) = 0 \quad \text{for} \quad j = 0, 1, \cdots, m-1.$$

On the other hand, for each $u \in C^{2m}[0,\pi]$, there exists a polynomial $p(x)$ of degree $2m-1$ such that $v(x) = u(x) - p(x)$ satisfies the conditions displayed above. It follows from the discussion of the periodic case above that the k-th coefficient of the sine series of $v(x)$ will be $O(k^{-2m})$. Therefore, the *modified* sine series

$$u(x) = p(x) + \sum_{k=1}^{\infty} \beta_k \sin kx$$

will converge to $u(x)$ much more rapidly than the ordinary sine series expansion of $u(x)$. It is a corollary that the approximating subspace spanned by (say) 1, x, x^2, x^3, and the $\sin kx$ with $k = 1, \cdots, N$ will be superior to that spanned by these $\sin kx$ alone, for smooth functions on $[0, \pi]$ satisfying general boundary conditions.

Weierstrass approximation theorem. Likewise, for given $f \in C[-1,1]$ and $\epsilon > 0$, there always exist an $n = n(\epsilon)$ and a polynomial $p_n(x)$ of degree n or less such that $\|f(x) - p_n(x)\|_\infty < \epsilon$.[6] Indeed, by restricting attention to even periodic functions and using the Chebyshev polynomials, one can derive this result from the corresponding fact about trigonometric approximation.[7] However, this very special method sheds little light on the following, very much stronger result.[8]

THEOREM 3. *Let* $u(\mathbf{x}) \in C^k(\bar{\Omega})$ *be given, where* $\bar{\Omega}$ *is a compact domain in* $\mathbb{R}^n$. *Then for any* $\epsilon > 0$, *there exists a polynomial* $p(\mathbf{x})$ *such that, for all partial derivatives of order* $j \leqslant k$, $|p^{(j)}(\mathbf{x}) - u^{(j)}(\mathbf{x})| \leqslant \epsilon$.

Although this *global, multivariate* approximation theorem is very attractive theoretically, no efficient method is known for implementing it practically, and we will now resume our review of *univariate* approximation theory.

Chebyshev approximation. Approximations which minimize $\|\phi - u\|_\infty$ are usually called Chebyshev approximations. Their constructive theory is largely restricted to polynomial and rational functions of *one* variable; it is not of much direct use for elliptic problems in $\mathbb{R}^n$ when $n > 1$. For $n = 1$ they have the following remarkable property.[9]

THEOREM 4. (Chebyshev equioscillation theorem). *For any function* $u \in C[a,b]$, *there is a unique polynomial* $p(x)$ *of degree* n *which minimizes* $\|p - u\|_\infty$ *on* $[a,b]$. *For this polynomial there are* $n+2$ *points* $x_0 < x_1 < \cdots < x_{n+1}$ *in* $[a,b]$ *where the error* $e(x) = p(x) - u(x)$ *has extrema* $e(x_i)$. *The* $e(x_i)$ *alternate in sign and all* $|e(x_i)| = \|p - u\|_\infty$.

THEOREM 4′. *In the (nonlinear) subspace* $\mathbb{P}_n^m$ *of rational functions* $g(x) = p(x)/q(x)$ *with polynomials* $p(x)$ *of degree* $k \leqslant m$, $q(x)$ *of degree* $j \leqslant n$, *and* p, q *relatively prime, there is a unique* $g(x)$ *which minimizes*

---

[6] This is the Weierstrass approximation theorem; see Davis [63, p. 107].

[7] D. V. Widder, *Advanced Calculus*, Prentice-Hall, 1947, §7.2; Davis [63, Chap. 6].

[8] Davis [63, Chap. 6]. We use the convenient multi-index notation for $\mathbf{s} = (s_1 \cdots s_n)$ with $s_1 + \cdots + s_n = s$, according to which $u^{(\mathbf{s})} = \partial^s u / \partial x_1^{s_1} \cdots \partial x_n^{s_n}$.

[9] Davis [63], Rice [64], or Walsh [60].

$\|g-u\|_\infty$. *This* $g(x)$ *has the Chebyshev equioscillation property with* $m+n+1-\min\{m-k, n-j\}$ *oscillations.*

Remez algorithm. In 1934, the Russian mathematician E. Remez used the Chebyshev equioscillation theorem to *compute* best approximations. For the relevant procedure see Fike [68, Chaps. 5 and 9], Hart et al. [68], or Rice [64, Chap. 6]. Most computer programs for function evaluation today use (discontinuous) piecewise polynomial or rational functions; the Remez algorithm is used to compute the coefficients for the pieces.

That is, the Remez algorithm[10] provides an effective means for computing the coefficients of the polynomial $p_n(x)$ of given degree n or less which minimizes $\|f(x)-p_n(x)\|_\infty$ or, equivalently, finding the a_k giving

$$\min_{p_n}\{\max_x |f(x)-p_n(x)|\} \qquad \text{for } p_n(x)=\sum_{k=0}^{n} a_k x^k. \tag{3.4}$$

One can also use other (less efficient) exchange methods or linear programming.[11]

4. Typical finite elements. As was stated in Chapter 6, §3, the 'finite element method' (FEM) originated as a variant of the Rayleigh-Ritz method, in which a variational integral is *minimized* over some 'approximating subspace' Φ of (typically) *piecewise* differentiable, globally continuous functions. Moreover, its rate of convergence, as a function of the element diameter h (hence of the number of unknowns) depends basically on the order of accuracy with which a locally smooth function (the unknown exact solution $u(\mathbf{x})$) can be approximated by functions $U_h \in \Phi$.

In Chapter 6, §§3 and 4, we estimated the errors in piecewise linear and piecewise bilinear approximating functions. The order of accuracy of these finite element approximations is only $O(h^2)$; hence to achieve four-digit accuracy, one might need a 100×100 mesh with 10^4 unknowns. In this section, we will describe two families of finite elements which approximate general 'smooth' functions with $O(h^3)$ accuracy. With such schemes, a 22×22 mesh with 484 unknowns could give four-digit accuracy, which is enough for most engineering purposes. This reduces the number of unknowns by a factor of twenty; the 'computational complexity' is reduced even more. Thus with band elimination, the number of multiplications is reduced by a factor of about 400 by the use of these 'higher-order' finite elements.

Example 1. Consider the (unique) *quadratic* interpolant

$$q(x,y)=a_0+a_1x+b_1y+a_2x^2+2cxy+b_2y^2 \tag{4.1}$$

[10] See E. L. Stiefel in SIAM J. Numer. Anal. 1 (1964), 164–176, and in Garabedian [65, pp. 68–82]; see also Meinardus [67, §7].

[11] J. B. Rosen, SIAM J. Numer. Anal. 7 (1970), 80–103.

to given values at the vertices and midpoints of the sides of a triangle. By an affine transformation, one can reduce this to the 'standard' triangle with vertices at $X=(h,0)$, $Y=(0,h)$, and $Z=(0,0)$. We let $z=h-x-y$, so that x/h, y/h, and z/h are *barycentric* coordinates: the point (x,y) is the center of mass of 'weights' x,y,z at X, Y, Z, respectively. Writing $u_0=u(0,0)$, $u_1=u(h,0)$, $u_2=u(0,h)$, $u_3=u(h/2,0)$, $u_4=u(h/2,h/2)$, and $u_5=u(0,h/2)$, a simple calculation gives

$$a_0=u_0,\ a_1=\frac{1}{h}(4u_3-3u_0-u_1),\ a_2=\frac{1}{h^2}(2u_1-4u_3+2u_0),\ c=\frac{4}{h^2}u_4;$$

the b_i are given by like formulas, completing the construction.

Let π be any subdivision of a polygon into T triangles. Then on each side of any of these triangles, $q(x,y)$ is the unique univariate quadratic interpolant to the three 'nodal values' at its ends and midpoint. Hence *any* choice of nodal values at the mesh-points gives a globally *continuous*, piecewise quadratic interpolant. The 'approximating subspace' for Dirichlet boundary conditions in this case is clearly $(V+S)$-dimensional, where V is the number of interior vertices, and S is the number of interior sides. But by Euler's formula, $S=V+T-1$; hence its dimensionality is also $2V+T-1$.

Error bounds. To prove that q approximates u, u_x, u_y, u_{xx}, u_{xy}, and u_{yy} well requires more ingenuity. By definition, the *interpolation error* $e=u-q$ vanishes at the six interpolation points; moreover, since q is quadratic, e has the same derivatives of all orders $j>2$ as u. Now expand e in a Taylor series with remainder $O(h^3)$ about any interior point $P=(\xi,\eta)$. Since $e=0$ at the interpolation points, (x_i,y_i), $i=1,\cdots,6$, we will have six equations

$$e(P)+(x_i-\xi)e_x(P)+(y_i-\eta)e_y(P)+\frac{1}{2}(x_i-\xi)^2e_{xx}(P)$$
$$+(x_i-\xi)(y_i-\eta)e_{xy}(P)+\frac{1}{2}(y_i-\eta)^2e_{yy}(P)=M_ih^3,$$

where h is the distance of (x_i,y_i) from P (hence at most the diameter of the triangle), and $|M_i|\leqslant\max|\partial^3u/\partial s^3|/6$, s being arc length.

The determinant of the coefficient matrix of these equations is independent of ξ and η. Solving for $e,e_x,\cdots,e_{yy}$ and taking due account of the fact that the first derivatives are multiplied by factors $O(h)$ and the second by factors $O(h^2)$, we obtain the following result.[12]

THEOREM 5. *Piecewise quadratic interpolation in a triangulated polygon gives a globally continuous function that approximates any* $u\in C^3(\bar{\Omega})$ *with*

[12] We define $u\in C^r$ to mean that all $\partial^{i+j}u/\partial x^i\partial y^j$ with $i+j\leqslant r$ exist and are continuous, and $u\in C^{r,s}$ to mean that these derivatives all exist and are continuous for $i=0,\cdots,r$, $j=0,\cdots,s$.

$O(h^3)$ accuracy, ∇u with $O(h^2)$ accuracy, and the second derivatives of u with $O(h)$ accuracy.

Example 2. Likewise, given u on a rectangle R with sides parallel to the axes, there is one and only one *biquadratic* function

$$q(x,y) = \sum_{i=0}^{2} \sum_{j=0}^{2} a_{ij} x^i y^j = a_{00} + a_{10}x + \cdots + a_{22}x^2y^2 \tag{4.2}$$

that can be interpolated to the $u(x_i, y_j)$ at the 4 vertices, the 4 midpoints of the sides, and the center of R. Again, the interpolant on each side is the unique univariate quadratic interpolant to the 3 nodal values on that side. Hence the piecewise biquadratic interpolant given by this scheme is also globally continuous.

By a standard error analysis which assumes that $u \in C^{3,3}$ (see Steffensen [50, pp. 204–206]), one obtains $u(x,y) = q(x,y) + e(x,y)$ where

$$e(x,y) = \frac{1}{6}x(x^2 - h^2)\, u_{xxx}(\xi_1, \eta_1) + \frac{1}{6}y(y^2 - k^2)\, u_{yyy}(\xi_2, \eta_2)$$
$$- x(x^2 - h^2)y(y^2 - k^2)\, u_{xxxyyy}(\xi_3, \eta_3)$$

and all $(\xi_i, \eta_i) \in$ R. Thus, the biquadratic interpolant has $O(h^3)$ error.

Here we derive a sharper error bound in terms of u_{xxx} and u_{yyy} alone, which only assumes that $u \in C^3$; see also §7.

THEOREM 6. *The error $e(x,y)$ arising from piecewise biquadratic interpolation to $u(x,y) \in C^3$ in a square mesh of side h satisfies the inequality*

$$|e(x,y)| \leqslant \frac{h^3}{64\sqrt{3}} [|u_{xxx}|_{\max} + |u_{yyy}|_{\max}]. \tag{4.3}$$

Proof. Let $q(x,y)$ be the biquadratic interpolant to $u \in C^3(\mathrm{R})$, and let its interpolation error be $e(x,y) = u(x,y) - q(x,y)$. Then clearly $e_{xxx} \equiv u_{xxx}$ and $e_{yyy} \equiv u_{yyy}$. Now for $y \in [-h,h]$, let $x_0 = -h$, $x_1 = 0$, $x_2 = h$, and consider the $e(x_i, y)$ as functions of y. By the Cauchy remainder formula, (2.1), for quadratic interpolation in y, we have

$$e(x_i, y) = \frac{1}{6}y(h^2 - y^2)\, u_{yyy}(x_i, \eta_i), \qquad \eta_i \in (-h,h).$$

Next, for each $y \in (-h,h)$, let $b(x,y)$ be the univariate quadratic interpolant (in x) to the $e(x_i, y_j)$:

$$b(x,y) = \frac{1}{h^2}[e(-h,y)x(x-h) + 2e(0,y)(h^2 - x^2) + e(h,y)x(x+h)]. \tag{4.4}$$

For any $x \in (0,h)$, we have $x(x-h) < 0$, while $2(h^2 - x^2)$ and $x(h+x)$ are positive. The sum $(h^2 + hx - x^2)/h^2$ of the absolute values of the coefficients of the $e(x_i, y)$ in (4.4) has the maximum value $5/4$.

Finally, we consider $f(x,y) = e(x,y) + b(x,y)$. By construction, $f(-h,y) = f(0,y) = f(h,y) = 0$. Hence, again by the Cauchy remainder formula for quadratic interpolation (this time in x), we have

$$f(x,y) = \frac{1}{6}\, x(h^2 - x^2) f_{xxx}(\xi, y),$$

for some $\xi \in (-h,h)$. Moreover, since $b(x,y)$ is quadratic in x, $f_{xxx} = e_{xxx} = u_{xxx}$. Also, $x(h^2 - x^2)$ has its maximum absolute value when $h^2 = 3x^2$, and this maximum is then $2h^3/3\sqrt{3}$. Therefore

$$(4.5) \quad |e(x,y)| = |f(x,y) - b(x,y)| \leqslant \frac{h^3}{9\sqrt{3}} |u_{xxx}|_{\max} + \frac{5h^3}{36\sqrt{3}} |u_{yyy}|_{\max}.$$

Halving the mesh, interchanging x and y and averaging, we obtain (4.3).

General convex quadrilaterals. The preceding biquadratic interpolation scheme can be extended to partitions of polygons into arbitrary convex quadrilaterals, as follows.[13]

LEMMA. *Let* $P_0 = (0,0)$, $P_1 = (1,0)$, $P_2 = (1,1)$, $P_3 = (0,1)$ *be the four corners of the unit square* $S = [0,1]^2$. *Define* α *and* β *as the bilinear interpolants on* S *to the corner values* $\alpha(P_k) = a_k$ *and* $\beta(P_k) = b_k$, *where the points* (a_k, b_k) *span a convex quadrilateral* Q. *Then the function* $\mathbf{f} : (s,t) \mapsto (\alpha(s,t), \beta(s,t))$ *is bianalytic from* S *to* Q, *and linear on edges.*

More precisely, α and β are bilinear functions with analytic *algebraic* inverses. They are linear if the quadrilateral is a parallelogram, and irrational otherwise. Thus $x = a(1 - st)$ and $y = bs - ct$ map S onto the quadrilateral with corners at $(\pm a, 0)$, $(0,b)$, and $(0,-c)$. Since $2bs = y + \sqrt{4 - 4ax + y^2}$, the inverse map is not rational.

Since the orders of accuracy of approximations to smooth functions and their derivatives are invariant under any bianalytic transformation, this lemma has the following consequence.

COROLLARY. *Let* ϕ *be any globally continuous piecewise polynomial interpolation scheme for a square which approximates all* j*-th derivatives,* $j = 0, 1, \cdots, k < n$, *with* $O(h^{n-j})$ *accuracy. Then for any subdivision of a polygon into quadrilaterals, the interpolant* $u(\mathbf{f}(\mathbf{x}))$ *defined in each such quadrilateral by the preceding lemma is also globally continuous and has the same order of accuracy.*

5. Hermite interpolants; univariate splines. We recall (Chapter 2, §4) that solutions of the Laplace equation $\nabla^2 u = 0$ minimize the Dirichlet integral $\mathbf{D}\langle u,u \rangle$, in the (affine) subspace of all *continuous*, piecewise smooth functions assuming the same boundary values. Similarly, among all *continuously differentiable* functions $U \in C^1(\Omega)$ that satisfy the boundary conditions

[13] This is a result of Tague and Kerr, proved in de Veubeke [64]. Convergence proofs for this scheme have been given by P. G. Ciarlet and P.-A. Raviart in Aziz [72, pp. 409–474] and in Comp. Methods Appl. Mech. Eng. 1 (1972), 217–249. See also Stephan-Wendland [82].

$$(5.1) \qquad U(\mathbf{y}) = g(\mathbf{y}) \quad \text{and} \quad \partial U(\mathbf{y})/\partial n = h(\mathbf{y}) \qquad \text{on } \Gamma = \partial\Omega,$$

the one which satisfies the biharmonic DE

$$(5.2) \qquad \nabla^4 u = \nabla^2[\nabla^2 u] = 0$$

minimizes the integral of the squared Laplacian[14]

$$(5.2') \qquad \mathbf{B}\langle U, U\rangle = \iint |\nabla^2 U|^2\, dR.$$

Since jumps in the gradient ∇U make an infinite contribution to $\mathbf{B}\langle U,U\rangle$, this suggests that in applying the Rayleigh-Ritz method to obtain approximate solutions of the 'clamped plate' problem (5.1)–(5.2), one should limit one's attention to conforming elements having no jumps in ∇U. That is, one should consider only local approximations which produce a function $U \in C^1(\Omega)$ when fitted together.

This fact makes it desirable to know how to construct so-called *conforming* finite element approximations, which are *globally continuously differentiable*.

Bicubic Hermite interpolation. The simplest continuously differentiable, piecewise polynomial approximants are constructed by Hermite interpolation as follows.

Example 3. Let the values of a smooth (class $C^{1,1}$) function *u and its partial derivatives* u_x, u_y, u_{xy} be given at the vertices of a rectangle with sides parallel to the coordinate axes. Then one and only one *bicubic* function

$$(5.3) \qquad \sum_{i=0}^{3}\sum_{j=0}^{3} a_{ij}x^i y^j = a_{00} + a_{10}x + \cdots + a_{33}x^3y^3$$

can be interpolated to the 16 given mesh values and derivatives. Moreover, its restriction to any edge is the unique (univariate, cubic) Hermite interpolant on that edge to the given values and slopes of u at its endpoints.

Hence, given any subdivision into rectangles of a *rectangular polygon* with sides parallel to the axes, by a rectangular mesh, there is a unique *piecewise bicubic Hermite* function which can be interpolated to the values of u and its derivatives u_x, u_y, u_{xy} at mesh-points. We will call these 'nodal values' and 'nodal derivatives'. A careful analysis proves[15]

THEOREM 7. *For piecewise bicubic Hermite interpolation to any* $u \in C^4(\bar{\Omega})$, *the interpolation errors are* $O(h^4)$ *in* u, $O(h^3)$ *in* ∇u, *and* $O(h^2)$ *in* u_{xx}, u_{xy}, u_{yy}.

---

[14] See Bergman-Schiffer [53, p. 235]. The fact that jumps in ∇u make an infinite contribution to $\mathbf{B}\langle U,U\rangle$ is obvious, since $\int (u'')^2\,dx = \int u''\,du'$, and, u'' is infinite where u' has a jump.

[15] Birkhoff-Schultz-Varga [68]; see also §7 below.

The bicubic Hermite elements of Example 3 can be generalized to $\mathbb{R}^n$ and to polynomials of higher (odd) degree (see §7 below). They have the practical advantage of giving 'conforming' global approximations in $C^1(\Omega)$ (see §9).

We begin with *univariate Hermite interpolation* of general odd degree $2r-1$ on a subdivided interval;[16] the case of a single interval was discussed in §2. Given $u \in C^{r-1}[a,b]$, let $\pi : a = x_0 < x_1 < \cdots < x_J = b$ be any subdivision of $[a,b]$ into J subintervals. We can define a *piecewise* polynomial interpolant $U \in C^{r-1}[a,b]$ to $u(x)$ of degree $2r-1$ on each j-th subinterval ('piece') $[x_{j-1},x_j]$ of $[a,b]$, as the unique polynomial $U(x) = c_0 + c_1 x + \cdots + c_{2r-1}x^{2r-1}$ that satisfies

$$(5.4) \qquad U^{(m)}(x_s) = u^{(m)}(x_s), \qquad m = 0,1,\cdots,r-1, \quad s = j-1, j.$$

In other words, U has the same values and first $r-1$ derivatives as $u(x)$ at the endpoints of each of the J subintervals.

Hermite *interpolants* of high degree to any smooth function give very accurate *approximations*. The values of $U(x)$ and error bounds are determined *locally* and independently in each interval $[x_{j-1},x_j]$. Replacing $[0,h]$ in Theorem 2 of §2 with $[x_{j-1},x_j]$ and setting $h_j = x_j - x_{j-1}$, we obtain for any $u \in C^{2r}[a,b]$ and $k = 0,\cdots,2r-1$,

$$(5.5) \qquad |u^{(k)}(x) - U^{(k)}(x)| \leqslant M_{r,k} h_j^{2r-k}, \qquad x \in [x_{j-1},x_j],$$

where $M_{r,k} = \|u^{(2r)}\|_{\max}/4^{r-k}(2r-2)!k!$ is as in (2.4′) with $u^{(2r)}$ in place of $f^{(2r)}$. For $k > r$, the bound (5.5) refers to *one-sided* derivatives at the mesh-points x_{j-1}, x_j.

Spline interpolation. Spline interpolation provides an intriguing and often useful alternative to piecewise Hermite interpolation of the same (odd) degree for solving elliptic problems, because it gives nearly the same accuracy with many fewer unknowns. Although spline interpolation is *not* local, it gives much smoother interpolants (many authors refer to what we call 'splines' as 'smooth splines', for this reason). Splines are therefore appropriate for geometrical design problems.[17]

A *spline function* of degree $2r-1$ on a subdivided interval $[a,b]$ is defined as a function $U \in C^{2r-2}[a,b]$ which is a polynomial of degree $2r-1$ or less on each subdivision. For a given partition $\pi : a = x_0 < x_1 < \cdots < x_J = b$, the *spline interpolant* to a given function $u \in C^{r-1}[a,b]$ is that spline function which satisfies

---

[16] For the theory of univariate Hermite interpolation and approximation, see Davis [63]; the basic formulas are due to Hermite.

[17] For applications to surface fitting, see the article by G. Birkhoff and C. de Boor in H. L. Garabedian [65, pp. 164–190]; also R. E. Barnhill in Rice [77, pp. 69–120].

$$(5.6) \qquad \begin{aligned} U(x_j) &= u(x_j) && \text{for } j=0,\cdots,J, \quad \text{and} \\ U^{(k)}(x_j) &= u^{(k)}(x_j) && \text{for } j=0, J \text{ and } k=1,\cdots,r-1. \end{aligned}$$

Obviously, $U \in C^{2r-2}[a,b]$.

The monomials x^s ($s = 0, \cdots, 2k-1$) and the functions

$$(x-x_j)_+^{2k-1} = \begin{cases} 0 & \text{if } x < x_j, \\ (x-x_j)^{2k-1} & \text{if } x \geqslant x_j, \end{cases}$$

form an obvious *basis* of spline functions on the interval $[a,b]$ subdivided by the partition π; with them, one can first match arbitrary jumps in $U^{(2k-1)}$ at x_j, and then match arbitrary Hermite-type endpoint conditions. However, this basis of spline functions is extremely ill-conditioned.

The effective computation of spline interpolants proceeds differently. For *cubic* splines, one can take $U'_j = U'(x_j)$, $j = 1, \cdots, J-1$, as unknowns, assuming U'_0 and U'_J to be given, and solve the linear system (with $h_j = x_{j+1} - x_j$)

$$(5.7) \qquad \begin{aligned} &h_j U'_{j-1} + 2(h_{j-1} + h_j)\, U'_j + h_{j-1}\, U'_{j+1} \\ &\qquad = 3[h_j(U_j - U_{j-1})/h_{j-1} + h_{j-1}(U_{j+1} - U_j)/h_j] \end{aligned}$$

with diagonally dominant tridiagonal coefficient-matrix, and then use cubic Hermite interpolation to the given U_j and the computed U'_j in each subinterval. On a uniform mesh, (5.7) reduces to

$$(5.7') \qquad h(U'_{j-1} + 4U'_j + U'_{j+1}) = 3(U_{j+1} - U_{j-1}).$$

Cubic B-splines. Alternatively, one can take as basis functions the so-called *B-splines* of Curry and Schoenberg [66].[18] These are univariate splines of degree $2k-1$, with the property that their support is confined to $2k$ subintervals. This property of a B-spline basis greatly reduces the labor of numerical quadrature in applying Rayleigh-Ritz-Galerkin methods. For a *uniform* mesh with mesh-length h, the univariate *cubic* B-spline centered at 0 has support $(-2h, 2h)$ and is given explicitly by the formula

$$(5.8) \qquad B_h(x) = \begin{cases} (x+2h)^3/6h^4 & \text{on } [-2h,-h], \\ (4h^3 - 6hx^2 - 3x^3)/6h^4 & \text{on } [-h,0], \\ B_h(-x) & \text{on } [0,2h]. \end{cases}$$

The Rayleigh-Ritz method can be applied to approximating subspaces of either Hermite or spline functions. Spline approximation of degree $2k-1$ has the same order of accuracy, $O(h^{2k})$ in u, as Hermite approximation

[18] For a complete treatment of B-splines, see de Boor [78, Chaps. IX–XI].

of the same degree. Spline approximation requires many fewer (about $1/k$ times as many) unknowns as Hermite approximation. However, it requires much more programming effort.[19]

In concluding our brief discussion of univariate splines, we emphasize the fact that their use for solving elliptic problems constitutes only a very small fraction of their interest. Besides their importance for computer-aided design and automated die production, they have many fascinating analytical properties. For these, we refer the reader to Schoenberg [69], Karlin [76], de Boor [78], and other references in the bibliographies of all three books. The book by de Boor also contains in its final chapter a 47 page analysis of tensor products of splines; cf. §7.

6. Triangular elements. The approximation techniques of (continuous) piecewise linear (Chapter 6, §3) and quadratic (Example 1 in §4 above) interpolation to nodal values in triangulated polygons can be generalized in two ways. First, one can use polynomials of any degree r, and second, one can use *simplicial* subdivisions of any polyhedron in $\mathbb{R}^n$. Such simplicial approximations with $n>2$ are not often used in practice; even tetrahedral elements (the case $n=3$) are awkward to program. In this section, we will derive some general properties of piecewise polynomial interpolation in triangles.

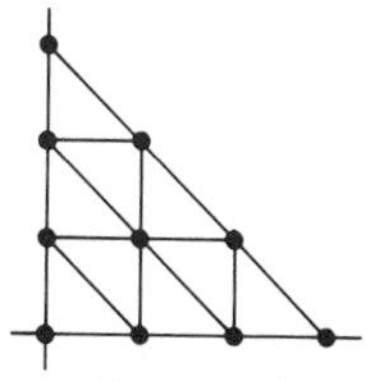

FIGURE 1

For simplicity, we will first consider the isosceles right mesh triangle T_h defined by $x \geqslant 0$, $y \geqslant 0$, $x+y \leqslant h$ (see Fig. 1). We can interpolate a *cubic* polynomial to the values $u(x_i,y_j)=u(ih/3,jh/3)$ with $i,j=0,1,2,3$ and $i+j \leqslant 3$ as follows. First match the seven values $u_{i,0}$ and $u_{0,j}$ by *univariate* cubic interpolants, to get

$$U_0(x,y)=a_{00}+a_{10}x+a_{20}x^2+a_{30}x^3+a_{01}y+a_{02}y^2+a_{03}y^3. \tag{6.1}$$

Then write the remainder

$$r(x,y)=u(x,y)-U_0(x,y)=xyq(x,y). \tag{6.2}$$

We can match the three remaining values u_{ij} by linear interpolation

$$R(x,y)=xy(a_{11}+a_{21}x+a_{12}y), \tag{6.2'}$$

since $R(x_i,0)=R(0,y_j)=0$ for $i,j=0,1,2,3$. Evidently, $U=U_0+R$ is a univariate cubic interpolant to u at each of four points on all three sides of the mesh triangle T_h; hence the interpolation error is $O(h^4)$ there. More generally, we have the following result.

---

[19] This labor can be avoided by using the B-spline package developed by Carl de Boor and available through IMSL.

THEOREM 8. *The scheme of cubic interpolation in the triangle* T_h *defined by* (6.1), (6.2), *and* (6.2′) *approximates* u *with* $O(h^4)$ *accuracy, its first derivatives (gradient) with* $O(h^3)$ *accuracy, its second derivatives with* $O(h^2)$ *accuracy, and its third derivatives with* $O(h)$ *accuracy*:

$$\partial^{r+s}e/\partial x^r\,\partial y^s = O(h^{4-r-s}), \qquad r+s \leqslant 3. \tag{6.3}$$

Sketch of proof. As in the proof of Theorem 5, expand the error $e(x,y)=U(x,y)-u(x,y)$ in Taylor series with remainder about any point $(\xi,\eta)\in \mathrm{T}_h$:

$$e(x,y) = e(\xi,\eta) + \sum_{r+s\leqslant 3} \frac{\partial^{r+s}e}{\partial x^r\,\partial y^s}(\xi,\eta)\,\Delta_{rs} + R_4(x-\xi,y-\eta), \tag{6.4}$$

where $\Delta_{rs} = (x-\xi)^r(y-\eta)^s/(r!)(s!)$. By construction, $e(x_i,y_j)=0$ at each of the 10 vertices (x_i,y_j); moreover, since $U(x,y)$ is a cubic polynomial, $e(x,y)$, $R_4(x,y)$, and $-u(x,y)$ have the *same fourth derivatives*, everywhere.

Hence, if M_4 is the largest fourth derivative of u, $|R_4| \leqslant KM_4h^4$ identically in T_h, and so the identity (6.4) gives 10 equations of the form

$$e(\xi,\eta) + \sum_{r+s\leqslant 3} \frac{\partial^{r+s}e}{\partial x^r\,\partial y^s}(\xi,\eta)\,\Delta_{rs} = -r_4(x_i-\xi,y_j-\eta), \tag{6.4′}$$

where $\Delta_{rs} = O(h^{r+s})$. Solving by Cramer's rule, we get the conclusion of the theorem.

General triangles. Any triangle can be reduced to the isosceles right triangle T_h by an affine transformation

$$(x,y) \longmapsto (b_{11}x+b_{12}y+c_1,\ b_{21}x+b_{22}y+c_2)\,; \tag{6.5}$$

moreover, such transformations carry cubic polynomials into cubic polynomials and k-th derivatives into linear combinations of k-th derivatives. Hence Theorem 8 holds in any mesh triangle, and *uniformly* if all interior angles are bounded uniformly away from 0° and 180°. Finally, since affine transformations (6.5) preserve ratios of lengths in any direction, cubic interpolation on the two sides of any edge of a triangulation gives consistent values — i.e., a *globally continuous*, piecewise cubic interpolant. Thus we have:

THEOREM 9. *Let* $u\in C^4(\overline{\Omega})$, *and consider triangulations of* $\overline{\Omega}$ *into triangles of diameter* h *or less, whose anlges are uniformily bounded away from* 0° *and* 180°. *Then piecewise cubic interpolation to values of* u *gives a globally continuous interpolant with errors as in Theorem* 8.

More generally, for any positive integer r, we can interpolate a unique polynomial of degree r in the triangle T_h discussed above, by matching nodal values $u_{ij} = u(ih/r, jh/r)$ at $r(r+1)/2$ points. This will give a globally continuous interpolant that approximates k-th partial derivatives with $O(h^{r+1-k})$ accuracy.

Simplicial elements.[20] Tetrahedral approximation is awkward to program and rarely used in practice; moreover, simplicial subdivisions of domains in $\mathbb{R}^n$ $(n > 3)$ are even less practical. Nevertheless, for theoretical completeness we will next sketch generalizations of Theorems 8 and 9 to arbitrary simplices.

Let $u(\mathbf{x}) \in C^r(\mathrm{P})$ be given in a closed polyhedron $\mathrm{P} \subset \mathbb{R}^n$, and let σ_h be any 'simplicial subdivision' of P into simplices of diameter at most h, with all vertex angles uniformly bounded away from 0° and 180°. Then we can approximate u with error $O(h^r)$, by using the the following technique of piecewise polynomial interpolation.

For $h = 1/r$, the 'lattice' points

$$(6.6)\quad h_{\mathrm{j}} = (j_0 h, \cdots, j_n h), \qquad \mathbf{j} = (j_0, \cdots, j_n), \quad j_i \geqslant 0, \quad \textstyle\sum_i j_i = 1$$

lie on faces of the standard $(n+1)$-simplex with vertices at $\mathbf{0}$ and the standard unit vectors $\mathbf{e}_i = (\delta_{i1}, \delta_{i2}, \cdots, \delta_{in})$. This is affine-equivalent to the standard *barycentric* simplex with *vertices* at $\mathbf{e}_0, \mathbf{e}_1, \cdots, \mathbf{e}_n$, consisting of all points (vectors) $\mathbf{x} = (x_0, x_1, \cdots, x_n)$ with

$$(6.7)\qquad \mathbf{x} = w_0\mathbf{e}_0 + w_1\mathbf{e}_1 + \cdots + w_n\mathbf{e}_n, \qquad w_i \geqslant 0, \quad \textstyle\sum_i w_i = 1,$$

and to every other 'solid' $(n+1)$-simplex.

THEOREM 10. *One can interpolate a unique polynomial $p(x_0, x_1, \cdots, x_n)$ of degree s to nodal values in any standard n-simplex uniformly subdivided into s^n congruent subsimplices with $h = 1/s$.*

The proof by induction on $n + s$ is straightforward. By induction on n, there exists a unique polynomial $q_0(x_0, x_1, \cdots, x_n)$ of degree s interpolating to the nodal values on the 'face' ($(n-1)$-simplex) $x_0 = 0$. Consider the remainder $r_0(\mathbf{x}) = u(\mathbf{x}) - q_0(\mathbf{x})$ at the smaller simplex of lattice points h_{j} with $j_0 > 0$, and write

$$(6.8)\qquad r_0(\mathbf{x}) = x_0\, p(\mathbf{x}) + r_1(\mathbf{x}).$$

Since $r_0(\mathbf{x}) = 0$ at the nodes on $x_0 = j_0 h = 0$, substitution of any polynomial $p(\mathbf{x})$ of degree $s-1$ in (6.8) gives $u(\mathbf{x}) - q_1(\mathbf{x}) = r_1(\mathbf{x})$, where $q_1(\mathbf{x}) = q_0(\mathbf{x}) + x_0\, p(\mathbf{x})$ is also a unique polynomial $p(x)$ of degree s. By induction on s, there *is* a polynomial $p(\mathbf{x})$ of degree $s-1$ such that $j_0\, hp(h_{\mathrm{j}}) = r_0(h_{\mathrm{j}})$ at the lattice points with $x_0 \neq 0$. Because $x_0\, p(\mathbf{x})$ is of degree s and $x_0\, p(\mathbf{x}) = 0$ when $x_0 = 0$, $q_1(\mathbf{x})$ interpolates to $u(\mathbf{x})$ at all the lattice points in the given simplex.

[20] See M. Zlamal, Numer. Math. 12 (1968), 394–409; R. A. Nicolaides, SIAM J. Numer. Anal. 9 (1972), 435–445 and 10 (1973), 182–189; A. Zenisek, J. Approx. Theory 7 (1973), 344–351; L. Mansfield, J. Math. Anal. Appl. 56 (1976), 137–164. P. J. Ciarlet and P.-A. Raviart, Archiv Rat. Mech. (1972), 177–199, and Ciarlet [78, Chap. 3] refer to the use of such elements as "Lagrange interpolation".

7. Tensor product approximations. As in Chapter 3, §7, and Chapter 4, §6, if Φ and Ψ are two vector spaces of univariate functions on intervals $[a,b]$ and $[c,d]$ with bases $\phi_r(x)$ and $\psi_s(y)$, respectively, then the products $\phi_r(x)\psi_s(y)$ form a basis for the 'tensor product' space of their linear combinations on the rectangle $[a,b]\times[c,d]$. And as in Chapter 4, §6, one can solve efficiently for the coefficients c_{rs} of the expansion $\sum c_{rs}\phi_r(x_i)\psi_s(y_j)$ on a product set of nodal values $U_{i,j}=u(x_i,y_j)$, by utilizing univariate interpolation formulas. The scheme of bicubic Hermite interpolation (§5) is the tensor product of two univariate cubic Hermite interpolation schemes, in this sense; we now describe an analogous scheme of bicubic *spline interpolation*.

Bicubic splines. Let the rectangle $\mathrm{R}=[0,a]\times[0,b]$ be subdivided by the partition

$$\pi: 0=x_0<x_1<\cdots<x_I=a, \qquad 0=y_0<y_1<\cdots<y_J=b.$$

By definition a *bicubic spline function* for the preceding rectangular subdivision is a function $s(x,y)\in C^{2,2}(\mathrm{R})$ that is *bicubic* in each mesh rectangle $[x_{i-1},x_i]\times[y_{j-1},y_j]$.

Besides providing convenient multivariate interpolants on cartesian products of intervals, the tensor product construction leads to algebraic existence and uniqueness theorems, of which we shall give only the original (and still the most important!) sample.[21]

THEOREM 11 (de Boor). *For each rectangular subdivision of a rectangle* R, *there is one and only one bicubic spline function which assumes given values at all mesh-points, has given normal derivatives* (u_x *or* u_y) *at all edge mesh-points, and given cross-derivatives* u_{xy} *at the four corner mesh-points.*

Equally essential, bicubic Hermite and spline interpolation give good *approximations*: $O(h^4)$ for values and $O(h^{4-j})$ for partial derivatives of order j if $u\in C^4(\mathrm{R})$, uniformly and regardless of the number of mesh-points.[22] In any one rectangle, the error satisfies[23]

$$\|e^{(r,s)}\|\leqslant M(\|u^{(4,0)}\|_\infty+\|u^{(0,4)}\|_\infty)h^{4-j}, \qquad j=\max\{r,s\},$$

provided $r,s\leqslant 4$; the exponents are best possible.

Thus spline approximating subspaces achieve the same order of accuracy as Hermite approximating subspaces of the same degree (e.g., cubic), with bases about one quarter as large. However, when one uses bases of 'patch

[21] C. de Boor, J. Math. and Phys. 41 (1962), 212–218; G. Birkhoff and C. de Boor in Garabedian [65, p. 173].

[22] For these and other results, see G. Birkhoff in Schoenberg [69, pp. 185–221]; also Birkhoff-Schultz-Varga [68]; D. D. Stancu, SIAM J. Numer. Anal. 1 (1964), 127–163; and W. Simonsen, Skand. Aktuarietidskr. 42 (1959), 73–89.

[23] See Bramble-Hilbert [71]. Their proof is nonconstructive; see Dupont-Scott [80].

functions' with minimum support (to get sparse matrices), one finds that the proportion of nonzero entries in the resulting matrices is about three times as great for splines as for Hermite interpolants. As a result, the net computational advantage of using spline as contrasted with Hermite approximating subspaces for solving elliptic problems is small, even in rectangles.

Multivariate Hermite approximation. By taking the tensor product of univariate *Hermite* interpolation schemes of degree $2r-1$, local multivariate Hermite interpolation schemes having $O(h^{2r})$ accuracy are easily constructed, for any r, in any number of dimensions. For example, *triquintic* Hermite interpolation in the box $\mathrm{B}: [0,a]\times[0,b]\times[0,c]$ matches all nodal values $u(x_i,y_j,z_k)$ and derivatives

$$\partial^{\alpha+\beta+\gamma}u(x_i,y_j,z_k)/\partial x^{\alpha}\,\partial y^{\beta}\,\partial z^{\gamma}, \qquad \alpha,\beta,\gamma=0,1,2,$$

for $x_i=0$ or a, $y_j=0$ or b, $z_k=0$ or c.

This gives an $8\times 27=216$-parameter family of biquintic Hermite 'finite elements' in a box. However, the *compatibility* condition that $u\in C^{2,2}(\mathrm{B})$ brings this down to 27 per mesh-point.

Approximation errors. Let S_1 and S_2 be two approximating subspaces of univariate functions, and let $\mathbf{P}_i: u_i \mapsto \mathbf{P}_i[u_i]=U_i$ be associated linear projectors. Then $u_i=U_i+e_i=\mathbf{P}_i[u_i]+(\mathbf{I}-\mathbf{P}_i)[u_i]$, and the error $e=u-U$ in the tensor product approximant $U=\mathbf{P}_1\otimes\mathbf{P}_2[u]\in S_1\otimes S_2$ is

$$e=(\mathbf{I}-\mathbf{P}_1)\otimes\mathbf{P}_2[u]+\mathbf{P}_1\otimes(\mathbf{I}-\mathbf{P}_2)[u]+(\mathbf{I}-\mathbf{P}_1)\otimes(\mathbf{I}-\mathbf{P}_2)[u].$$

Example 4. Let $\mathbf{P}_1$ and $\mathbf{P}_2$ denote quadratic interpolation on uniform meshes with mesh lengths h and k. Then $\mathbf{P}_1\otimes\mathbf{P}_2$ gives the biquadratic interpolant (Example 2, §4) to u at nine points: the corners, midpoints of sides, and center of the rectangle $[-h,h]\times[-k,k]$. The term $(\mathbf{I}-\mathbf{P}_1)\otimes\mathbf{P}_2[u]$ in the error is the quadratic interpolant in y to $x(h^2-x^2)u_{xxx}(\xi(x,y),y)/6$, while the term $(\mathbf{I}-\mathbf{P}_1)\otimes(\mathbf{I}-\mathbf{P}_2)[u]$ is

$$\frac{1}{36}x(h^2-x^2)y(k^2-y^2)u_{xxxyyy}(\xi'(x,y),\eta'(x,y)),$$

and so on. This gives the bound (see also Steffensen [50, pp. 204–206])

$$|e(x,y)|\leqslant\frac{1}{9\sqrt{3}}h^3[\|u_{xxx}\|_\infty+\|u_{yyy}\|_\infty+\frac{1}{9\sqrt{3}}h^3\|u_{xxxyyy}\|_\infty]. \tag{7.1}$$

Although (7.1) is less sharp than the inequality (4.3), its 'tensor product' method of proof is more general.

8. Rectangular polygons. Because Hermite interpolation is *local*, multivariate Hermite interpolation and approximation to sufficiently smooth functions in rectangular polygons, and more generally in any *rectangular polytope* (union of boxes whose edges are parallel to the coordinate axes), can be used without additional complications or loss of accuracy.

Thus, let (R,π) be any *subdivided rectangular polygon*, that is, any polygon with sides parallel to the coordinate axes, cut up into rectangles by a 'rectangular mesh' of straight 'mesh lines' (also parallel to the axes) joining (opposite) edges of R. We assume that every edge of R lies on a mesh line. Then, given u, u_x, u_y, and u_{xy} at all mesh-points (intersections of mesh lines), there is a unique piecewise bicubic Hermite interpolant $U \in C^{1,1}$ that matches these values and derivatives. Moreover, if $u \in C^4(\mathrm{R})$, then by Theorem 7 this interpolant approximates u with $O(h^4)$ accuracy.

Quasi-interpolants. The *global* dependence of spline *interpolants* on the nodal values has an unpredicable effect on the errors when a very nonuniform mesh is used. This can be avoided by using an alternative *local* method of spline *approximation* by 'quasi-interpolants' invented by de Boor and Fix,[24] which uses B-splines having local support. For *any* partition π, the resulting errors are $O(|\pi|^4)$. The one-dimensional case of this method, motivated by the 'thin beam' interpretation and matching 'moments' of $u^{(2m)}$, had been proposed earlier by one of us.[25]

Polygons. More generally, many nonrectangular polygons can be represented as unions of rectangular polygons and boundary triangles, while almost all domains can be approximated by polygons. In §9, we will describe a technique of 'blending' that makes it possible to approximate smooth functions in such polygons accurately by piecewise polynomial approximants.

Bivariate splines. Bicubic spline interpolation (and approximation) in rectangular polygons is much more complicated, and not to be recommended to the inexperienced. We shall indicate here only a few striking results; for details and generalizations, see the references cited below, and Birkhoff in Schoenberg [69, pp. 187–222]. For deeper theoretical properties and thoughtful observations, see Mansfield [74].[26]

Define a bicubic spline function $f \in Sp(\mathrm{R},\pi,2)$ as a function $f(x,y)$ which is: (i) a bicubic polynomial in each mesh rectangle, and (ii) a function of class $C^2(\mathrm{R})$. It follows from (i) and (ii) that $f \in C^{2,2}(\mathrm{R})$.[27] Also, for a rectangular polygon with p reentrant corners and hence $4+2p$ sides, the dimension of the bicubic spline subspace $Sp(\mathrm{R},\pi,2)$ is $M+(E-p)+4$, where M is the total number of mesh-points and E the number of mesh-points on the edges of R. Moreover, Carlson and Hall[27] have constructed an 'analytically well-set' interpolation scheme for

---

[24] de Boor-Fix [73].

[25] G. Birkhoff, J. Math. Mech. 16 (1967), 987–990; see also Schoenberg [69, pp. 208–209].

[26] The authoritative monograph by de Boor [78, pp. 360–362] explicitly disclaims covering this challenging subject.

[27] Ralph Carlson and C. A. Hall, J. Approx. Theory 4 (1971), 37–53.

interpolating to appropriate values at interior mesh-points and derivatives at boundary mesh-points.

Variational conditions. Cubic splines and piecewise cubic Hermite interpolants can be characterized variationally as 'thin beams' which minimize $\int [u'']^2 \, dx$ subject to interpolation and 'clamped' constraints, respectively (see Example 1, Chapter 6, §1). Similarly, multivariate tensor products of these piecewise cubic polynomials are characterized variationally in terms of multiple integrals. Namely, they minimize

$$\mathbf{J}[u] = \iint [\partial^4 u / \partial x^2 \partial y^2]^2 \, dx \, dy \tag{8.1}$$

in the class of all smooth functions interpolating to the same conditions (see Ahlberg-Nilson-Walsh [67, p. 242 and §8.6]). Hence, relative to the quadratic semi-norm $||u|| = \mathbf{J}[u]^{1/2}$, bicubic spline interpolation gives a 'best approximation' (see §3).

Unfortunately, they do not *uniquely* minimize $\mathbf{J}$: if one adds to u any function $v(x)$ which vanishes together with $v'(x)$ and $v''(x)$ at all mesh-points, then $\mathbf{J}[u+v] = \mathbf{J}[u]$ also minimizes $\mathbf{J}$ and solves the same interpolation problem, without in general being a bicubic spline.[28] However, we do have the following result.

THEOREM 12. *In any rectangle, bicubic spline functions minimize the inner product* $\mathbf{B}\langle f,f \rangle$, *where*

$$\mathbf{B}\langle f,g \rangle = \iint_{\Omega} f_{xxyy} \, g_{xxyy} \, dx \, dy + \int_{\Gamma} f_{ss} \, g_{ss} \, ds. \tag{8.2}$$

These results based on tensor products are limited to rectangles (and boxes). To find an analogous variational characterization of spline functions on rectangular polytopes subdivided into box-shaped cells is an outstanding unsolved problem. The best results known are those in Mansfield [74]; see also her paper in Numer. Math. 20 (1972), 99–114.

Direct variational methods do not establish the existence of bicubic spline interpolants and approximants to functions on general plane domains, because the greatest lower bound of $\mathbf{J}$ in (8.1) need not be attained. To avoid this theoretical difficulty, it is simplest to use the following result to extend functions smoothly to a 'box' in which such approximants can be proved to exist.

THEOREM 13 (Whitney's extension theorem).[29] *Any 'smooth' function can be extended to all space without losing differentiability.*

We do not know of any good algorithm for actually *computing* such an extension.

---

[28] See Mansfield [74]; our next theorem is her Theorem 1. See also her earlier papers in SIAM J. Numer. Anal. 8 (1971), 115–126, and Numer. Math. 20 (1972), 99–114.

[29] H. Whitney, Trans. AMS 36 (1934), 63–89; C. Coatmelec, Ann. Ec. Norm. Sup. 83 (1966), and Schoenberg [69, pp. 29–49].

As Schultz[30] and Varga [71] have pointed out, this result makes it possible to derive *a priori* estimates of the order of accuracy of 'best' spline approximations to general 'smooth' functions in general regions, from the corresponding 'tensor product' formulas for box-shaped regions.

In particular, Schultz has shown that, for a significant class of elliptic problems, including the Poisson equation with natural (Neumann) boundary conditions, bicubic spline functions on 'quasi-uniform' meshes with uniformly bounded mesh-ratios give accuracy of the maximum possible order, $O(h^4)$.

9. Blending. A very flexible method of constructing bivariate and multivariate 'finite elements' from univariate functions is provided by 'blending' techniques. Although the idea is certainly very old, its systematic application to approximation schemes seems to have been first studied explicitly in depth by W. J. Gordon and one of us.[31]

Blending techniques can be regarded as schemes for *interpolating* to values on the boundary and on mesh lines, so as to give globally continuous functions. They provide tentative answers to questions like the following: "Given the values $g(\mathbf{y})$ of a smooth function $u(\mathbf{x})$ on the boundary Γ of a domain Ω, what is the best *a priori* approximation to $u(\mathbf{x})$ in Ω, and what is its probable error?"

Example 5. In the unit square $S = [0,1]^2$, suppose we know that

$$u(x,k) = \gamma_k(x), \qquad u(k,y) = \beta_k(y), \qquad k = 0, 1. \tag{9.1}$$

In the absence of other information, the best guess as to the interpolant is

$$\begin{aligned} U(x,y) = {} & (1-x)\beta_0(y) + x\beta_1(y) + (1-y)\gamma_0(x) + y\gamma_1(x) \\ & - \alpha_{00}(1-x)(1-y) - \alpha_{01}(1-x)y - \alpha_{10}x(1-y) - \alpha_{11}xy, \end{aligned} \tag{9.2}$$

where $\alpha_{jk} = u(j,k) = \beta_j(k) = \gamma_k(j)$.

As was first proved by Mangeron, this is the unique interpolant to the given boundary values that satisfies the hyperbolic 'Mangeron' DE $U_{xxyy} = 0$; thus we have a curious example of a well-set *hyperbolic* DE problem, with boundary conditions of the kind usually associated with an *elliptic* DE. More relevant, the interpolant obtained by this blending scheme approximates any function $u(x,y) \in C^4$ in a square of side h with $O(h^4)$ accuracy, as we will show in Chapter 8, §2.

By an affine transformation, the preceding scheme can be extended to any rectangle, as follows. First form the bilinear interpolant

---

[30] M. H. Schultz, SIAM J. Numer. Anal. 6 (1969), 161–183 and 184–209. Schultz analyzes both L_∞ and L_2 approximations.

[31] G. Birkhoff and W. J. Gordon, J. Approx. Theory 1 (1968), 199–208. Many of the first results proved there had been derived earlier by D. Mangeron, D. D. Stancu, S. A. Coons, and others.

$b(x,y) = b_0 + b_1x + b_2y + b_3xy$ to the four corner values. After this has been subtracted from the given $g(\mathbf{y}) = u(\mathbf{y})$ on Γ, the remainder $h(\mathbf{y})$ will clearly vanish at all four corners. This reduces the α_{jk} in (9.2) to zero, whence $u(x,y) = b(x,y) + B(x,y)$, where $B(x,y)$ is the sum of the *linear* interpolants to edge values of the remainder along lines parallel to the coordinate edges. Still more generally, the 'blending' scheme of Example 5 can be applied to any convex quadrilateral by the parametrization described at the end of §4 above.

Example 6. The technique outlined in Example 4 of Chapter 2, §1, gives a series representation of the solution of $\nabla^2 u = 0$ in the unit square S with $u = g(\mathbf{y})$ on ∂S *except* at the four corners of S. The solution at *all* points of S $\cup$ ∂S can be obtained by writing $u = v + b$, where b is the (harmonic) bilinear interpolant to u at corners of S. One then applies the method of Example 4 in Chapter 2 to solve $\nabla^2 v = 0$ in S, $v = g(\mathbf{y}) - b(\mathbf{y})$ on ∂S.

Not only does one obtain the solution at all points in S $\cup$ ∂S, but one also obtains a series which converges *faster* on the boundary.

Mesh lines. Most important, the bilinear blending scheme of Example 5 can be combined with *univariate* interpolation and approximation along mesh lines, to provide accurate *globally continuous* interpolants to arbitrary boundary data.

For example, as applied to *cubic* Hermite or spline interpolants on mesh lines, or other piecewise cubic approximants having $O(h^4)$ *univariate* accuracy, linear blending gives 12-parameter Adini element approximants (in fact, interpolants to the corner u, u_x, u_y if univariate Hermite interpolants are used), which are exact for all cubic polynomials (see Zienkiewicz, [77, p. 234]). This suggests that the *bivariate* error of Adini elements is also $O(h^4)$ for functions $u \in C^4$, and this is the case.

More generally, interpolation by *multilinear blending* to the values of any $u \in C^{2k}$ on the boundary $\Gamma = \partial\Omega$ of any 'box' $\prod [0,h_i]$ in $\mathbb{R}^n$ is accurate to order $O(h^{2k})$,[32] as was shown by W. J. Gordon in two important papers.[33] His results can be extended (by suitable parametrization) to arbitrary convex hexahedra with plane quadrilateral faces, and give globally continuous, highly accurate approximations. Gordon also showed how to interpolate by blending to (compatible) normal derivatives as well as to values on the boundary. On the unit square S of Example 5, the error is then a multiple of $[xy(1-x)(1-y)]^2$ and $O(h^8)$ for S $= [0,h]^2$.

---

[32] This follows from the fact that for polynomials, it is an 'ideal' interpolation scheme with principal ideal $\prod x_i(h_i - x_i)$; see G. Birkhoff in Coffman-Fix [79].

[33] W. J. Gordon, J. Math. Mech. 18 (1969), 931–952, and SIAM J. Numer. Anal. 8 (1971), 158–177. The compatibility conditions for normal derivatives amount to requiring that $\partial^2 u/\partial x\, \partial y = \partial^2 u/\partial y\, \partial x$ at corners.

Most interesting for design purposes, Gordon introduced in these papers the concept of 'spline-blending' through networks of curves.

As the diagrams of Fig. 2 illustrate, bilinear blending is compatible with *local* mesh refinement; hence it provides a simple and very practical way of getting reasonable accuracy near singularities.[34] In this sense, it can be regarded as competitive with the multigrid method discussed in Chapter 5, §10, as well as with the 'singular elements' of §10.

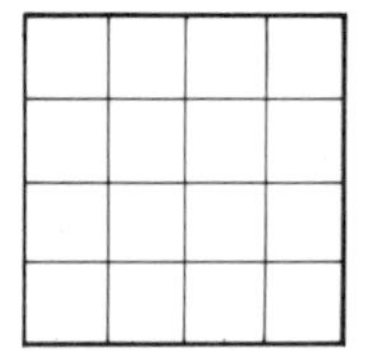
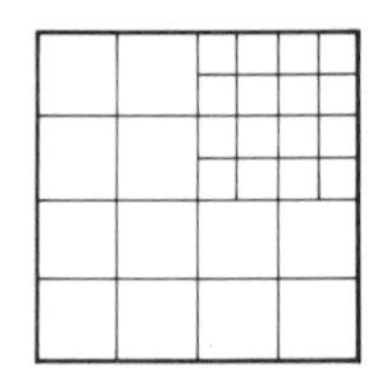
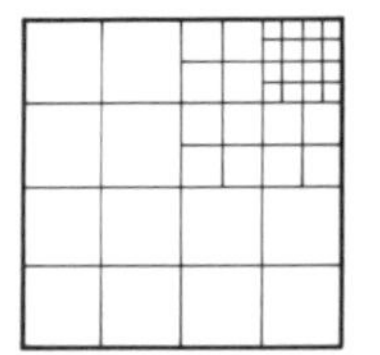

Figure 2

Interpolation in triangles.[35] Next, we discuss a somewhat analogous 'trilinear blending' scheme for interpolating to given boundary values on the edges of a triangle. This can be defined as follows, in the 'standard triangle' with vertices (0,0), (1,0), and (0,1); see Fig. 3. Set

$$u(x,y) = L(x,y) + g(x,y) = L(x,y) + G(x,y) + e(x,y),$$

where L is the linear interpolant to corner values,

$$L(x,y) = u(0,0)z + u(1,0)x + u(0,1)y,$$

where $z = 1 - x - y$, and

$$2\,G(x,y) = \frac{z}{1-y}g(Q_1) + \frac{x}{1-y}g(Q_2) + \frac{z}{1-x}g(R_1) + \frac{y}{1-x}g(R_2) + \frac{x}{1-z}g(S_1) + \frac{y}{1-z}g(S_2).$$

Its error is $e(x,y) = O(h^3)$; for any polynomial u, the error is a multiple of xyz. Conversely, the interpolant to any multiple of xyz is zero; hence the interpolation scheme is 'ideal' in the sense defined by G. Birkhoff in Coffman-Fix [79].

This scheme was applied to construct higher-order conforming finite elements in triangulated polygons by G. Birkhoff in Aziz [72, pp. 363–385], and sharp error bounds were derived by R. E. Barnhill and L. Mansfield in J. Approx. Theory 11 (1974), 306–318. It was extended to a

[34] G. Birkhoff, J. C. Cavendish, and W. J. Gordon, Proc. Nat. Acad. Sci. 71 (1974), 3423–3425; and J. C. Cavendish, J. Comp. Phys. 19 (1975), 211–238.

[35] R. E. Barnhill, G. Birkhoff, and W. J. Gordon, J. Approx. Theory 8 (1973), 114–128; and G. Birkhoff, J. Math. Anal. Appl. 42 (1973), 474–484.

tricubic blending scheme for constructing compatible triangular elements by G. Birkhoff and L. Mansfield in J. Math. Anal. Appl. 47 (1974), 531–553.

Conforming triangular elements. Although Hermite functions provide satisfactory 'conforming' elements for elasticity problems, their use is limited to rectangular polygons. Moreover, to construct 'conforming' *triangular* finite elements turns out to be surprisingly difficult.

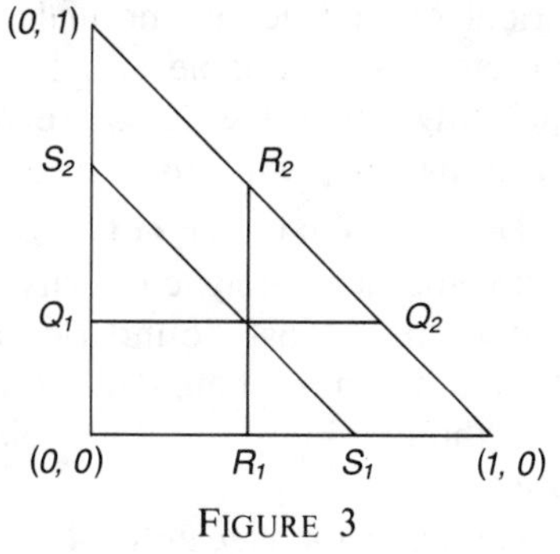

FIGURE 3

An early solution to the problem of constructing conforming triangular elements was by the Clough-Tocher[36] method of subdividing each mesh triangle into three subtriangles, using quadratic interpolation on each outer edge, and then applying 'C^1-matching' to their interfaces. A mathematically more satisfactory solution consists in using piecewise *quintic* polynomial elements invented by Argyris and Bell, whose normal derivative is constrained to be a *cubic* polynomial.[37] This gives an 18-parameter family of 'elements' in each triangle (because $21-3=18$), which become *compatible* if one fits them together by interpolating to the same values of u, u_x, u_y, u_{xx}, u_{xy}, u_{yy} at the corners of each triangle, and of $\partial u/\partial n$ at the midpoints of each side (see C. A. Felippa and R. W. Clough in Birkhoff-Varga [70, pp. 210–253]).

Alternatively, as was first shown by Irons and Zlamal, one can construct a 15-parameter family of 'quartic' *rational* functions: interpolate one of these to specified edge and vertex values in each triangle of a triangulation, and piece these together compatibly so as to get a family of globally C^1-approximants. Another 15-parameter family of conforming 'rational' triangular elements, and a 12-parameter family, have also been constructed.[38]

Actually, one can often get accurate solutions to plate bending problems (5.1)–(5.2) with *non*-conforming elements.[39] 'Tricubic' polynomials are

[36] See L. Mansfield, Math. Comp. 32 (1978), 135–142.

[37] See Zienkiewicz [77, §10.15].

[38] See G. Birkhoff and L. Mansfield, J. Math. Anal. Appl. 47 (1974), 531–553; here references to earlier work may be found. For 'compatible' tetrahedral elements, see L. Mansfield, J. Math. Anal. Appl. 56 (1976), 137–164.

[39] Zienkiewicz [77, §10.7] writes: ". . . in most cases, the accuracy obtained by the non-conforming triangle gives . . . results superior to those attainable with equivalent 'conforming' elements". See also Strang's amusing comments on 'variational crimes' in Aziz [72] and in Strang-Fix [73].

among the simplest such non-conforming triangular elements. By definition, they are (at most) *quartic* polynomial functions whose variation is (at most) cubic along lines that are parallel to any edge. In the 'standard' triangle with vertices at (0,0), (1,0), and (0,1), the functions

$$1,\ x,\ y,\ x^2,\ xy,\ y^2,\ x^3,\ x^2y,\ xy^2,\ y^3,\ x^3y-x^2y^2,\ x^2y^2-xy^3,$$

provide a convenient basis of tricubic polynomials. From these, a basis of 'tricubic' quartic polynomials for any other triangle can easily be constructed by a suitable affine transformation parametrization. Such 'tricubic' polynomial elements are *nearly* conforming and very simple; we surmise that they will prove accurate and efficient for many plate problems.

10. Boundary elements. A very troublesome aspect of solving elliptic problems accurately concerns the high-order approximation of boundary conditions. Most boundaries are not polygonal, and even on polygonal boundaries most boundary conditions are not of polynomial form, and so unorthodox kinds of finite elements or other special methods must be used.

Isoparametric elements. The standard finite element recipe for handling curvilinear boundaries is provided by 'isoparametric' elements (see Zienkiewicz [77, Chap. 8]). These are, typically, triangular or quadrilateral elements having one or more curved sides, and given by simple parametrizations as diffeomorphic images of ordinary triangles and rectangles.[40] Figure 4 indicates how to decompose quite general plane domains into interior rectangles and 'isoparametric' boundary elements.

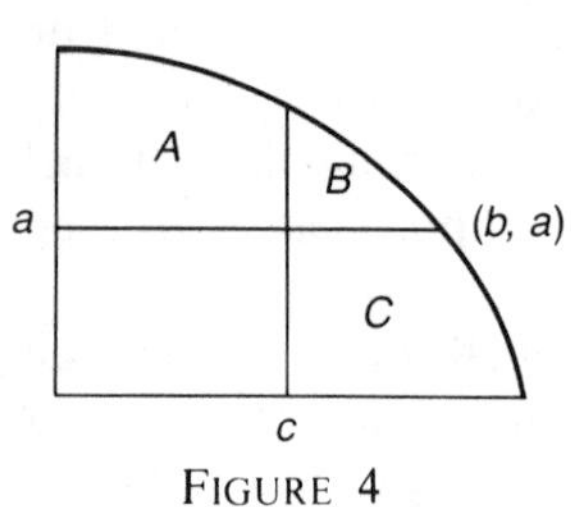

FIGURE 4

Thus in the quadrilateral A, one can take as parameters x and $\eta=(y-a)/\phi(x)$, where the curved side is $y=a+\phi(x)$. Likewise, in the quadrilateral C, one can take y and $\eta=(x-c)/\psi(y)$, where the curved side is $x=c+\psi(y)$. Finally, in the curvilinear triangle B, one can use x and $\eta=(y-a)(x-b)/\sigma(x)(c-b)$, where the curved side is $y=a+\sigma(x)$.

Sometimes one can map the entire domain Ω diffeomorphically onto a rectangle, and make a rectangular subdivision of the parameter plane. In such cases the 'transfinite elements' of Gordon and Hall[41] may be very convenient. However, this is only possible (with a diffeomorphic inverse), when Ω has precisely four corners.

[40] A *diffeomorphism* is a continuously differentiable transformation with nonvanishing Jacobian, hence equally smooth inverse.

[41] W. J. Gordon and C. A. Hall, Numer. Math. 21 (1973), 109–129.

When boundary curves are approximated by functions which are not piecewise polynomials, another problem can arise, which is less serious. They too can be accurately treated by blending techniques, as was first shown by MacLeod and Mitchell.[42] These blending techniques give rise to globally continuous functions, and also permit one to match exactly Dirichlet-type boundary conditions. However, Rayleigh-Ritz-Galerkin methods then require one to evaluate double integrals (presumably numerically). This gives rise to another source of error.

Polygonal approximation. To avoid these complications, it is often most convenient to approximate the domain by a triangulated *polygon*. For domain constants which depend monotonically on Ω, the resulting error is then only $O(h^2)$. But in general, the effect of variations in the boundary on solutions of the Laplace and other elliptic DE's is very hard to predict, as is illustrated by the large classical literature concerned with this subject (and with isoperimetric problems),[43] and also by the Babuska paradox of Chapter 6, §8.

Penalty methods. Whether it is the boundary shape or the boundary values that are to be approximated, failure to satisfy the boundary data can always be partially compensated for by 'penalty' methods. For variational problems, these add to the variational quantity to be minimized a large multiple of the error made in matching boundary conditions, as a 'penalty'.

Bramble has recently written a review article which summarizes the ingenious contributions that various authors, including especially himself and Babuska, have made to the problem of analyzing the orders of accuracy (for given operation and storage counts) that are achieved by such 'penalty' methods.[44]

Singular elements.[45] The assertions about the orders of accuracy of the approximation schemes discussed so far in this chapter (and in Chapter 3) have referred to 'smooth' functions. This was appropriate because solutions of elliptic DE's with smooth coefficients *are* smooth in the *interior* of Ω.

Nevertheless, one can still achieve higher orders of accuracy near corners in *two* dimensions with finite element approximations, by including so-called 'singular elements' in the basis of approximating functions. These singular elements are designed to match the leading terms in the

[42] R. MacLeod and A. R. Mitchell, J. Inst. Math. Appl. 10 (1972), 382–393.

[43] See M. Schiffer, Am. J. Math. 68 (1946), 417-450.

[44] J. H. Bramble and A. H. Schatz, Comm. Pure Appl. Math. 23 (1970), 653–675; J. H. Bramble, Advances in Math. 16 (1975), 187–196; see also I. Babuska, Math. Comp. 27 (1973), 221–228; J. E. Dendy, Jr., SIAM J. Numer. Anal. 11 (1974), 604–636; and J. Pasciak, SIAM J. Numer. Anal. 16 (1979), 1046–1059.

[45] See G. J. Fix, J. Math. Mech. 18 (1969), 645–658; and Strang-Fix [73, Chap. 8].

local power series expansions of solutions, as determined by Lehman's theorem (Chapter 2, §8). These are determined by the vertex angle at the corner, and the coefficients of the (elliptic) DE. Specifically, one can reduce linear source problems defined by

$$\text{(10.1)} \qquad \mathbf{L}[u] = -(pu_x)_x - (pu_y)_y + qu = f(x,y)$$

to the form assumed in Lehman's theorem, by dividing through by $p(x,y)$, provided that p and q are analytic functions.

Eigenfunction approximation. Accurate approximations to eigenfunctions of second-order elliptic DE's can also be obtained by including in the basis 'singular' elements which reproduce the behavior near 'singular' points at corners of the boundary Γ. We consider the eigenfunction problem defined by the equations

$$\text{(10.2)} \qquad \mathbf{L}[u] = \lambda u \quad \text{in } \Omega, \qquad u \equiv 0 \quad \text{on } \Gamma,$$

near an analytic corner (i.e., a corner where two analytic arcs meet), with interior angle π/α. By a translation and a rotation we can reduce to the case that the corner is at the origin and the arcs are tangent to $\theta = 0$, $\theta = \pi/\alpha$, respectively. In real form, Lehman's theorem states that u can be expanded near such a corner in a series of the form

$$\text{(10.3)} \qquad u = a_1 r^{\alpha} \sin \alpha\theta + a_2 r^{2\alpha} \sin 2\alpha\theta + \cdots,$$

plus perhaps similar terms having an additional $\log r$ factor.

Extensive numerical experiments on eigenfunctions of the Helmholtz equation, reported elsewhere by Birkhoff and Fix[46] suggest that these last logarithmic factors are superfluous in approximating eigenfunctions. We note further that, in the important case of a convex 90° corner, $\alpha = 2$, and the series (10.3) is part of an ordinary Taylor series; hence one need not use singular elements at all! However, in the case of reentrant 90° corners, such as arise with L-shaped and U-shaped regions, $\alpha = 2/3$ and singular elements with $\phi = r^{2/3} \sin(2\theta/3)$ and $\phi = r^{4/3} \sin(4\theta/3)$ are needed to obtain accurate results.

Interfaces.[47] Similar techniques can be applied to corners of interfaces separating two regions consisting of different materials. Here the physically natural interface conditions for the one-group diffusion equation

---

[46] G. Birkhoff and G. J. Fix in Birkhoff-Varga [70, pp. 111–151]. The case of an L-shaped region was treated by other methods with spectacular results, in Fox-Henrici-Moler [67] (see also Chapter 8, §4), 89–102, and with singular elements by G. J. Fix, J. Math. Mech. 18 (1969), 645–658.

[47] See R. B. Kellogg in Hubbard [71, pp. 351–400]; G. Birkhoff, J. Approx. Theory 6 (1972), 215–230; M. V. K. Chari and P. P. Silvester, *Finite Elements in Electrical and Magnetic Field Problems*, Wiley, 1980.

(10.4) $$\nabla \cdot [p(x,y)\nabla u] + [\lambda\rho(x,y) - q(x,y)]u = s(x,y)$$

are continuity of u and $p\partial u/\partial n$ across interfaces, as in Chapter 2, §8.

For example, consider the case of a corner near which, in suitable polar coordinates, $p(x,y)$ is constant in prescribed angular sectors — e.g., a piecewise Helmholtz equation. The substitution $u = r^\nu g(\theta)$ then leads to a periodic Sturm-Liouville system of the form

(10.5) $$g''(\theta) + \nu^2 g(\theta) = 0 \quad \text{in } (\alpha_{i-1}, \alpha_i).$$

The *characteristic exponents* ν_i are the eigenvalues; using them, one can construct singular elements appropriate to the interface. These same singular elements can be used more generally, to increase the order of accuracy of finite element approximations to solutions of DE's whose coefficients have limiting values as $r \downarrow 0$ which are piecewise constant in angular sectors.

This technique was first used by Fix, Gulati, and Wakoff,[48] who successfully applied it to the multigroup diffusion equation. They again found it unnecessary to use terms with logarithmic factors. An alternative method of 'homogenation' for treating interfaces with corners has been developed by Babuska; for the method and its applicability, see his paper in de Boor [74, pp. 213–277].

The Model Problem. On the other hand, singular elements having a logarithmic factor are definitely needed to achieve higher-order accuracy with inhomogeneous DE's and/or inhomogeneous boundary conditions.

Consider the Model Problem of the Poisson DE $-\nabla^2 u = 1$ in the unit square $[0,1]^2$ for the boundary condition $u \equiv 0$.

The double Fourier sine series for u is given in (7.5) on page 74. In this series, $c_{k,l} \approx 2 \times 10^{-10}$ for $k = 1001$, $l = 1$; hence one needs many more than 1000 terms of the series to evaluate u to $O(10^{-10})$ accuracy.

Since $u(x,0) \equiv u(0,y) \equiv 0$, clearly $u_{xx}(x,0) \equiv u_{yy}(0,y) \equiv 0$ on the axes; hence $\nabla^2 u(0,0) = 0$. Because $\nabla^2 u = -1$ at all interior points, it follows that u cannot have continuous second derivatives at the corner (0,0).

Indeed, by Lehman's theorem (p. 44), the singularity there is like that of the harmonic function $\mathrm{Im}\{z^2 \log(z)\}$ which satisfies the same boundary conditions on the axes.

Set $w = u + [(x-1/2)^2 + (y-1/2)^2]/4$, so that w is harmonic and has fourfold symmetry. For any harmonic approximation to w, the error in the interior is bounded by the maximum error on the boundary (see Chapter 2, §7, Corollary 3). By subtracting the singular element

[48] G. J. Fix, S. Gulati, and G. I. Wakoff, J. Comp. Phys. 13 (1973), 209–228. Similar methods were also successfully used by K. F. Hansen and C. M. Kang, Nucl. Sci. Eng. 51 (1973), 456–495.

$c \operatorname{Im}\{(z-z_i)^2 \log[(z-z_i)e^{i\phi_i}]\}$ at each of the four corners together with

$$\sum_{k=0}^{N} c_{4k}\,[(x-1/2)^2+(y-1/2)^2]^{2k} \cos 4k\theta,$$

S. C. Eisenstat used least squares approximation (Chapter 8, §3) on the boundary to find coefficients that reduced the error in w to $O(10^{-12})$ with $N=6$; specifically:[49]

$$\begin{aligned}
c &= \pm 0.3183099062144\times 10^{0}, & c_0 &= 0.1469642607553\times 10^{0},\\
c_4 &= 0.2000000125856\times 10^{1}, & c_8 &= 0.1000000062928\times 10^{1},\\
c_{12} &= -0.2983663620768\times 10^{-1}, & c_{16} &= -0.9430879889679\times 10^{-3},\\
c_{20} &= 0.4398951975566\times 10^{-5}, & c_{24} &= 0.2394115460351\times 10^{-5}.
\end{aligned}$$

Club Modulef. A large number of boundary and finite element computer programs based on finite element methods has been collected by the Club Modulef, a group of scientists from Western Europe solving elliptic problems by the methods of this chapter. For further information, see D. Bégis, F. Hecht, and M. Vidrascu in Birkhoff-Schoenstadt [84, pp. 23–34].

[49] We thank Professor Eisenstat for permission to list his coefficients.

Chapter 8

Integral Equation Methods

1. Introduction. Not only are some elliptic problems best treated by analytical methods like those described in Chapter 2, but others can best be solved numerically after extensive analytical work. Such solution methods may be called '*quasi-analytic*'.

In particular, many special problems, especially those involving the Laplace, Poisson, and Helmholtz equations in two dimensions, can be most effectively treated using integral equations. The theory of integral equations also sheds a great deal of light on other methods for solving elliptic problems, including their discretizations. Indeed, the analysis of elliptic boundary value problems and eigenproblems was revolutionized around 1900 by two major developments: the creation of a general theory of integral equations, and the invention of the Lebesgue integral. The Lebesgue integral made it possible to interpret 'points' in $L^2(\Omega)$ and other 'function spaces' (e.g., Sobolev spaces) as meaningful functions, and so to apply the functional-analytic methods mentioned in Chapters 6 and 7.[1] Integral equations will be the main theme of this chapter.

Integral equations arise in many contexts, of which the most basic involve the *Green's function* $G(\mathbf{x};\boldsymbol{\xi})$ of a linear differential operator $\mathbf{L}$ in a domain Ω, for specified homogeneous linear boundary conditions on $\Gamma = \partial\Omega$. By definition, this is the 'kernel' $G(\mathbf{x};\boldsymbol{\xi})$ of a linear *integral operator*

$$\mathbf{G}[f](\mathbf{x}) = \int_\Omega G(\mathbf{x};\boldsymbol{\xi})f(\boldsymbol{\xi})\,dR(\boldsymbol{\xi}), \tag{1.1}$$

such that $\mathbf{L}[\mathbf{G}[f]] = f$ (i.e., the operator $\mathbf{G}$ is a right-inverse of $\mathbf{L}$), and that $\mathbf{G}[f]$ satisfies the boundary conditions for all smooth f. For this it is necessary that G be smooth, satisfy $\mathbf{L}_\mathbf{x}[G] = 0$, except for a singularity at $\mathbf{x} = \boldsymbol{\xi}$, and itself satisfy the boundary conditions. (As is customary, $\mathbf{L}_\mathbf{x}[G]$ signifies that $\mathbf{L}$ is applied to $G(\mathbf{x};\boldsymbol{\xi})$ as a function of $\mathbf{x}$ with $\boldsymbol{\xi}$ fixed.)

The significance of formula (1.1) is illustrated most simply in the one-dimensional case of ordinary DE's.

[1] For their systematic application to elliptic problems, see Ciarlet [78].

Example 1. Consider the following two-endpoint problem:

$$-u'' = f(x), \qquad u(-1) = u(1) = 0, \tag{1.2}$$

already discussed in Chapter 1, §4. Its Green's function is piecewise bilinear in x and ξ:

$$G(x;\xi) = \frac{1}{2}(1 - x\xi - |x-\xi|). \tag{1.3}$$

Similarly, for a clamped beam,

$$u^{iv} = f(x) \quad \text{with} \quad u(0) = u'(0) = u(1) = u'(1) = 0,$$

the Green's function is given by

$$12G(x;\xi) = |x-\xi|^3 - (x+\xi)^3 + 6x\xi(x+\xi)(1+x\xi) - 4x^2\xi^2(3+x\xi)$$

(see Courant-Hilbert [53, p. 376]). For any fixed ξ, it is a (cubic) *spline* with a single joint at $x = \xi$ (Chapter 1, §7).

Cubic polynomials are the exact solutions of the DE $u^{iv} = 0$; moreover, the error $e(x) = u(x) - h(x)$ in the cubic Hermite interpolant $h(x)$ to any $u \in C^4[0,1]$ satisfies

$$e(0) = e'(0) = 0, \qquad e(1) = e'(1) = 0, \tag{1.4}$$

and $e^{iv}(x) = u^{iv}(x)$. Hence

$$e(x) = \int_0^1 G(x;\xi)u^{iv}(\xi)\,d\xi, \tag{1.5}$$

where $G(x;\xi)$ is the Green's function for the differential operator d^4/dx^4 and the boundary conditions (1.4). One formula for this Green's function has been displayed above. Since

$$(x-\xi)_+^3 = \frac{1}{2}[|x-\xi|^3 + (x-\xi)^3],$$

where $x_+^k = x^k$ if $x \geqslant 0$, and 0 otherwise, this formula is equivalent to

$$G(x;\xi) = \frac{1}{6}[(x-\xi)_+^3 - x^2(1-\xi^2)(x+2x\xi-3\xi)]. \tag{1.6}$$

We will give some familiar examples of Green's functions of partial DE's in §2, and apply them to to get precise *error bounds* of the form

$$e(\mathbf{x}) = \int_\Omega G(\mathbf{x};\boldsymbol{\xi})\mathbf{M}[u](\boldsymbol{\xi})\,dR(\boldsymbol{\xi}), \tag{1.7}$$

where $\mathbf{M}$ is an appropriate differential operator. In exact arithmetic, (1.7) has as a corollary the asymptotic formula

$$e_h(\mathbf{x}) \sim \int_\Omega G(\mathbf{x};\boldsymbol{\xi})r_h(\boldsymbol{\xi})\,dR(\boldsymbol{\xi}), \tag{1.7'}$$

where $r_h(\mathbf{x}) = \mathbf{L}_h[u] - f$ is the truncation error.

In §3, we will explain the least squares methods due to Bergman, Picone, Fichera, and Vekua. These are projection methods related to the 'penalty methods' of Chapter 7, §10, on the one hand, and on the other to Green's functions. Next, in §4, we will discuss 'boundary element' methods and describe their relation to the integral equations of potential theory and Green's third identity (already introduced in Chapter 2, §4). In turn, these equations are the basis of numerical methods of *conformal mapping*, which will be the theme of §§5–7, beginning in §5 with a discussion of the computation of Green's functions of the Laplace operator.

In §8, we will turn our attention to applications of integral equation methods to *three*-dimensional problems; and in §9 we describe the capacitance matrix method.

In §10, we will review some of the principal methods that have been proposed for solving the Helmholtz equation. The methods described there presumably have analogues applicable to other elliptic problems that arise from (parabolic) heat conduction and (hyperbolic) linear wave equations. Then finally, in §§11–12, we will give rigorous formulations of some general theorems. In particular, we will indicate how Poincaré applied Poisson's integral formula (Chapter 2, §3) in its full generality, to obtain an existence proof for solutions of the Dirichlet problem.

2. Green's functions; Peano-Sard kernels. The concept of the Green's function G of a linear differential operator $\mathbf{L}$ (for specified linear homogeneous boundary conditions) has turned out to be extraordinarily fruitful. Intuitively, $G(\mathbf{x};\boldsymbol{\xi})$ is that function which (for fixed $\boldsymbol{\xi}$) satisfies $\mathbf{L}[G]=\delta(\mathbf{x}-\boldsymbol{\xi})$. Here δ stands for the Dirac delta-function of n variables, and so $\delta(\mathbf{x}-\boldsymbol{\xi})$ represents the 'density' of a unit point-mass at $\boldsymbol{\xi}$ in empty space. It follows (from the theory of distributions) that the solution of $\mathbf{L}[u]=f$ for the specified boundary conditions is given as in (1.1), by the integral formula

$$u(\mathbf{x})=\mathbf{G}[f](\mathbf{x})=\int G(\mathbf{x};\xi)f(\xi)\,d\Omega, \qquad d\Omega=d\xi_1\cdots d\xi_n. \tag{2.1}$$

The oldest example of a Green's function is $1/4\pi r$, the Green's function for the Poisson equation $-\nabla^2 u=f$ in $\mathbb{R}^3$ ('free space'). This example was discussed in Chapter 2, §3; there the 'boundary condition' amounted to what Kellogg calls "regularity at infinity" [K, Chap. VIII, §3], i.e., $G(\mathbf{x};\boldsymbol{\xi})\rightarrow 0$ as $|\mathbf{x}|\rightarrow\infty$.

Historically, Poisson proceeded in the reverse order; he considered the 'Newtonian potential' [K, p. 150] of a charge (or mass) distribution of density $\rho(\boldsymbol{\xi})$ as the integral

$$u(\mathbf{x})=\int\frac{\rho(\boldsymbol{\xi})}{|\mathbf{x}-\boldsymbol{\xi}|}\,d\Omega=\int\frac{de(\boldsymbol{\xi})}{4\pi|\mathbf{x}-\boldsymbol{\xi}|}, \tag{2.2}$$

where $de(\boldsymbol{\xi})=4\pi\rho(\boldsymbol{\xi})\,d\Omega$. Poisson then derived from this definition the DE $-\nabla^2 u = 4\pi\rho$. Hence in $\mathbb{R}^3$,

$$G(\mathbf{x};\boldsymbol{\xi}) = \frac{1}{4\pi\,|\mathbf{x}-\boldsymbol{\xi}|} \tag{2.2'}$$

is the Green's function of this DE.

In the plane, 'fundamental solutions' of the Laplace and biharmonic equations, $-\nabla^2 u = 0$ and $\nabla^4 u = 0$, are

$$-\frac{1}{2\pi}\log r \quad \text{and} \quad \frac{1}{8\pi}\,r^2 \log r,$$

respectively. In bounded plane regions, Green's functions for $-\nabla^2 u = f$ and $\nabla^4 u = f$ behave like these near $\mathbf{x} = \boldsymbol{\xi}$, differing from them by smooth solutions which depend on the specific region and boundary conditions.

In the plane, it is often convenient to use complex variables. For example, consider the Green's function for the Poisson equation in the unit disk $\Omega : x^2 + y^2 < 1$, and $u \equiv 0$ on $\Gamma = \partial\Omega$. Let

$$\mathbf{x} = x + iy = r\,e^{i\theta}, \qquad \boldsymbol{\xi} = \zeta + i\eta = \rho\,e^{i\phi}, \qquad \boldsymbol{\xi}' = \zeta' + i\eta' = \frac{1}{\rho}\,e^{i\phi},$$

so that $\boldsymbol{\xi}$ and $\boldsymbol{\xi}'$ are inverse to each other in the unit circle Γ, i.e., $\zeta' = \zeta/\rho^2$, $\eta' = \eta/\rho^2$. Suppose $\boldsymbol{\xi}$ is in Ω, and let r_1 and r_2 be the distances $|\mathbf{x}-\boldsymbol{\xi}|$ and $|\mathbf{x}-\boldsymbol{\xi}'|$:

$$r_1 = \sqrt{r^2+\rho^2-2r\rho\cos(\theta-\phi)}, \qquad r_2 = \frac{1}{\rho}\sqrt{r^2\rho^2+1-2r\rho\cos(\theta-\phi)}.$$

Clearly on $\mathbf{x} \in \Gamma$, r_1/r_2 is the constant ρ; hence

$$G(x,y;\zeta,\eta) = \frac{1}{2\pi}\log(\rho r_2/r_1),$$

which is zero on $\mathbf{x} \in \Gamma$, is the Green's function for the problem. (For analogous Green's function of the Poisson equation and zero boundary conditions on a sphere, see Chapter 2, §3.)

Note that whereas the Green's functions of *ordinary* DE's (two-endpoint problems) are *bounded*, those for second-order elliptic *partial* DE's are usually unbounded[2] as in (2.2'). Note also that all of the above Green's functions are *symmetric* in $\mathbf{x}$ and $\boldsymbol{\xi}$: $G(\mathbf{x};\boldsymbol{\xi}) = G(\boldsymbol{\xi};\mathbf{x})$. This symmetry is an important general property of the Green's functions of self-adjoint DE's.

Self-adjoint DE's. As in Chapter 1, §4, a second-order linear differential operator $\mathbf{L}$ is called 'self-adjoint' when it is of the Sturm-Liouville form[3]

---

[2] However, the Green's function of the biharmonic equation in free space is $-(r^2\log r)/8\pi$, which is bounded near $r = 0$ (Courant-Hilbert [53, p. 370]).

[3] See Chapter 1, §1, and Birkhoff-Rota [78], p. 45 and Chaps. 9–10. For a general definition of self-adjoint differential operators, see Courant-Hilbert [62, p. 235]; see also Stakgold [79].

(2.3) $$\mathbf{L}[u] = -[p(x)u']' + q(x)u.$$

Similarly, second-order partial differential operators of the form

(2.3′) $$\mathbf{L}[u] = -\nabla \cdot [p(\mathbf{x})\nabla u] + q(\mathbf{x})u, \qquad p > 0,$$

are called self-adjoint, as in Chapter 1, §1.

For homogeneous boundary conditions of the form $u = 0$, $\partial u/\partial n = 0$, or more generally $\partial u/\partial n + h(\mathbf{y})u = 0$, the Green's function of any self-adjoint operator (2.3′) is also *symmetric*. Because of this, boundary conditions of the above forms are also called 'self-adjoint'.

Example 2. Consider the (elliptic) partial DE

(2.4) $$\begin{aligned} \mathbf{L}[u] = &-[p(x,y,z)u_x]_x - [p(x,y,z)u_y]_y - [p(x,y,z)u_z]_z \\ &+ q(x,y,z)u = f(x,y,z), \qquad p > 0, \quad q \geqslant 0, \end{aligned}$$

in a bounded three-dimensional domain Ω, with the homogeneous boundary conditions

(2.4′) $$\alpha u + \beta \partial u/\partial n = 0, \quad \text{with} \quad \alpha \geqslant 0, \quad \beta \geqslant 0, \quad \alpha + \beta > 0.$$

We now state a theorem asserting the existence of a Green's function for this three-dimensional linear source problem; a very general proof for the Dirichlet problem will be presented at the end of §11. Similar theorems hold for n-dimensional domains $\Omega \subset \mathbb{R}^n$.

THEOREM 1. *The Green's function for the problem of Example* 2 *exists and is that positive function*[4] $G(\mathbf{x};\boldsymbol{\xi})$ *on* $\bar{\Omega} \times \bar{\Omega}$ *which, for fixed* $\boldsymbol{\xi}$, (a) *satisfies the DE* (2.4) *with* $f = 0$ *except at* $\mathbf{x} = \boldsymbol{\xi}$, (b) *satisfies the boundary conditions* (2.4′), (c) *and satisfies*

(2.5) $$G(\mathbf{x};\boldsymbol{\xi}) \sim \frac{1}{4\pi p(\boldsymbol{\xi})\,|\mathbf{x} - \boldsymbol{\xi}|},$$

as $\mathbf{x} \to \boldsymbol{\xi}$.

As was mentioned in Chapter 2, §7, Green's functions of many second-order elliptic DE's are positive for boundary conditions of the form (2.4′). Moreover, the quadratic functional $\langle f, \mathbf{G}[f] \rangle$ is positive for all $f \neq 0$ (it is an 'inner product') even more generally. For the physical significance of these properties, see the remarks of Kurt Friedrichs in Aziz [69, pp. 53–62].

Eigenfunctions. The symmetry and square-integrability of the Green's function of a linear self-adjoint elliptic operator $\mathbf{L}$, for self-adjoint boundary conditions, can be used to prove the completeness of the *eigenfunctions* of elliptic problems in compact domains. By definition (cf. §1, below (1.1)), the DE $\mathbf{L}[\phi] = \lambda\phi$ is equivalent to the Fredholm *integral equation*

(2.6) $$\mathbf{G}[\psi](\mathbf{x}) = \int_\Omega G(\mathbf{x};\boldsymbol{\xi})\psi(\boldsymbol{\xi})\,d\boldsymbol{\xi} = \mu\,\psi(\mathbf{x}).$$

[4] See Frank-von Mises [30, vol. 1, p. 579] or Courant-Hilbert [53, p. 370].

The eigenvalues of the Green's function are clearly the reciprocals $\mu_n = 1/\lambda_n$ of those of **L**, but the eigen*functions* ϕ_n and ψ_n are the same. For this reason, matrix approximations to **L** and **G** have the same condition number: $\lambda_{\max}/\lambda_{\min} = \mu_{\max}/\mu_{\min}$.

For *self-adjoint* problems, the integral equation (2.6) has a symmetric kernel, which is *bounded* for *ordinary* DE's (two-endpoint problems), and square-integrable for linear source problems in $\mathbb{R}^2$ and $\mathbb{R}^3$. Hilbert and E. Schmidt applied this fact to prove the existence of a complete, orthonormal *basis of eigenfunctions* for such problems. A very readable version of their methods and results is presented in Courant-Hilbert [53, Chaps. III–VI].

One can easily express the Poisson integral kernel of Chapter 2, (3.9′), in terms of the Green's function, as its normal derivative $\partial/\partial\nu$ with respect to $\boldsymbol{\xi}$ on the boundary. Specifically, in three dimensions:

$$u(\mathbf{x}) = \int_{\partial\Omega} \frac{\partial G}{\partial \nu}(\mathbf{x};\boldsymbol{\xi}) g(\boldsymbol{\xi})\, d\sigma, \tag{2.7}$$

where u is the harmonic function assuming the values $u = g(\mathbf{y})$. For disks, spheres, and other special domains, the Poisson kernels $\partial G/\partial\nu$ can be expressed as elementary functions in closed form. Thus, for the sphere of radius a we have [K, p. 241]

$$u(\rho,\phi,\theta) = \frac{a^2-\rho^2}{4\pi a}\int_{\partial\Omega} \frac{u(a,\phi',\theta')}{(\rho^2+a^2-2a\rho\cos\gamma)^{3/2}}\, a^2 \sin\theta'\, d\phi'\, d\theta',$$

where γ is the angle between lines from the origin to (ρ,ϕ,θ) and to (a,ϕ',θ').

Peano-Sard kernels. The Green's function in (1.5) gives an exact formula for the error in cubic Hermite interpolation; integral kernels with this property are called *Peano-Sard kernels*. Peano-Sard kernels can be used to obtain optimal bounds on the errors in cubic Hermite and spline interpolation.

Thus, for *Hermite* interpolation of degree $2m$ on an interval of length h, repeated differentiation of a generalization of (1.5) gives

$$e^{(j)}(x) = \int_0^h G_{2m}^{(j,0)}(x;\xi)\, u^{(2m)}\, d\xi, \tag{2.8}$$

for $j = 0, 1, \cdots, m-1$. From this, setting $U_{2m} = \max|u^{(2m)}(x)|$, we can derive 'best possible' error bounds of the form[5]

$$\max|e^{(j)}(x)| \leqslant K_j U_{2m} h^j [x(h-x)]^{m-j}. \tag{2.9}$$

For *cubic* Hermite interpolation, we have $K_1 = \sqrt{3}/216$, $K_2 = 1/12$, $K_3 = 1/2$. For *quintic* Hermite interpolation, similarly,

[5] G. Birkhoff and A. Priver, J. Math. Phys. 46 (1967), 440–447.

$$K_1 = \frac{\sqrt{5}}{30000}, \quad K_2 = \frac{1}{1920}, \quad K_3 = \frac{1}{120}, \quad K_4 = \frac{1}{10}, \quad K_5 = \frac{1}{2}.$$

For any partition $\pi : 0 = x_0 < x_1 < \cdots < x_n = 1$ of $[0,1]$, we now define the *cardinal functions* for cubic spline interpolation as those cubic spline functions which satisfy $C_i'(0) = C_i'(1) = 0$ and $C_i(x_j) = \delta_{i,j}$ for $j = 0, \cdots, n$. The *error* $e_\pi(x)$ in the cubic spline interpolant on π to a given function $f \in C^4[0,1]$ can be shown to satisfy[6]

$$e_\pi(x) = \int_0^1 [G(x,\xi) - \textstyle\sum_i C_i(x) G(x_i,\xi)]\, df'''(\xi). \tag{2.10}$$

By examining more closely the oscillations of the $C_i(x)$, one can then show that *if* the ratios $h_i/h_j = (x_i - x_{i-1})/(x_j - x_{j-1})$ of mesh-lengths are kept uniformly bounded as $n \to \infty$, then $e_\pi^{(j)}(x) = O(|\pi|^{4-j})$ for $j = 0, 1, 2$.

A more complicated analysis based on the same ideas (specifically, on Sard's extensions of the Peano kernel technique to higher dimensions), gives error bounds for *multivariate* Hermite interpolation.[7] Obtaining analogous error bounds for multivariate *spline* interpolation is much more difficult; a useful survey article on this subject was published by Martin Schultz in Schoenberg [69, pp. 279–347].[8] Alternative error bounds for spline approximations on *uniform* meshes can be based on ideas from Fourier analysis.[9]

The same technique can be applied to the bilinear blending scheme of Chapter 7, §9. The error e in bilinear blending to the boundary values of u satisfies the inhomogeneous Mangeron DE $e_{xxyy} = u_{xxyy}$. Since the Mangeron differential operator $\partial^4/\partial x^2 \partial y^2$ is the tensor product of the univariate operators $\partial^2/\partial x^2$ and $\partial^2/\partial y^2$, its Green's function in a rectangle is the product of two Green's functions for intervals given by (1.3). These being positive, the biggest error for $|u_{xxyy}| \leq M$ is given by setting $u_{xxyy} = M$. An easy calculation gives the error in this case as

$$e(x,y) = M \iint G(x,\xi) G(y,\eta)\, d\xi\, d\eta. \tag{2.11}$$

For a square of side h, this gives the optimal error bound[10]

$$\|e\|_\infty \leq \frac{3}{16} h^2 \|u_{xxyy}\|_\infty. \tag{2.12}$$

---

[6] See G. Birkhoff and C. de Boor, J. Math. Mech. 13 (1964), 827–836, esp. p. 828, (6). See also C. A. Hall, J. Approx. Theory 1 (1968), 209–218, and Comm. ACM 12 (1969), 450–452; C. A. Hall and W. W. Meyer, J. Approx. Theory 16 (1976), 105–122.

[7] See Sard [63] and Birkhoff-Schultz-Varga [68].

[8] This gives many other references; see also Schumaker [81].

[9] Cf. G. J. Fix and G. Strang, Studies in Applied Math. 48 (1969), 265–273.

[10] G. Birkhoff and W. J. Gordon, J. Approx. Theory 1 (1968), 199–208.

Triangles. Finally, improved error bounds for 'smooth interpolation in triangles' have been obtained by Barnhill and Mansfield.[11]

3. Least squares method. Conceptually, one of the simplest schemes for solving linear elliptic problems is the *least squares* method. This was originally called the 'kernel function' method because of its relation to a larger theory based on properties of Green's, Neumann's, and kernel functions.[12] It has been described by Bergman and Herriot as follows:[13]

> The method of the kernel function for solving elliptic-boundary value problems . . . consists essentially of three steps:
>
> 1. A procedure for generating a (complete) set $\{h_\nu\}$, $\nu = 1, 2, \cdots$, of particular solutions of the partial differential equation.
> 2. A procedure for deriving from this system a set of particular solutions $\{\psi_\nu\}$, $\nu = 1, 2, \cdots$, which are orthonormal in a suitable sense over a given domain.
> 3. A procedure for determining from the prescribed boundary values a linear combination
>
> $$U = \sum_{\nu=1}^{N} a_\nu \psi_\nu \tag{3.1}$$
>
> [closely] approximating the desired solution.

Clearly, one can "approximate the desired solution" in any of a variety of norms. For the classical Dirichlet problem, approximation in the max ('uniform') norm is attractive theoretically, because the completeness of *harmonic polynomials* in this space is assured by classic theorems of Bergman and Walsh, which assert that every function harmonic in a compact simply connected domain can be uniformly approximated by a harmonic polynomial, to within any $\epsilon > 0$.[14] These theorems generalize an earlier theorem of Runge which asserts that any complex analytic function in $\bar{\Omega}$ can be uniformly approximated by a complex polynomial. This result, in turn, generalizes the classic Weierstrass approximation theorem stated in Chapter 7, §3. On the other hand, since $\bar{\Omega}$ is compact, uniform convergence in Ω (or on Γ) implies mean square convergence with respect to any reasonable integral such as (3.5a), (3.5b), or (3.5c) below.

---

[11] R. E. Barnhill and L. Mansfield, J. Approx. Theory 11 (1974), 306–318.

[12] S. Bergman, Quart. Appl. Math. 4 (1946), 233–245, and 5 (1947), 69–81. See also S. Bergman, *The Kernel Function and Conformal Mapping*, Am. Math. Soc., 1950.

[13] Bergman-Herriot [61]. See also S. Bergman and J. G. Herriot, Numer. Math. 7 (1965), 42–65. K. T. Hahn, Pacific J. Math. 14 (1964), 944–955; S. Bergman and B. Chalmers, Math. Comp. 21 (1967), 527–542.

[14] C. Runge, Acta Math. 6 (1885), 229–244; S. Bergman, Math. Ann. 86 (1922), 238–271; Walsh [69, esp. pp. 19, 35].

It is simpler computationally to use *least squares* approximation on the boundary, i.e., to minimize

$$\int_{\Gamma} [U(x,y) - g(x,y)]^2 \, dx \, dy.$$

A third alternative is offered by least squares approximation in the Dirichlet semi-norm, i.e., minimizing

$$\mathbf{D}\langle u,u\rangle = \int_{\Omega} (\nabla u \cdot \nabla u) \, dR = \int_{\Gamma} u \frac{\partial u}{\partial n} \, ds,$$

a formula that generalizes to any number of dimensions. (Of course, in practice, one must use suitably *discretized* analogues of the above formulas.)

Polynomial least squares. In particular, given $g(\mathbf{x})$ on the boundary Γ of a plane region Ω, one can look for that harmonic polynomial $h_n(\mathbf{x})$ of degree n which gives the best mean square (or a best uniform) approximation to $g(\mathbf{x})$ on Γ, and use $h_n(\mathbf{x})$ to approximate the solution of the Dirichlet problem in Ω for the specified boundary conditions. One basis $\{ b_k \}$ for the space of harmonic polynomials is 1, x, y, x^2-y^2, $2xy$, x^3-3xy^2, $\cdots$, with

$$b_{2n} = \mathrm{Re}\{(x+iy)^n\} \quad \text{and} \quad b_{2n+1} = \mathrm{Im}\{(x+iy)^n\}.$$

The approximate solution of Dirichlet problems, using least squares approximation by harmonic polynomials, was already automated by Bergman in 1946–7 using IBM punch-card computers. It was further developed by Davis-Rabinowitz [61, §10],[15] who observed that any one of three different procedures can be followed for finding a least squares ('closest') approximation to the solution of an elliptic problem. Namely, the basis functions ψ_ν of the approximating subspace Ψ can be chosen so that either

(a) the ψ_ν satisfy the differential equation but not the boundary condition,

(b) ψ_0 satisfies the boundary condition, and $\psi_1, \psi_2, \cdots$, satisfy homogeneous boundary conditions, but the ψ_ν do not satisfy the differential equation,

(c) the ψ_ν satisfy neither the differential equation nor the boundary condition.

Sometimes it is preferable to *homogenize* the DE. Thus, suppose we wish to solve

$$\nabla^2 u = x^2 + y^2 \quad \text{in} \quad \mathrm{S} = [-1,1] \times [-1,1], \tag{3.2}$$

---

[15] See also P. J. Davis and P. Rabinowitz, J. Assoc. Comput. Mach. 1 (1954), 183–191, and 13 (1966), 296–303.

for the boundary values $u = 0$ on $\Gamma = \partial S$. Let $\psi_0 = (x^4+y^4)/12$; then $v = u - \psi_0$ will be harmonic in S and satisfy $v = (x^4+y^4)/12$ on Γ.

By symmetry, we can approximate v uniformly closely on Γ, and hence (by the maximum principle of Chapter 2, §7) in S, by a linear combination of the $\psi_k = \mathrm{Re}\{z^{4k-4}\}$ for $k = 1, 2, \cdots$. To do this accurately and efficiently, one should first *orthonormalize* the ψ_k on the perimeter $\Gamma = \partial S$, by the Gram-Schmidt process.[16] Then make an orthonormal expansion of v, to get the desired approximation.

In general, for an elliptic DE in two or more dimensions

$$\mathbf{L}[u](\mathbf{x}) = 0, \qquad \mathbf{x} \in \Omega, \tag{3.3}$$

with Dirichlet, Neumann, or 'mixed' boundary conditions

$$\mathbf{B}[u](\mathbf{y}) = a(\mathbf{y})u + b(\mathbf{y})\partial u/\partial n = g(\mathbf{y}), \qquad \mathbf{y} \in \Gamma, \tag{3.4}$$

we may set up the following integrals corresponding to (a), (b), and (c) above, respectively:

$$I = \int_\Gamma w(\mathbf{y})\,\{g(\mathbf{y}) - \mathbf{B}[U](\mathbf{y})\}^2\,dS, \tag{3.5a}$$

$$I = \int_\Omega w(\mathbf{x})\,\{\mathbf{L}[U](\mathbf{x})\}^2\,dR, \tag{3.5b}$$

$$I = \int_\Omega w_1(\mathbf{x})\{\mathbf{L}[U](\mathbf{x})\}^2\,dR + \int_\Gamma w_2(\mathbf{y})\{g(\mathbf{y}) - \mathbf{B}[U](\mathbf{y})\}^2\,dS, \tag{3.5c}$$

and seek the values of $a_1, a_2, \cdots, a_N$ in (3.1) which render these integral measures of discrepancy as small as possible.

Davis-Rabinowitz [61] applied the method successfully to solve the Dirichlet problem for $\mathbf{L}[u] = \nabla^2 u = 0$ in a cube, a sphere, an elliptic annulus, and a square with its corners cut out, as well as for $\mathbf{L}[u] = \nabla^4 u = 0$ in various domains.[17]

The 'least squares' method is also applicable to some DE's with variable coefficients. Thus, for the DE

$$\nabla^2 u + (2r^2 - 4)\,u = 0, \qquad r^2 = x^2 + y^2, \tag{3.6}$$

one can calculate a basis of solutions ψ_k having the symmetry of the square by power series expansions. Using these, Bergman-Herriot [61] solved (3.6) in the square $|x| \leqslant 1, |y| \leqslant 1$ with circularly rounded corners, for boundary values having the 8-fold symmetry of the domain. Their Step 2 greatly reduced the amplification of roundoff errors, by orthonormalizing the basis of polynomial solutions.

---

[16] The *QR* factorization, with Householder transformations, of the matrix of (column) basis vectors is more stable; see Stewart [73] or C. L. Lawson and R. J. Hanson, *Solving Least Squares Problems*, Prentice-Hall, 1974. See also P. J. Davis in Todd [62, pp. 347–379].

[17] See also D. Levin et al., J. Inst. Maths. Appl. 22 (1978), 171–178.

The 'least squares' methods described above are especially well-suited to solving the Laplace, Helmholtz, and modified Helmholtz equations. They are less well-adapted to nonlinear DE's or to most DE's with variable coefficients; hence they are much less generally applicable than difference and finite element methods.

Vekua's ideas. For the similar method of Vekua, we refer the reader to I. N. Vekua, *New Methods for Solving Elliptic Equations*, North Holland, 1967; P. Henrici, ZaMP 8 (1957), 168–203; and S. C. Eisenstat, SIAM J. Numer. Anal. 11 (1974), 654–680. For related analytical considerations, see R. P. Gilbert et al. in Springer Lecture Notes in Math. No. 430, 1974, 184-260; also M. Kracht, E. Kreyszig, and G. Schroder, SIAM Rev. 26 (1982), 52-71.

L-shaped membrane. The techniques described above can be adapted to eigenproblems. In an outstanding paper, Fox-Henrici-Moler [67] exploited them to to compute eigenvalues of the Laplace operator for an L-shaped membrane. By Lehman's theorem (Chapter 2, §8), the corresponding eigenfunctions are analytic except at the reentrant corner of this membrane. There they are asymptotic to linear combinations of

$$J_{2n/3}(k_j r)\sin(2n\theta/3), \qquad k_j{}^2 = \lambda_j. \tag{3.7}$$

To exploit this fact, the authors proved the following new approximation theorem.

THEOREM 2. *For p_i and q positive functions, let λ^* and u^* be an approximate eigenvalue and eigenfunction which satisfy*

$$\sum_{i=1}^{n} \frac{\partial}{\partial x_i}\left[p_i(\mathbf{x})\frac{\partial u}{\partial x_i}\right] + \lambda q(\mathbf{x})u = 0 \qquad \text{on } \Omega,$$

$$\int_\Omega u^2(\mathbf{x})q(\mathbf{x})\,dR = \int_\Omega q(\mathbf{x})\,dR,$$

but not necessarily $u = 0$ on $\Gamma = \partial\Omega$. Let $\epsilon = \max_{\mathbf{y}\in\Gamma}|u^*(\mathbf{y})|$. If $\epsilon < 1$, then there exists an eigenvalue λ satisfying

$$\frac{|\lambda - \lambda^*|}{\lambda^*} \leqslant \frac{\sqrt{2}\epsilon + \epsilon^2}{1+\epsilon^2}.$$

They then computed u^* and λ^* by *boundary collocation* (Chapter 6, §6) using successive approximation. In their basis they included functions of the form (3.7), with λ_j a parameter to be determined.

Using this procedure, extremely accurate eigenvalues for the L-shaped membrane were obtained in the paper cited. Only later were equally accurate results obtained using 'singular' finite elements (Chapter 7, §10).[18]

[18] G. J. Fix, S. Gulati, and G. I. Wakoff, J. Comp. Phys. 13 (1973), 209–228.

4. Boundary element methods.[19] The harmonic polynomials used in §4 to approximate solutions to *interior* Dirichlet problems are superpositions of potentials due to multipoles at ∞. Using *inversion*, one can approximate solutions of *exterior* Dirichlet problems similarly by superpositions of multipoles located at the origin, provided that this is in the complement of the closure of the domain Ω in which the problem is to be solved. Clearly, greater flexibility in approximating such solutions is had if 'sources and sinks'[20] (including dipoles and multipoles as limiting cases) are allowed to be located at more than one point.

Rankine bodies. A classic application of the source-and-sink method is to the approximation of potential flows around axially symmetric solids by superpositions of poles and axial dipoles on the axis of symmetry. When airships (also called 'dirigibles') were still used in air transport, Fuhrmann and von Kármán calculated potential flows around them by such superpositions.[21]

Mathematically, in terms of the stream function of the flow, this amounts to finding a distribution of charges on the axis such that

$$\Psi(x,r) = Ur^2 + \int \frac{(x-\xi)}{\sqrt{(x-\xi)^2+r^2}}\, de(\xi) \tag{4.1}$$

vanishes identically on Γ, where $r^2 = y^2+z^2$. Thus, it is classic that a *linear* source distribution with density $\rho = x$ on the interval $[-1,1]$, superposed on a uniform flow, approximates the flow around a slender ellipse with foci at $(\pm 1,0)$. In space, the velocity potential

$$\phi = A\mu\left[\frac{1}{2}\zeta \ln \frac{\zeta+1}{\zeta-1} - 1\right], \tag{4.2}$$

in elliptic coordinates (Lamb [32, §105]), approximates the flow around a slender prolate spheroid.

Unfortunately, it is only exceptionally that potential flows around axially symmetric cylinders and solids have analytic continuations without singularities all the way to the axis of symmetry. Hence not all axially

[19] We use here the suggestive name of C. Brebbia, ed., *Recent Advances in Boundary Element Methods*, Pentech Press, 1978; *New Developments in Boundary Element Methods*, Springer, 1980; *Boundary Element Methods*, Springer, 1981. See also F. Kang's article in *Proceedings of China-France Symposium on Finite Element Methods*, F. Kang and J. L. Lions, eds., Gordon and Breach, 1983; G. C. Hsiao and R. C. McCamy, SIAM Rev. 15 (1973), 687–705; G. C. Hsiao, P. Kopp, and W. L. Wendland, Computing 25 (1980), 89–130; and E. Stephan and W. L. Wendland, Computer Methods in Appl. Mech. Engin. 36 (1983), 331–358.

[20] 'Sources' and 'sinks' are, mathematically, just 'poles' of opposite sign (in empty space).

[21] G. Fuhrmann, Dissertation, Göttinger, 1912. Th. von Kármán, Abh. Aer. Inst. Aachen 6 (1927), p. 1; see Prandtl-Tietjens, *Applied Hyrdo- and Aeromechanics*, Dover, 1957, p. 187; see also F. Kenne and K. Burg, *Singularitätens Verfahren der Strömungslehre*, Karlsruhe, 1975.

symmetric bodies are 'Rankine bodies', flows around which can be exactly expressed in this way by distributions of finite total charge along the axis.

The 'boundary element method' to be described below is free from this limitation. It approximates exact solutions of linear elliptic problems by superpositions of *Green's functions* having poles (or dipoles) on the boundary of the domain Ω of interest. It utilizes integral equations and integral identities to express functions of $n = 2$ or 3 variables satisfying some differential equation in terms of a function of one less variable on the boundary. Since only *boundary* values and derivatives need be considered, it is usually easier to compute with and visualize approximations constructed in this way, than approximations obtained by difference or ordinary finite element methods.

Integral equations of potential theory. To approximate *general* harmonic functions in a given domain Ω, one must use distributions of sources and sinks (or dipoles) over its entire boundary. This amounts to a classic reformulation of the Dirichlet and Neumann problems of potential theory in terms of the so-called 'integral equations of potential theory'. These can be derived easily from Green's third identity (Chapter 2, Theorem 7) as our starting point. In the plane, this identity is given by the formula

$$u(x_0, y_0) = \frac{1}{2\pi} \int_\Gamma \frac{\partial}{\partial \nu} \log r \, ds - \frac{1}{2\pi} \int_\Gamma \log r \frac{\partial u}{\partial \nu} \, ds, \tag{4.3}$$

where $r^2 = (x - x_0)^2 + (y - y_0)^2$. This formula is valid at any interior point (x_0, y_0) of *any* plane domain Ω bounded by a smooth curve Γ, for any function $u(x, y)$ which is harmonic in the compact domain $\bar{\Omega}$.

Green's third identity, whether in the plane or in space, asserts that (under suitable differentiability hypotheses) every function harmonic on the closure $\bar{\Omega} = \Omega \cup \partial\Omega$ of Ω is the potential of a distribution or 'spread' on $\partial\Omega = \Gamma$ of poles *and* dipoles normal to Γ. However, just as a function u harmonic in a compact region $\bar{\Omega}$ is determined by *either* its values u *or* its normal derivatives on Γ, so u can be represented as the potential of a spread of poles on Γ (a 'single layer'), *or* as the potential of a spread of normal dipoles (a 'double layer'). If Γ is smooth, moreover, the respective *densities*, $\rho(\mathbf{y})$ and $\mu(\mathbf{y})$, of these spreads vary smoothly with position.

If one takes the density $\mu(\mathbf{y})$ of the *double* distribution as an unknown function, then the resulting potential assumes given values $u(\mathbf{y}) = g(\mathbf{y})$ on Γ (i.e., the Dirichlet problem) if and only if $\mu(\mathbf{y})$ satisfies the following *Fredholm integral equation* of the 'second kind' in the plane[22]

$$g(\mathbf{y}) = \pm\mu(\mathbf{y}) + \frac{1}{\pi} \int \frac{\partial}{\partial \nu} \log \frac{1}{r} \mu(\xi) \, ds(\xi). \tag{4.4}$$

[22] For their derivation, see [K, pp. 286—287] and Garabedian [64, §9.3]. By the 'Fredholm alternative' (Garabedian [64, §10.1]), uniqueness implies existence, and vice versa.

The + sign corresponds to the solution of the *exterior* problem, and the − sign to that of the *interior* Dirichlet problem [K, Chap. XI, §2].

Likewise, the *Neumann* problem can be solved by taking as unknown the density $\rho(\mathbf{y})$ of a *single* layer on Γ. The normal derivative $s(\mathbf{y}) = \partial u/\partial n$ of such a spread is given in $\mathbb{R}^2$ (the plane) by

$$s(\mathbf{y}) = \pm\rho(\mathbf{y}) + \frac{1}{\pi}\int \frac{\partial}{\partial n} \log \frac{1}{r} \rho(\xi)\, ds(\xi). \tag{4.5}$$

The preceding integral equations, and their analogues in $\mathbb{R}^3$, are called 'the integral equations of potential theory'; they have been studied in great depth.

The solutions of

$$f(\mathbf{y}) = \mu(\mathbf{y}) + \lambda \int K(\mathbf{y};\xi)\mu(\xi)\, dS(\xi) \tag{4.6}$$

for intermediate values of λ, with $-1 < \lambda < 1$, are also of physical interest as 'induced potentials'.[23] For example, they describe the initial acceleration potential of a liquid globule in an incompressible fluid suddenly accelerated from rest.

5. Relation to conformal mapping. The Green's function of the *two-dimensional* Laplace operator on a domain Ω is associated with a *conformal map* of Ω onto the unit disk. To see the connection, let $\mathbf{0} \in \Omega$, and let

$$G(x,y;0,0) = -\frac{1}{2\pi} U = -\frac{1}{2\pi}[\log r + \phi(x,y)], \quad r = \sqrt{x^2+y^2}, \tag{5.1}$$

be the Green's function of ∇^2 for a source at $\mathbf{0}$.

The harmonic *conjugate* of $U = \log r + \phi$ is clearly the function (unique up to a constant addend),

$$V = \theta + \psi(x,y) = \theta + \int (\phi_x\, dy - \phi_y\, dx) \tag{5.2}$$

satisfying the Cauchy-Riemann equations $V_y = U_x$, $V_x = -U_y$. Moreover, the function

$$W = U + iV = \log r + i\theta + \phi + i\psi \tag{5.3}$$

maps the punctured domain $\Omega - \mathbf{0}$ onto the cylinder defined by the W-plane (mod $2\pi i$). Hence

$$w = e^W = r\, e^{i\theta}\, e^{\phi + i\psi} = z\, e^{\phi+i\psi} = \rho\, e^{i\sigma} \tag{5.4}$$

maps Ω onto the interior of the unit disk, $|w| < 1$, in the w-plane. The origin $z = 0$ is mapped onto $w = 0$, and the constant of integration in

[23] See G. Birkhoff, *Studies in Math. and Mech. Honoring Richard von Mises*, Academic Press, 1954, 88–96. Here reference is made to Frank-von Mises [35, vol. 2. Chap. 17, §2], and in Bergman-Schiffer [53].

(5.2) can be chosen so that $z_1 = r_1 \in \Gamma$, with $\arg\{z_1\} = 0$, is mapped onto $w_1 = 1$. Then $\sigma = \theta + \psi = V$, and ψ is the 'angular distortion'. defined as the difference between the polar angle σ in the unit disk in the W-plane and the corresponding angle θ in the z-plane.

In particular, if Γ is the unit circle $z = e^{i\theta}$, then on Γ the formula

$$v(\theta) = v_0 + \frac{1}{2\pi} \oint \cot[(\theta - \sigma)/2]\, u(\sigma)\, d\sigma \tag{5.5}$$

expresses v in terms of u. Formula (5.5) is valid if the Cauchy principal value is taken and u is analytic in $|z| < 1$ and continuous in $|z| \leqslant 1$.

It is easy to derive the singular convolution kernel[24] $\cot[(\theta - \sigma)/2]$ in (5.5) formally. By a Fourier series expansion, we get

$$v(\theta) = v_0 + \frac{1}{\pi} \sum_{k=1}^{\infty} \int [\cos k\sigma \sin k\theta + \sin k\sigma \cos k\theta]\, u(\sigma)\, d\sigma .$$

Setting $\phi = \sigma - \theta$ and $\cos\phi + \cos 2\theta + \cdots = [\cot\phi/2]/2$, we obtain (5.5). Actually, when solving problems in circular domains, it is usually most effective to calculate directly with the Fourier coefficients.

The efficient use of interpolation formulas to evaluate the singular integral (5.5) numerically has been analyzed in Gaier [64, p. 74ff]; by P. J. Davis in Todd [62, pp. 468–484]; by R. Kress in Computing 10 (1972), 177–187; and elsewhere.

Computing Green's functions. Since the Green's function satisfies (5.1), it can be computed by determining ϕ as the solution the Dirichlet problem in Ω, for the boundary condition $\phi = -\log|r|$, by any of the methods described in Chapters 3–7.

There are also many ways of computing approximate values $\Phi_{i,j}$ and $\Psi_{i,j}$, of $\phi(x_i, y_j)$ and $\psi(x_i, y_j)$ on a square mesh. For example, we can use the standard 5-point approximation

$$\Phi_{i,j} = \frac{1}{4}[\Phi_{i+1,j} + \Phi_{i,j+1} + \Phi_{i-1,j} + \Phi_{i,j-1}].$$

We can then compute the conjugate function Ψ of Φ approximately (up to an additive constant) at the *centers* $(i'h, j'h)$ of the previous mesh squares by the formulas

$$\Psi_{i',j'} - \Psi_{i',j''} = \Phi_{i+1,j} - \Phi_{i,j}, \qquad \Phi_{i',j'} - \Phi_{i'',j'} = \Phi_{i,j} - \Phi_{i,j+1}.$$

Here we adopt the convenient notation $i' = i + 1/2$, $i'' = i - 1/2$, $j' = j + 1/2$, $j'' = j - 1/2$.

From the two *discrete harmonic functions* thus computed on a pair of *staggered* square meshes, we can obtain, by local (Bessel) cubic or other

[24] The kernel $K(s,t)$ of an integral equation is called a 'convolution kernel' when it is of the form $K(s-t)$ (on $\mathbb{R}$ or the unit circle).

interpolation schemes, a continuously defined *complex*-valued approximation to an analytic function $\omega(x+iy)=\Phi+i\Psi$.

Stated another way, the constructions (5.3) and (5.4) enable one to approximate the *Green's function* $G(z;\zeta)=G(x,y;\xi,\eta)$ for the Laplace DE in Ω and the boundary conditions $u\equiv 0$ on Γ.

In the limit as $h\downarrow 0$, the preceding constructions give functions u and v as limits of the approximate Φ and Ψ given by (5.3) and (5.4) which are *differentiable* (i.e., that their *gradients* also converge). It is easy to show that the function $w(z)$ maps the domain Ω *conformally* onto the *unit disk* $|w|<1$ in the w-plane. That is, the preceding formulas give a *constructive* implementation of Riemann's mapping theorem with $w(0)=0$. More precisely, each of the computational schemes described in Chapters 3–7 gives such an implementation (see also §6).

Therefore, by using inverse interpolation, we can construct a one-one conformal transformation from the unit disk in the w-plane onto the given domain Ω in the z-plane. (Note that, by a translation of coordinates in the z-plane, we can map $\mathbf{0}$ onto *any* point ζ_0 interior to Γ, but that the mapping is only unique up to a rotation $\theta\mapsto\theta+\theta_0$ of the unit disk.)

Discrete Green's functions. The inverses of matrices A associated with elliptic difference operators for Dirichlet boundary conditions give 'discrete Green's functions'. For a plane domain and $k=(i,j)$, $k'=(i',j')$:

$$G(x_i,y_j;\xi_{i'},\eta_{j'})=(A^{-1})_{k,k'}.$$

Thus, A^{-1} should be a good approximation to the Green's function of the given problem, especially away from the diagonal $k=k'$.

6. Conformal mapping. As we showed in Chapter 2, §§1–2, it is easy to solve the Laplace and Poisson equations in the unit disk. This fact, and the many physical applications of these DE's,[25] give enormous importance to implementing practically the Riemann mapping theorem stated there. We next state another version of this so-called fundamental theorem of conformal mapping.

THEOREM 3. *Let Ω be any bounded, simply connected*[26] *domain of the complex z-plane with smooth boundary Γ, and let z_0, z_1, be points inside Ω and on Γ, respectively. Then there exists a unique one-one conformal map of Ω onto the unit disk $r<1$ in the complex t-plane, which maps z_0 into* 0 *and z_1 into* 1.

[25] Several such applications are carefully explained in [KK, Chap. VI]. For the analytical theory, outstanding is Z. Nehari, *Conformal Mapping*, McGraw-Hill, 1952; see also Bergman-Schiffer [53].

[26] I.e., that every closed curve in Ω can be shrunk (in Ω) to a point.

The truth of this theorem follows from the existence of a Green's function in Ω with its 'pole' at z_0, for the boundary conditions $G(z;z_0) \equiv 0$ on Γ.[27] Set $\log \rho = -2\pi G(z;z_0)$, and define σ as the harmonic conjugate of the function $\log \rho$:

$$\sigma(z) = 2\pi \int_{z_1}^{z} \left[\frac{\partial G}{\partial y} dx - \frac{\partial G}{\partial x} dy \right]. \tag{6.1}$$

The function $t = \rho e^{i\sigma}$ achieves the desired mapping (see §5).

The existence of such Green's functions seemed "physically obvious" to Green, because $G(z;z_0)$ represents the electrostatic potential at z of a unit (line) charge at $z = z_0$, in the presence of a grounded cylindrical conductor having Γ as its profile (cf. [K, pp. 176 and 236] for the three-dimensional analogue).

An analogous theorem states that there exists a conformal map from any *doubly* connected plane region onto the annulus $1 < \rho < \gamma$, for some $\gamma > 1$. This theorem is plausible from the following physical interpretation. Consider the equilibrium temperature $\tau_1(x,y)$ for boundary values 0 on the inner boundary and 1 on the outer boundary. Now set $\tau = \lambda \tau_1$, choosing the constant λ (i.e., rescaling τ_1) so as to make

$$\oint \left[\frac{\partial \tau}{\partial y} dx - \frac{\partial \tau}{\partial x} dy \right] = 2\pi,$$

for any closed path winding once counterclockwise around the annulus. Let θ be the conjugate function of τ. By the strict maximum principle, τ can have no critical points in the domain; hence for $\tau = \log \rho$, and the function $\zeta = \rho e^{i\theta}$ should again achieve the desired mapping.

Riemann's mapping theorem, and its development into an effective *analytical* technique by Schwarz, Christoffel, and others, made possible the explicit solution of a large number of two-dimensional problems involving the Laplace equation.[28] We will describe next a few key methods and results which facilitate its efficient *numerical* implementation.

Lichtenstein's method. We have seen (§5) that the Green's function of a domain Ω defines a conformal map of Ω onto the unit disk. As Lichtenstein observed, the function mapping the boundary Γ of the domain Ω onto the boundary γ of the unit disk satisfies an integral equation. We now specify this integral equation, following Birkhoff-Young-Zarantonello [53].[29] Suppose Γ is given parametrically by $z = z(q)$, where q is a

[27] Unless the contrary is stated, Green's functions are for the boundary condition $G(\mathbf{x};\boldsymbol{\xi}) \equiv 0$ for $\mathbf{x}$ or $\boldsymbol{\xi}$ on Γ.

[28] See the atlas of analytical conformal maps compiled by H. Kober [52].

[29] See also Todd [55], reviewed in Math. Revs. 17 (1956), p. 540; and Gaier [64, pp. 6–8].

periodic variable with period 2π, and set, for any z_j, $z \in \Gamma$,

$$A(z_j;z) = \frac{1}{\pi\rho_j}\frac{\partial\rho_j}{\partial n}\frac{ds}{dq}, \qquad \rho_j = |z - z_j|, \quad ds = |dz|. \tag{6.2}$$

The mapping is then determined by the 'angular distortion' $\psi(z)$, defined as the difference $\psi = \sigma - \theta$ between the angle σ on the unit circle $\Gamma: t = e^{i\sigma}$ and the polar angle $\theta = \theta(q) = \arctan[y(q)/x(q)]$ (see §5). This function $\psi(z)$ satisfies the following Fredholm integral equation of the 'second kind'

$$\psi(z_j) = \lambda \oint A(z_j;z)\psi(z)\, dq + \Phi(z_j), \tag{6.3}$$

where $\lambda = +1$ for interior and $\lambda = -1$ for exterior mappings, and

$$\Phi(z_j) = \frac{1}{\pi}\oint \frac{\log r}{\rho_j}\, d\rho_j = \frac{1}{\pi}\oint (\theta - 2\alpha_j)\, d\alpha_j + \theta_j + \pi, \tag{6.3$'$}$$

where the second equality follows from Lemma 2 in Birkhoff-Young-Zarantonello [53].

Numerical implementation. Now let n be any positive integer, and set $z_j = z(q_j) = z(2\pi j/n)$ (i.e., adopt a uniform mesh on the parametric circle). Using trapezoidal quadrature for maximum accuracy (since the contour integrals are analytic and periodic),[30] we can compute approximate values $\psi_n(z_j)$ of the $\psi(z_j)$, as follows.

Let $a_{j,k} = A(z_j, z_k)/n$; then trapezoidal quadrature of (6.3) amounts algebraically (for any fixed j) to setting

$$\psi_j = \lambda \sum a_{j,k}\psi_k + \Phi_j, \qquad \Phi_j = \Phi(z_j). \tag{6.4}$$

The system (6.4) is singular for interior mappings: $\psi = 1$ is an eigenfunction for $\lambda = 1$. However, as was first shown by Poincaré,[31] it can always be solved by setting $\psi_j^{(0)} = 0$, and iterating:

$$\psi_j^{(m+1)} = \sum a_{j,k}\psi_k^{(m)} + \Phi_j. \tag{6.5}$$

Since $w(z)/z = \rho\, e^{i\psi}/r$ reduces to $e^{i\psi(\zeta)}/r(\zeta)$ for $\zeta \in \partial\Omega$, once ζ has determined on the boundary, one can evaluate w at interior points z of Ω by Cauchy's integral formula:

$$\log(w(z)/z) = \frac{1}{2\pi}\oint \frac{-\log|z-\zeta| - i\psi(\zeta)}{z-\zeta}\, d\zeta, \tag{6.6}$$

where $|z-\zeta|$ is the distance between $z \in \Omega$ and $\zeta \in \partial\Omega$.

[30] Birkhoff-Young-Zarantonello [53, §4], and P. J. Davis in Langer [59, pp. 45–61].

[31] H. Poincaré, Acta Math. 20 (1897), 59–142. For estimates of the rate of convergence, see L. V. Ahlfors, Pacific J. Math. 2 (1952), 271–280, and H. Royden, ibid, pp. 385–394.

Gerschgorin's method.[32] Instead of using (6.3), whose solution is the angular distortion, one can compute the angle σ at the boundary of the unit disk. This satisfies the integral equations

$$\sigma(z_j) = \lambda \oint A(z_j;z)\sigma(z)\,dz - \beta(z_j), \qquad j = 1,\cdots,n, \tag{6.7}$$

where $\beta(z_j)$ is the angle subtended at z_j by $z_1 - z_0$, and $w(z_0) = 0$, $w(z_1) = 1$. As with Lichtenstein's method, one can modify this so as to make $A(z_j;z)$ and $\beta(z)$ periodic and analytic. One then obtains maximum accuracy by trapezoidal quadrature; i.e., by iterating

$$\sigma^{(m+1)}(z_j) = \sum_{k=1}^{N} a_{j,k}(\sigma^{(m)}(z_k) - 2\alpha_{j,k}) + 2\theta_j + \pi. \tag{6.8}$$

Other methods. Many other numerical procedures have been proposed for solving Fredholm integral equations of the second kind like (4.4), (4.5), and (6.3). Thus Atkinson recommends using Simpson's rule or Gauss quadrature instead of trapezoidal quadrature,[33] while Birkhoff and de Boor assert (in Garabedian [65, p. 180]) that "Spline functions seem ideally suited to the approximate solution of Fredholm integral equations with *smooth kernels* $K(x,y)$". Chebyshev and Gauss collocation also give good results for 'smooth' (i.e., highly differentiable) kernels. However, very few authors distinguish the 'discretization' stage from the 'solution' stage, or discuss the 'condition number' of the matrices used to approximate the kernel. We now make this distinction and discuss dependence of the accuracy of numerical techniques on the domain being mapped.

Obviously, trapezoidal quadrature, though most efficient for smooth boundaries, will generally not be accurate in the case of domains with corners such as rectangles. For rectangles, it seems reasonable to locate mesh-points at the zeros of the Chebyshev or Legendre polynomials, for the reasons explained in Chapter 7, §2, and perhaps at the corners as well, although this is dubious since ρ may well be infinite there.

Kantorovich and Krylov [KK, pp. 130–140] describe a numerical technique due to Krylov and Bogolioubov (1929) for solving the integral equations (4.4) and (4.5) of *two*-dimensional potential theory.[34] In both cases, the exact solutions for an ellipse (ellipsoid) can be given explicitly. More generally, because the kernel of the integral equation of potential theory is nonsingular, one can apply similar methods to solve (in principle) Dirichlet and Neumann problems for smooth bodies of general shape.[35]

---

[32] S. Gerschgorin, Rec. Math. (Mat. Sbornik) 40 (1933), 48–58. See also [KK, Chap. V, §9]; Gaier [64, pp. 8–9].

[33] See Atkinson [76] and also his article in TOMS 2 (1976), 154–171.

[34] See also E. J. Nystrom, Acta Math. 54 (1930), 185–204, and M. S. Lynn and W. P. Timlake, Numer. Math. 11 (1968), 77–98.

[35] L. Landweber, Taylor Model Basin Rep. 761 (1951); J. P. Moran, J. Fluid Mech. 17 (1964), 285–304; R. J. Riddell, Jr., J. Comp. Phys. 31 (1979), 21–41 and 42–59; Fairweather, et al. [79]. For a survey of recent work, see Atkinson [76].

7. Theodorsen's method. For applications to two-dimensional airfoil theory, one wants to map the exterior Ω' of the unit circle $r = e^{i\sigma}$ conformally onto the exterior of some specified airfoil profile γ having a sharp *trailing edge*. For such applications, a very successful numerical method of conformal mapping is that of Theodorsen [31].[36] It may be described as follows (Birkhoff-Young-Zarantonello [53, §6] or Gaier [64, Kap. II]).

First apply a Joukowsky, Kármán-Trefftz, or other conformal transformation to map the exterior of γ onto the exterior of some *nearly circular*, simple closed curve Γ of class C^1 with center 0, as in Gaier [64, Kap. II, §4]. Let this Γ be specified by $r = \exp(F(\theta))$.

The conjugate of any periodic function $g(\sigma)$ is

$$\mathbf{C}[g](\sigma) = \frac{1}{2\pi}\oint \cot[(\sigma - \sigma')/2]\, g(\sigma')\, d\sigma'. \tag{7.1}$$

This may be computed as follows. Take again as an unknown function the angular distortion $u(\sigma) = \sigma - \theta$ with conjugate $v = \log r - \log \rho$, to determine the desired conformal map of $r\,e^{i\sigma}$ onto Γ. Since $v = -F(\sigma - u(\sigma))$ on Γ, we clearly have

$$u(\sigma) = \pm\, \mathbf{C}[F](\sigma - u(\sigma)), \tag{7.2}$$

with the plus sign for the exterior mapping and the minus sign for the interior mapping. For a well-chosen preliminary transformation of γ onto Γ, $F(\theta)$ will be reasonably smooth and nearly constant. In such cases, the nonlinear integral equations (7.1)–(7.2) can usually be solved by iteration.[37]

Analytic domain perturbation. Theodorsen's method is most effective when Γ is a small perturbation of the unit circle $r = e^{i\sigma}$. In the case of *analytic* domain perturbation, it is worth trying a very different *perturbed power series* technique due to Kantorovich.[38] For example, consider the family of coaxial ellipses

$$x^2 + y^2 - 2\lambda(x^2 - y^2) = 1. \tag{7.3}$$

For each λ, the appropriate mapping is given by

[36] See also T. Theodorsen and I. E. Garrick, NACA Tech. Rep. 452, 1934; Warschawski [45]; I. E. Garrick, J. Nat. Bur. Standards 19 (1952), 137–147; and M. Gutknecht, Numer. Math. 36 (1981), 405–429.

[37] For general discussions of numerical methods for solving integral equations, see H. Bueckner in Todd [62, pp. 439–484]; Ben Noble's articles in Anselone [64] and in *Conference on Applications of Numerical Analysis*, Lecture Notes in Mathematics, 228, E. J. Morris, ed., Springer 1971, 137–154. See also Delves-Walsh [74]; Atkinson [76]; and C. T. H. Baker, *The Numerical Treatment of Integral Equations*, Oxford Univ. Press, (1978).

[38] [KK, Chap. V, §6] and Gaier [64, pp. 166–168]. For the original presentation, see L. V. Kantorovich, Mat. Sbornik 40 (1933), 294–325.

$$(7.4) \qquad z = f(\zeta;\lambda) = \sum_{k=1}^{\infty} a_k(\lambda)\zeta^k,$$

where power series in λ for the $a_k(\lambda)$ can be computed recursively. Up to an error of $O(\lambda^6)$, one finds

$$(7.5) \qquad \begin{aligned} z = (1-\tfrac{1}{2}\lambda^2+\tfrac{3}{8}\lambda^4)[\zeta+(\lambda-2\lambda^3+3\lambda^5)\zeta^3+(2\lambda^2-9\lambda^4)\zeta^5 \\ +(5\lambda^3-36\lambda^5)\zeta^7+14\lambda^4\zeta^9+42\zeta^{11}]. \end{aligned}$$

Other methods. Before terminating this brief survey of numerical conformal mapping, a few other techniques should at least be mentioned. First, we note a technique due to Carrier,[39] selected by Gaier [64, Kap. I, §2.3] along with Gerschgorin's and Theodorsen's methods, for detailed analysis.

Gaier [64] also reviews many other successful numerical procedures for constructing conformal maps, and his bibliography is very complete through 1963. An incisive but less comprehensive survey may be found in C. Gram (ed.), *Selected Numerical Methods for . . . Conformal Mapping*, Regencentralen, 1962. Moreover, [KK, Chap. V] gives an authoritative discussion of what was known about the subject a decade earlier.

For applications of the Gerschgorin method to domain perturbation, see Gaier [64, pp. 59–60] and Warschawski.[40] Even if Ω has corners, one can adapt Gerschgorin's method; see Gaier [64, pp. 16–21]. For further results, see Warschawski [56] and, for early numerical experiments, Beckenbach [52] and Todd [55].

Newer methods. Next, we mention a method of Symm that has been recommended by a group of workers at the Los Alamos Scientific Laboratories,[41] in preference to those discussed above. This involves solving a Fredholm integral equation of the *first* kind, a procedure that is ordinarily less stable numerically than solving Fredholm integral equations of the second kind, such as arise in the Lichtenstein-Gerschgorin method of §6.

Finally, we call attention to a method recently developed by Menikoff and Zemack [80] for treating highly distorted domains having a periodic free boundary, such as naturally arise in Taylor instability. This has been successfully applied to periodic waves by Meiron, Orszag, and Israeli.[42]

All these methods can be greatly speeded up by using the FFT.[43] A survey reviewing these and other methods has been written by Fornberg [80], as a prelude to a description of still another new method.

---

[39] G. F. Carrier, Quar. Appl. Math. 5 (1947), 101–104.

[40] S. Warschawski, J. Math. Mech. 19 (1970), 1131–1153.

[41] G. T. Symm, Numer. Math. 9 (1966), 250–258, and 10 (1967), 437–457; J. K. Hayes, D. K. Kahaner, and R. G. Kellner, Math. Comp. 26 (1972), 327–334.

[42] D. I. Meiron, S. A. Orszag, and M. Israeli, J. Comp. Phys. 40 (1981), 345–360.

[43] See P. Henrici, SIAM Rev. 21 (1975), 481–527.

8. Three-dimensional problems. Integral equations similar to (4.4) and (4.5), for solving three-dimensional Dirichlet and Neumann problems of potential theory, can also be constructed. These integral equations are[44]

$$g(\mathbf{y}) = \pm\mu(\mathbf{y}) + \int K(\mathbf{y};\boldsymbol{\xi})\mu(\boldsymbol{\xi})\,dS(\boldsymbol{\xi}), \tag{8.1}$$

and

$$s(\mathbf{y}) = \pm\rho(\mathbf{y}) + \int K(\boldsymbol{\xi};\mathbf{y})\rho(\boldsymbol{\xi})\,dS(\boldsymbol{\xi}). \tag{8.2}$$

The kernel $K(\mathbf{y};\boldsymbol{\xi})$ in (8.1) is $(1/2\pi)\partial(1/r)/\partial\nu$; it is the same in (8.2), but its arguments are interchanged: $K(\boldsymbol{\xi};\mathbf{y}) = (1/2\pi)\,\partial(1/r)/\partial n$.

To solve truly three-dimensional Dirichlet or Neumann problems numerically, using the integral equations of potential theory, tends to be very expensive. Thus, even in the simplest case of a *cube*, a mesh length $h = 1/I$ gives $6I^2$ unknowns and a dense coefficient matrix. Hence about $100I^6$ arithmetic operations are required to determine the approximate density (ρ or σ), after which about $30I^5$ operations are needed ($30I^2$ per interior mesh-point) to determine the values of the potential.[45]

A good recent reference for the general principles involved is G. J. Symm's article in Delves-Walsh [74, Chap. 24]. The relevant formulas are analogous to those in the plane, and the existence and uniqueness theory for the solution of integral equations, apart from vastly greater difficulties in constructing Green's functions, is almost identical.[46] In particular, as was shown by Poincaré long ago (1896), the integral equations of potential theory for *convex* solids can be solved by simple iteration: the Neumann series $I + K + K^2 + \cdots$ always converges for the appropriate Fredholm integral equation of the *second* kind, (8.1).

Moreover, in two *or* three dimensions, whereas the solution of elliptic problems by difference *or* finite element methods yields tabulations from which values at specific points are either obtained in the course of the computation or can be computed very quickly from stored (tabulated) numbers, it takes a complete numerical quadrature to yield the value at any particular interior point, when an integral equation of potential theory is used. Thus in the plane, with an $I \times I$ mesh and $4I$ boundary mesh-points, although about $20I^2 \approx (4I)^3/3$ additions and multiplications may

---

[44] Computations based on these formulas have been performed by G. Fairweather et al., [79].

[45] J. F. Brophy used integral equation and other techniques to compute the added mass and other domain constants of a cube, in his Ph.D. Thesis, Purdue University, 1983.

[46] Symm's article is based on the papers by M. S. Lynn and W. P. Timlake, Numer. Math. 11 (1968), 77–98, and Y. Ikebe, M. S. Lynn, and W. P. Timlake, SIAM J. Num. Anal. 6 (1969), 334–346. See also Y. Ikebe, SIAM Rev. 14 (1972), 465–491, which is, however, largely theoretical.

suffice to determine the (charge or dipole) density distribution on the boundary, after that, another $80I^3$ will be needed to compute the potential at the I^2 interior points, and $160I^3$ to compute the two velocity components.

9. Capacitance matrix method. In the *capacitance matrix method*,[47] a Poisson (or Helmholtz) equation on a domain Ω is extended to a rectangle R. First a fast solver (Chapter 4, §§5–6) is used to solve $-\nabla_h^2 V = f$ (or $-\nabla_h^2 V + \lambda V = f$) for the extended problem. Next a dense linear system of equations (whose coefficients form the 'capacitance matrix') is solved to obtain the density of a spread of poles or dipoles at mesh-points on the boundary $\Gamma = \partial\Omega$. By construction, when the potential due to the spread is added to V, one gets the solution of the discrete problem both inside and on Γ.[48]

Consider the Poisson equation and Dirichlet boundary conditions

$$-\nabla^2 u = f(\mathbf{x}), \quad \mathbf{x} \in \Omega, \qquad u = g(\mathbf{y}), \quad \mathbf{y} \in \Gamma = \partial\Omega, \tag{9.1}$$

where Ω is a plane domain inside a rectangle R. If the boundary is smooth, the solution of (9.1) can be written as

$$u(\mathbf{x}) = F(\mathbf{x}) + M(\mathbf{x}), \tag{9.2}$$

where

$$\begin{aligned} F(\mathbf{x}) &= \int_\Omega \frac{1}{2\pi} \log \frac{1}{r} f(\xi)\, dR(\xi), \\ M(\mathbf{x}) &= \int_\Gamma \frac{1}{\pi} \frac{\partial}{\partial \nu} \log \frac{1}{r} \mu(\xi)\, dS(\xi). \end{aligned} \tag{9.2'}$$

Since $F(\mathbf{x})$ is expressed in terms of the given function $f(\mathbf{x})$, the solution of (9.1) is obtained by finding the dipole distribution $\mu(\mathbf{y})$ on Γ, for which $M(\mathbf{y}) = u(\mathbf{y}) - F(\mathbf{y})$. This satisfies an 'integral equation of potential theory', namely (see (4.4))

$$g(\mathbf{y}) - F(\mathbf{y}) = -\mu(\mathbf{y}) + \int_\Gamma \frac{1}{\pi} \frac{\partial}{\partial \nu} \log \frac{1}{r} \mu(\xi)\, ds(\xi). \tag{9.2''}$$

---

[47] See Buzbee-Dorr-George-Golub [71] and the references they cite. See also R. W. Hockney, J. Appl. Phys. 39 (1968), 4166–4170; Hockney [70]. Hockney attributes the method to O. Buneman.

[48] W. Proskurowski and O. Widlund, Math. Comp. 30 (1976), 433–468; and SIAM J. Sci. Stat. Comput. 1 (1980), 410–425. See also O. Widlund, *Lecture Notes in Mathematics*, vol. 631, Springer, 1978, 209–219. For higher order accuracy, see V. Pereyra, W. Proskurowski, and O. Widlund, Math. Comp. 31 (1977), 1–16; for the three-dimensional case, see D. P. O'Leary and O. Widlund, Math. Comp. 33 (1979), 849–879. See also W. Proskurowski, TOMS 5 (1979), 36–49; in Schultz [81, pp. 391–398]; in *Advances in Computer Methods for Solving Partial Differential Equations* IV, R. Vichnevetsky and R. S. Stepleman, eds., (1981), 274–280; and TOMS 9 (1983), 117–124.

Let there be a square mesh on R with mesh spacing h and mesh-points R_h. Let Ω_h denote the set of 'interior' mesh-points in Ω, whose four nearest neighbor mesh-points are also in Ω (see Fig. 1). At each of these, approximate the Poisson equation by $-\nabla_h^2 U = f$. Let $\partial\Omega_h$ denote the set of 'irregular' mesh-points, mesh-points in Ω which have at least one nearest neighbor mesh-point outside Ω (perhaps on Γ); and let p denote the number of these points. Let Γ_h denote the points on the boundary Γ intersected by mesh lines. On $\partial\Omega_h$ use the standard 5-point approximation, (4.2) of Chapter 3, with five nodes in $\Omega_h \cup \partial\Omega_h \cup \Gamma_h$; let $\mathbf{L}_h[U] = f$ denote the difference equation at these points. Finally, let $\Omega_h^* = R_h \cap (\Omega \cup \partial\Omega_h)'$ denote the 'exterior' mesh-points in the complement of $\Omega \cup \partial\Omega_h$. At these points use $-\nabla_h^2 U = f$, where f has been extended from Ω to R.

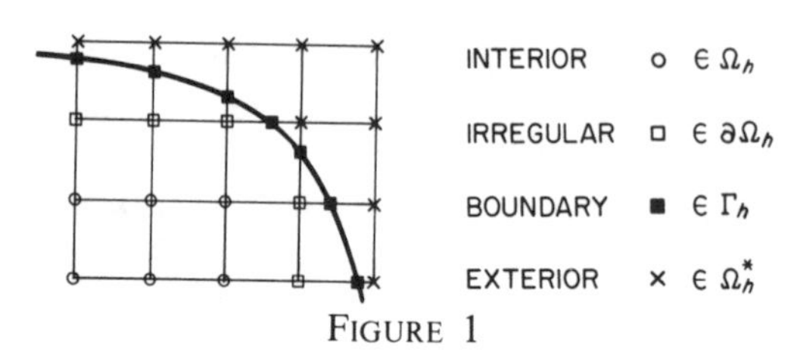

FIGURE 1

As usual, each ΔE is associated with the unknown at the central stencil point. Order the equations and unknowns so that nodal values in Ω_h come first, then those in $\partial\Omega_h$, and finally Ω_h^*. By transposing to the right side the terms involving the given boundary function g for points in Γ_h, we get the vector equation $A\mathbf{U} = \mathbf{b}$.

Now consider a second matrix B associated with the 5-point operator $-\nabla_h^2$ at all points of R_h (and for the same ordering as A). The entries of $A - B$ are zero except for the p rows for the equations associated with $\partial\Omega_h$. Following Proskurowski and Widlund, we write $A - B = WZ^T$, where W and Z have p columns; W has zero entries except in its p rows corresponding to the nodal values in $\partial\Omega_h$ and there it is the $p \times p$ identity, i.e., $W^T\mathbf{w}$ gives the restriction to $\partial\Omega_h$ of any mesh-function $\mathbf{w}$ on R_h and $W^T W = I$ is the $p \times p$ identity matrix.

In analogy to (9.2)–(9.2′), we write the solution of

$$A\mathbf{U} = (B + WZ^T)\mathbf{U} = \mathbf{b} \tag{9.3}$$

as $\mathbf{U} = B^{-1}\mathbf{b} + B^{-1}D\boldsymbol{\mu}$, where D has p columns, each representing a discrete dipole at a point of $\partial\Omega_h$, and $\boldsymbol{\mu}$ gives the (unknown) discrete dipole distribution at these points. Substitution, simplification, and multiplication by W^T give an equation for $\boldsymbol{\mu}$:

$$C\boldsymbol{\mu} = (W^T D + Z^T D)\boldsymbol{\mu} = -Z^T B^{-1}\mathbf{b}, \tag{9.4}$$

where the $p \times p$ matrix C is called the *capacitance matrix*. The capacitance matrix is dense; storage and cost limit the size of p when direct methods are used to determine $\boldsymbol{\mu}$. Proskurowski and Widlund suggest solving

$$C^T C\boldsymbol{\mu} = -C^T Z^T B^{-1}\mathbf{b}$$

by the conjugate gradient method (Chapter 5, §3) to determine $\boldsymbol{\mu}$.

10. Helmholtz equation. Another problem area in which classical methods continue to play a major role concerns the *Helmholtz* or 'reduced wave' equation

$$\nabla^2\phi + k^2\phi = 0. \tag{10.1}$$

This has the complex 'fundamental solution' $e^{ik|x|}$ in one space dimension, and e^{ikr}/r in three. It arises in the theories of sound, of electromagnetic waves, of 'seiches', and so on. Most of its applications stem from the fact that if ϕ satisfies Eq. (10.1), then 'standing waves' defined by $u(\mathbf{x},t) = e^{ickt}\phi(\mathbf{x})$ satisfy the wave equation $u_{tt} = c^2\nabla^2 u$; cf. Chapter 1, §5. The analytical approximation and numerical treatment of scattering and other problems involving (10.1) depend crucially on the ratio $d/\lambda = kd/2\pi$ of the diameter of the object to the wave length $\lambda = 2\pi/k$ under consideration. This ratio can easily vary from 0.01 (sound waves scattered by a raindrop) to 10^{10} (light waves reflected by a metal sphere).

The original applications were to sound waves in an organ pipe or room, with $\partial\phi/\partial n = 0$ as the physically appropriate boundary condition. This is an *eigenfunction* problem, whose first few eigenvalues $\lambda_j = k_j^2$ and corresponding eigenfunctions $\phi_j(\mathbf{x})$ can be accurately determined, but whose higher eigenvalues become increasingly dense, and the $\phi_j(\mathbf{x})$ correspondingly hard to isolate. We have already reported in Chapter 7, §10, on the use of FEM with singular elements to solve this problem and will not discuss it further here.[49]

We have also already discussed methods for solving the *modified* Helmholtz equation

$$\nabla^2\phi = k^2\phi + f(\mathbf{x}), \tag{10.2}$$

which can best be treated as a linear source problem with absorption. Especially relevant to this problem is the 'capacitance matrix' method, already mentioned in §9.

More interesting is the *radiation* problem. This is an *exterior* problem: one looks for a complex-valued solution $u(\mathbf{x})$ of (10.1) in the region $\Omega \subset \mathbb{R}^3$ *outside* a surface Γ, which satisfies Sommerfeld's radiation condition

$$\lim_{r\to\infty} r\,|\partial u/\partial r - i\,ku| = 0 \tag{10.3}$$

and a Neumann boundary condition

[49] See R. J. Riddell, Jr., J. Comp. Phys. 31 (1979), 21–59.

$$(10.3') \qquad \partial u/\partial n = g(\mathbf{y}) \quad \text{on} \quad \Gamma.$$

Integral equation methods have been used extensively in treating radiation problems. This approach requires knowing the Green's function; when it is not known, or impractical to compute, difference and finite element methods are of course still available.[50]

An area that has been intensively studied, because of the practical importance of radar, concerns microwave *transmission* in waveguides and related *scattering* phenomena. Here the Helmholtz DE (10.1) plays a central role. Thus it determines the TE-modes and TM-modes of transmission through cylindrical wave guides of general cross-section.[51]

Similar problems arise in sonar, which has important applications to underwater acoustics. Variational methods for treating these have been developed by Fix and Marin. C. I. Goldstein has analyzed an analogous variational method for treating wave guide problems.[52]

We also note the application by Goldstein and others of a 'weak element' approximation due originally to Milton Rose.[53]

Finally, an algorithm for solving the *three*-dimensional Helmholtz equation has been published by O'Leary and Widlund.[54] Solutions of the Helmholtz problem have been also obtained as limits of *time-dependent* solutions of parabolic and hyperbolic DE's (the latter with radiation conditions). We shall not discuss these approaches here.[55]

11. Rigorizing potential theory. It should be evident from the theoretical results and numerical evidence presented in this book, that one can solve a wide variety of elliptic boundary value problems with high accuracy, in very general regions, by applying simple arithmetic processes with sufficient perseverance. However, the *precise* generality with which these processes converge to a solution, and with which solutions exist or are unique, is very hard to determine. Thus Kellogg devotes an entire chapter [K, Chap. IV] to defining the class of 'regular regions' in which he claims to have proved the divergence theorem on which his later results

[50] A. K. Aziz and R. B. Kellogg, Math. Comp. 37 (1981), 261–272; C. I. Goldstein, Numer. Math. 38 (1981), 61–82.

[51] See D. S. Jones, *Methods in Electromagnetic Wave Propagation*, Clarendon Press, 1979. For the state of the art a decade earlier, see See IEEE Trans. on Microwave Theory and Techniques, vol. 17, no. 8.

[52] G. J. Fix and S. P. Marin, J. Comp. Phys. 26 (1978), 253–270; C. I. Goldstein, Math. Comp. 39 (1982), 309–324.

[53] M. E. Rose, Numer. Math. 24 (1975), 185–204. See also D. Greenspan and P. Werner, Archiv Rat. Mech. Anal. 23 (1966), 288–316, and R. E. Kleinman and G. F. Roach, SIAM Rev. 16 (1974), 214–236.

[54] D. P. O'Leary and O. Widlund, TOMS 7 (1981), 239–246.

[55] See, for example, G. A. Kriegsmann and C. S. Morawetz, SIAM J. Sci. Stat. Comp. 1 (1980), 371–385.

depend. Likewise, Gilbarg-Trudinger [77] discuss several *non*-existence theorems for mildly 'pathological' domains. In this book, we have tried to state only theorems which are plausibly motivated and for which rigorous proofs are available, even though we only sketch them. Generally speaking, such proofs become simpler in proportion as one is willing to make 'smoothness' assumptions. These assumptions progress from that of mere *continuity*, through those of continuous differentiability, of class C^k, of class C^∞, to that of *analyticity* (expandability in power series).

In these last two sections, we will try to give a clearer idea of the generality in which some of the basic theorems about elliptic boundary value and eigenproblems have been proved with 'Weierstrassian rigor', with references to other sources where detailed discussions can be found.

Riemann mapping theorem. We have stated versions of this in Chapter 2, §2, and in §6 of this chapter. Actually, it is sufficient that Ω be open, and that Ω be simply connected and omit two points.[56]

Korn's theorem. More generally, let S be *any* compact, simply connected Riemann surface (two-dimensional manifold) with parameters u, v, and distance differential

$$ds^2 = e(u,v)\,du^2 + 2f(u,v)\,du\,dv + g(u,v)\,dv^2, \qquad eg > f^2.$$

A deep classical result, Hilbert's 19th problem, first solved by Korn,[57] states that S can be mapped conformally onto a disk. That is, S can be reparametrized so that $e = g = 1$, $f = 0$, and the (u,v)-domain is $u^2 + v^2 \leqslant 1$.

Dirichlet principle. The classical result, that 'source problems' of the form

$$-\nabla \cdot [p(\mathbf{x}) \nabla u] + q(\mathbf{x}) u = f(\mathbf{x}), \qquad p(\mathbf{x}) > 0, \quad q(\mathbf{x}) \geqslant 0,$$

have one and only one solution for given continuous boundary values, has become so familiar that it is natural to accept this result as 'obvious' from the physical interpretations of Chapter 1, §§2–3. Actually, the existence of such solutions is very hard to prove rigorously, even in the simplest case ($p = 1$, $q = f = 0$) of the Dirichlet problem:

$$\nabla^2 u = 0 \quad \text{in } \Omega, \qquad u = g(\mathbf{y}) \quad \text{on } \partial\Omega. \tag{11.1}$$

Indeed, Dirichlet's 'proof' of existence was fallacious. It was based on the true Dirichlet principle which he discovered: that to satisfy (11.1) is equivalent (for 'smooth' functions) to minimizing the Dirichlet integral:

---

[56] The proof in Ahlfors [79, p. 230] is rigorous.

[57] See the article by Bers in Proc. Symp. Pure Math. XXVIII, F. Browder (ed.), Amer. Math. Soc., 1976, pp. 559-609.

$$\text{(11.2)} \qquad \mathbf{D}\langle u,u\rangle = \int_\Omega \nabla u \cdot \nabla u \, dR = \int_\Omega \Sigma_i \, (\partial u/\partial x_i)^2 \, dR,$$

subject to the specified boundary conditions $u = g(\mathbf{y})$ on $\partial\Omega$.[58] Dirichlet also observed that, since the integral (11.2) is positive (or zero in the trivial case $u = const.$), its values must have a greatest lower bound.

Dirichlet claimed that, in consequence, there must exist a function ϕ whose Dirichlet integral $\mathbf{D}\langle\phi,\phi\rangle$ attains this greatest lower bound m. He argued that, to find ϕ, it suffices to select a 'minimizing sequence' of functions $\phi_1, \phi_2, \cdots$ with $\mathbf{D}\langle\phi_k,\phi_k\rangle < m+2^{-k}$, and 'pass to the limit' as $k \to \infty$. This fallacious argument was accepted by Riemann, who used it to 'prove' the fundamental theorem of conformal mapping (§6) and to provide a basis for his general theory of integrals of algebraic functions (so-called 'Abelian integrals').

However, as Weierstrass pointed out, it is not clear why the ϕ_k should approach any limit at all, *or* that as $k \to \infty$, $\lim \mathbf{D}\langle\phi_k,\phi_k\rangle = \mathbf{D}\langle\phi,\phi\rangle$ just because $\phi_k \to \phi$. Hence Riemann's proof was fallacious, although his conclusions were essentially correct.

Also, even in the unit disk, any function $u(r,\theta)$ that assumes the boundary values $u(1,\theta) = 1$ on $(0,\pi)$ and $u(1,\theta) = -1$ on $(-\pi,0)$, must have an infinite Dirichlet integral; cf. Chapter 2, §1, Example 1a. This is because the sum $\pi\Sigma\, j(a_j^2 + b_j^2)/2$ is already infinite for the solution

$$\text{(11.3)} \qquad \frac{4}{\pi}\sum_{j=1}^{\infty} \frac{r^{2j-1}}{2j-1} \sin([2j-1]\theta).$$

Thus, for any other function converging as $r \uparrow 1$ to the same boundary values, it must be infinite *a fortiori*. Still another counterexample to the Dirichlet principle, which is physically less interesting but mathematically simpler, is provided by the sum of the series

$$\text{(11.3')} \qquad \phi = \sum_{k=1}^{\infty} k^{-2} r^{k^2} \sin k^2\theta = \mathrm{Im}\left\{\sum_{k=1}^{\infty} k^{-2} z^{k^2}\right\}.$$

This is a uniformly convergent series of orthogonal functions, whose sum is harmonic in Ω and continuous in $\overline{\Omega}$, yet whose Dirichlet integral is $(1+1+1+\cdots)\pi$.[59]

Hilbert tried to vindicate the Dirichlet principle as a so-called 'direct variational approach', in 1900. He did this by showing that, in the *plane*, one can construct solutions of the Dirichlet problem as limits of minimizing sequences. In spite of Courant's proud boast[60] that "Today Dirichlet's

[58] The true 'Dirichlet principle' was first derived by Kelvin (for the Neumann problem) and should presumably be called the 'Thomson-Dirichlet principle'.

[59] See J. Hadamard, *Oeuvres*, vol. III, 1245–1248.

[60] Courant [50, p. 3]. In general, Courant [50] provides an excellent exposition and historical analysis of the direct variational approach. The numerical implementation of this approach was discussed in Chapters 6 and 7, above.

principle has become a tool as flexible and almost as simple as that originally envisaged by Riemann", its rigorous justification is very difficult except in the case of plane regions; the same is true of various other related function-theoretic methods.[61]

The best proof of the existence of solutions of the Dirichlet problem, for a general region Ω in $n > 2$ dimensions, is based on the Poincaré-Perron 'méthode de balayage' to be explained in §12. For a more recent review of existence theorems, see M. V. Keldych, Uspehi Mat. Nauk 8 (1941), 171–231 (Math. Revs. 3, p. 123).

Such existence theorems obviously imply the existence (for each $\boldsymbol{\xi} \in \Omega$) of a *Green's function* in $\mathbb{R}^n$. The formula

$$G(\mathbf{x};\boldsymbol{\xi}) = g(\mathbf{x};\boldsymbol{\xi}) - V(\mathbf{x};\boldsymbol{\xi}), \qquad g(\mathbf{x};\boldsymbol{\xi}) = \frac{|\mathbf{x}-\boldsymbol{\xi}|^{2-n}}{(n-2)\omega_n}. \tag{11.4}$$

where ω_n is the surface area of the unit hypersphere in $\mathbb{R}^n$, obviously gives such a Green's function if V solves the Dirichlet problem for $\nabla^2 V \equiv 0$ in Ω and $V(\mathbf{y};\boldsymbol{\xi}) = g(\mathbf{y};\boldsymbol{\xi})$ on Γ (see Courant-Hilbert [62, p. 244]).

12. Méthode de balayage. Aware of the preceding facts (many of which had not been previously stated), Poincaré gave the first rigorous proof of an existence theorem for the Dirichlet problem of (11.1). He assumed only that Ω had a reasonably smooth boundary Γ (that it was a 'regular' solid),[62] and that $g(\mathbf{y})$ was continuous on Γ.

Then one can 'fill' Ω with a sequence of spheres $S_1, S_2, S_3, \cdots$, so that $\cup S_i = \Omega$, and construct a continuous interpolant $u_0(\mathbf{x})$ in Ω to $g(\mathbf{y})$ on Γ (such an interpolant always exists). Running through the S_i in the order S_1; S_1, S_2; S_1, S_2, S_3; $S_1, \cdots$, one can define $u_k(\mathbf{x})$ recursively by

$$u_k(\mathbf{x}) = \begin{cases} u_{k-1}(\mathbf{x}) & \text{outside } S_i, \\ P[u_{k-1}] & \text{in } S_i, \end{cases} \tag{12.1}$$

where P is the harmonic interpolant inside S_i to the values of u_{k-1} on ∂S_i.

Poisson integral formula. The harmonic interpolant used above is obtainable from the Poisson integral formula; to state and prove this rigorously in its full generality also requires sharp logical reasoning. We stated this formula for the unit disk in Chapter 2, §3, and for a sphere in

[61] See the articles by J. Serrin and E. Bombieri in Proc. Symp. Pure Math. XXVIII, Amer. Math. Soc., 1976, pp. 507–538. Contrary to what is stated in the Table of Contents, Serrin reviewed Problem 20 while Bombieri reviewed Problem 19.

[62] Technically, it suffices that each $\mathbf{y} \in \Gamma$ lie on a sphere S *exterior* to Ω, so that $S \cap \Omega = \varnothing$; see [K, pp. 326–329]. This hypothesis is not fulfilled at reentrant corners, and is unnecessarily strong.

§2 of this chapter. This formula is related to harmonic functions by *two* results: (a) any function $u(\mathbf{x})$ harmonic in $\bar{\Omega} = \Omega \cup \Gamma$, $\Gamma = \partial\Omega$, satisfies the Poisson integral formula, and (b) given $g(\mathbf{y})$ continuous on Γ, the surface of a 'ball' in $\mathbb{R}^n$ of radius a, the formula

$$u(\mathbf{x}) = \int_\Gamma P(\mathbf{x};\mathbf{y})g(\mathbf{y})\,dS(\mathbf{y}) = \mathbf{P}[g](\mathbf{x}), \qquad \mathbf{x} \in \Omega, \tag{12.2}$$

defines a function harmonic in Ω and continuous in $\bar{\Omega}$.

These results are *converses* of each other, in the sense that (a) *assumes* u to be harmonic and *deduces* (12.2), whereas (b) *assumes* (12.2) and *deduces* that $\nabla^2 u = 0$. They are very satisfactory, since only continuity is assumed in (b), and one can prove from (12.2) not only that $u \in C^2(\Omega)$, but even that $u(\mathbf{x})$ is *analytic* in Ω.

As a special case of (12.2), we have Gauss' theorem of the arithmetic mean: for $\mathbf{x}$ the *center* of Γ, $P(\mathbf{x};\mathbf{y})$ is the *constant* $1/\omega_n$, where ω_n is the hyperarea of the unit sphere. In this case we have a notable converse.

Converse of arithmetic mean theorem. Let $u(\mathbf{x})$ be continuous in $\bar{\Omega}$, and satisfy Gauss' theorem of the arithmetic mean there. Then $u(\mathbf{x})$ is harmonic ($\nabla^2 u \equiv 0$) and therefore *analytic* in Ω.

As $k \to \infty$, the $u_k(\mathbf{x})$ in (12.1) converge to a *limit* $u(\mathbf{x})$ which is continuous in $\bar{\Omega}$, takes on the given boundary values [K, pp. 311–319], and satisfies Gauss' theorem of the arithmetic mean in some neighborhood of every $\mathbf{x} \in \Omega$. It follows from the converse of this theorem, stated above, that $u(\mathbf{x})$ is harmonic and analytic in Ω, as well as continuous in $\bar{\Omega}$. This is a satisfactory existence theorem for the Dirichlet problem.

Generalizations. Although Gauss' theorem of the arithmetic mean has analogues for the Helmholtz, modified Helmholtz, and other constant-coefficient elliptic DE's, one must use very different methods to establish rigorously general existence theorems for variable-coefficient DE's such as the linear source problem.

Here we should again emphasize the extreme *difficulty* and *sophistication* of the subject, with which Hilbert's Problems 19, 20, and 23 were concerned. Specifically, "the scope of [Hilbert's] twentieth problem", concerned with the *existence* of solutions of [elliptic] boundary value problems, is "little less than encyclopaedic".[63] We conclude this chapter by mentioning a few more basic facts.

In the first place, the *linear* case is far better understood than the nonlinear case. Many general results may be found (perhaps with highly technical hypotheses) in the excellent books by Bers-John-Schechter [64], and Gilbarg-Trudinger [77]. However, the nonlinear case is a jungle; the

[63] From pp. 507–508 of J. Serrin's survey article in *Mathematical Developments Arising from Hilbert's Problems*, Felix Browder (ed.), Proc. Symp. Pure Math. XXVIII, Amer. Math. Soc., 1976.

first thing to realize is that very few problems (not even the Plateau problem) are fully understood.

A second important fact is the dependence of our knowledge on the number n of independent variables. The case $n = 1$ of two-endpoint problems for ordinary DE's is quite well understood. Moreover, many powerful function-theoretic tools are available for the case $n = 2$. Thus Hilbert's original (1900) vindication of the Dirichlet principle was for $n = 2$; the same is true of Serge Bernstein's existence theorem for quasilinear second-order elliptic DE's (1910–12), and its 1930 extension by Schauder. Third, Dirichlet-type boundary conditions are much easier to treat than those of Neumann or mixed type.

As a final indication of the unreliability of intuition and analogy in this area, we mention the fact that Hilbert's conjecture that "solutions of regular analytic variational problems must be analytic", turned out to be false.[64]

[64] G. Stampacchia, ibid., p. 620; see also pp. 26, 66.

Chapter 9

ELLPACK

1. Overview. In this chapter, we describe some capabilities of ELLPACK, a system for solving elliptic problems. This was constructed, beginning in 1976, with the help of many researchers (see Rice-Boisvert [84]) and coordinated by John R. Rice.[1] It consists of a high-level *language* to state the problem, a *preprocessor* to translate the user's description of the problem into Fortran, and a *library* of subroutines grouped into *modules* which carry out the tasks necessary to solve the problem. For a more complete descriptions of the ELLPACK language, modules, and many of its uses, see Rice-Boisvert [84].

ELLPACK is an extension of Fortran, designed to solve second-order elliptic DE's of the form

$$Au_{xx}+2Bu_{xy}+Cu_{yy}+Du_x+Eu_y+Fu=G \quad \text{on } \Omega, \tag{1.1}$$

for boundary conditions of the form

$$au+bu_x+cu_y=g \quad \text{on } \partial\Omega, \tag{1.1'}$$

where Ω is a bounded plane domain.[2] Here, as in Chapter 1, (1,1), $A, \cdots, G$, and $a, \cdots, g$, are given functions of position, and $AC > B^2$.

An ELLPACK *program* consists of a sequence of *statements* or 'segments', written in the simple, user-oriented ELLPACK *language*. These statements specify the problem, the discretization, the method of solution, and the type of output desired. Fortran statements can be intermixed with ELLPACK statements.

---

[1] ELLPACK is available for a license from The ELLPACK Project, Department of Computer Sciences, Purdue University, West Lafayette, IN 47907. Currently, the fee is $250 per year, which includes a tape and two copies of Rice-Boisvert [84].

[2] ELLPACK can also solve some systems of DE's and nonlinear problems; moreover it also has some three-dimensional capabilities.

For example, the problem $\nabla^2 u = -1$ on an L-shaped domain, with $u \equiv 0$ on the boundary, is described by

```
EQUATION.              UXX + UYY = - 1.
BOUNDARY. U = 0. ON LINE 0.,0. TO 2.,0. TO 2.,1. &
                 TO 1.,1. TO 1.,2. TO 0.,2. TO 0.,0.
```

The ampersand, &, at the end of a line signifies that the statement is continued on the next line. A square mesh with $h = 1/10$ and the 5-point approximation $\nabla_h^2 U = -1$ are specified by

```
GRID.  21 X POINTS 0. TO 2. $ 21 Y POINTS 0. TO 2.
DISCRETIZATION.         5 POINT STAR
```

The dollar sign, $, separates statements written on one line. To halve the mesh lengths, it suffices to replace the 21's above with 41's. The red-black ordering, solution by SOR, and a table of the values U_{ij} at mesh-points are obtained with

```
INDEXING.                 RED BLACK
SOLUTION.                    SOR
OUTPUT.                    TABLE(U)
```

These statements, followed by the terminator END, form a complete ELLPACK *program*. Note that a period must follow the keyword beginning each new segment.

The ELLPACK *system* is written in standard Fortran and is designed to be portable. The system processes an ELLPACK program by first translating it into a Fortran source program, called the ELLPACK *control program*. The ELLPACK *preprocessor*, which does the translation, determines the sizes of all arrays and inserts appropriate DIMENSION statements, named COMMON blocks, calls to ELLPACK subroutines, and numerous assignment statements which initialize variables, all as required to solve the problem stated in the user's ELLPACK program. This control program is compiled, linked to the precompiled ELLPACK subprograms in the ELLPACK library, and then executed to obtain the approximate solution of the user's problem.

The subprograms of the ELLPACK library are grouped into *modules* for specific kinds of tasks in the solution process. For example, a DISCRETIZATION module fills arrays with the values of the coefficients and the right side of a set of linear algebraic equations which approximate the DE and boundary conditions. An INDEXING module constructs arrays of pointers which, in effect, order the equations and the unknowns. A SOLUTION module solves the resulting linear algebraic system (vector equation). Several choices are available for each type of module, and one of them is selected in the ELLPACK user's program by giving its name, e.g., NESTED DISSECTION.

Uses of ELLPACK. ELLPACK was originally designed[3] as a research tool to help in evaluating algorithms for solving elliptic problems. In its present form, ELLPACK can be used to solve a large variety of problems with little programming (or debugging!) effort. The study of solutions of such problems can be helpful in designing special purpose software for solving large scale problems in engineering and physics.[4] In addition, it can be used as an educational tool which enables students to compare the accuracy and efficiency of various numerical methods.

2. Processing of an ELLPACK program. Consider $-\nabla^2 u = 4(r^2-1)$ in the unit disk $x^2+y^2 \leqslant 1$ and $u \equiv 0.75$ on $r=1$, with exact solution $u = r^2 - r^4/4$. The program below was used to generate the numbers listed in Table 2 of Chapter 3, §3.

```
* EXAMPLE 1
OPTIONS.      MAX X POINTS = 65 $ MAX Y POINTS = 65
EQUATION.     UXX + UYY = 4.*(1. - X**2 - Y**2)
BOUNDARY.     U = 0.75 &
     ON X = COS(T), Y = SIN(T) FOR T = 0. TO 2.*PI
FORTRAN.
         DO 20 I = 1, 5
            IGRID = 1 + 2**(I + 1)
            H     = 2./(IGRID - 1)
GRID.       IGRID X POINTS -1. TO 1.
            IGRID Y POINTS -1. TO 1.
DISCRETIZATION.   5 POINT STAR
INDEXING.         AS IS
SOLUTION.         BAND GE
OUTPUT.           MAX(ERROR)
FORTRAN.
          RATIO = R1NRMI/H**2
          PRINT 10, IGRID - 1, R1NRMI, RATIO
    10    FORMAT(/' N =',I5,', MAX ERROR ='F15.10,
     A       ', ERROR/H**2 =',F10.6)
    20 CONTINUE
SUBPROGRAMS.
        FUNCTION TRUE(X,Y)
        RSQR = X**2 + Y**2
        TRUE = RSQR - RSQR**2/4.
        RETURN
        END
END.
```

The first line is a comment, denoted by the * in column 1. The OPTIONS segment specifies that subgrids of a 65×65 grid will be used.

---

[3] See J. R. Rice in Rice [77, 319–341], and in Schultz [81, pp. 135–162].

[4] See P. Gherson, P. S. Lykoudis, and R. E. Lynch, Seventh Internat. Conf. on MHD Electrical Power Generation II, MIT, (1980), 590–594.

The boundary is given parametrically; PI is an ELLPACK reserved word for the number π. The lines between FORTRAN (or SUBPROGRAMS) and the next segment name are user-supplied Fortran statements.

In each pass through the DO-loop, a uniform grid is set up, a system of algebraic equations is constructed, and the system is solved by band Gauss elimination. Also, the maximum of the error at mesh-points is computed as requested in the OUTPUT segment; it is stored by ELLPACK in R1NRMI. In the second FORTRAN segment, $||e||_{max}/h^2$ is computed and the information in Table 2 of Chapter 3, §3, is printed.

Whenever ELLPACK is asked to compute an error, it evaluates U(X,Y) − TRUE(X,Y), where U(X,Y) is the ELLPACK generated approximation and the Fortran function TRUE(X,Y) must be user-supplied in the SUBPROGRAMS segment.

Execution. We outline how the program above is processed; the actual execution is more complicated than our simplified description of it. A similar procedure applies to any ELLPACK program.

The ELLPACK preprocessor analyzes the program line-by-line, and constructs a Fortran *source program* called the ELLPACK *control program*. It translates this information into Fortran, appending the Fortran statements onto one of several files to form pieces of the control program. One file, say 'file1', contains the Fortran declarations and COMMON blocks for the 'main' program of the control program. Another, say 'file2', contains the executable statements, and, say 'file3', contains subprograms.

In the control program, ELLPACK uses names with their first character a C, I, L, R, or Q, respectively, for common blocks, variables of type integer, logical, and real, and for subprogram names. The second character of these is a digit (0,1, ···,9). For example, R1GRDX and R1GRDY are the names of the arrays which contain the coordinates on the axes of the grid lines; I1GRDX and I1GRDY are the variables containing the sizes of these arrays. Such names should be avoided in user-supplied Fortran statements, unless an ELLPACK variable is needed in the user's program (as R1NRMI is in EXAMPLE 1).

When the OPTIONS segment of EXAMPLE 1 is analyzed, the preprocessor defines the arrays R1GRDX and R1GRDY and their sizes (each 65). Specifically, the preprocessor translates this information into two pieces of Fortran code: a pair of array and COMMON declarations and a pair of assignment statements. The preprocessor appends the Fortran statements

```
COMMON / C1GRDX / R1GRDX(65)
COMMON / C1GRDY / R1GRDY(65)
```

to the declarations file, 'file1', and the assignment statements

```
I1GRDX = 65
I1GRDY = 65
```

to the file of executable statements, 'file2'. These arrays and their dimensions are passed to those subprograms which need them, in the subprograms's argument lists.

When the EQUATION segment is analyzed, the preprocessor determines which coefficients are zero, constant, or variable. It assigns values to several logical switches that code this information about the DE, and these are available to the modules. In EXAMPLE 1, the DE is the Poisson equation, and the logical variables specifying 'constant coefficients', 'Poisson equation', and 'homogeneous DE' are assigned values; this is done by appending the statements

```
L1CSTC = .TRUE.
L1POIS = .TRUE.
L1HMEQ = .FALSE.
```

to 'file2', containing the executable statements of the main program.

To enable discretization modules to evaluate the right side of the DE, the preprocessor constructs the subprogram Q1RHSD. For EXAMPLE 1, it appends

```
FUNCTION Q1RHSD(X,Y)
Q1RHSD = 4.*(1. - X**2 - Y**2)
RETURN
END
```

to the subprograms file, 'file3'. The only thing which changes in the function Q1RHSD from run to run is the right side of its assignment statement. The preprocessor just copies whatever is on the right side of the equal sign in the user-supplied EQUATION segment to the right side of this assignment statement. Thus, the right side of the DE could have been F(X,Y), in which case the second line of function would be Q1RHSD = F(X,Y); the user must then supply F(X,Y) as a FUNCTION in the SUBPROGRAMS segment.

Similarly, the preprocessor constructs the subroutine Q1PCOE to evaluate the coefficients of the DE. When the user's BOUNDARY segment is analyzed, the preprocessor constructs subprograms which locate points on the boundary and evaluate the given boundary conditions there. These are appended to 'file3'.

The user-supplied Fortran statements following the first FORTRAN in EXAMPLE 1 are merely appended to 'file2', containing the executable main program statements. When the GRID segment is analyzed, the preprocessor appends to 'file2' a call to the domain processor in the ELLPACK library. Later, when the program executes, the domain processor constructs a grid and stores information about the location of the boundary with respect to the mesh lines. The DISCRETIZATION, INDEXING, SOLUTION, and OUTPUT segments are handled the same way. Also, more named COMMON blocks are appended to the declarations file,

'file1', with dimensions as appropriate for the modules requested. Next, some more of the user's Fortran statements (after the second FORTRAN in EXAMPLE 1) are copied onto 'file2'.

The SUBPROGRAMS segment must be the last one in the user's ELLPACK program, and when it is reached, the preprocessor completes the main program of the control program and appends the user's subprograms to 'file3', the subprogram file. At this point, all the necessary Fortran statements are in the three files. The files are concatenated to form the ELLPACK control program to solve the user's problem. This program is then compiled, linked to the ELLPACK library, and executed.

3. Problem definition. ELLPACK segment names, such as EQUATION, BOUNDARY, and so on, begin in column 1 of a line (or card) and end with a period. These names may be abbreviated, but at least the first two letters of a name must be used. A star * in column 1 denotes a comment. A dollar sign, $, is used as a separator when more than one statement is written on a line. An ampersand, &, at the end of a line denotes that the statement is continued on the next line. Except for column 1, segment names, and user-supplied Fortran statements, ELLPACK programs are written in free format (blanks are ignored). Column 1 of an ELLPACK statement must contain the first letter of a segment name, a *, or a blank; for Fortran statements, see §7.

Fortran arithmetic expressions, such as +3., EXP(X+Y)*SIN(X), etc., are included in some segments. Fortran statements can be freely mixed between ELLPACK segments; the segment names FORTRAN and SUBPROGRAMS indicate that Fortran statements follow up to the next segment name.

We now describe in more detail the segments which, as in the examples of §§1 and 2, define the problem and which allow user-supplied Fortran subprograms to be included in an ELLPACK program.

EQUATION segment. An ELLPACK program can have one EQUATION segment which specifies the elliptic DE to be solved; for example:

```
EQ. -3.*UXX -*UYY + 5.*UX + X**2*U = EXP(X+Y)+2.*X
```

The keywords U, UX, UY, UXX, UXY and UYY stand for the unknown and its first two derivatives; X, Y are used for the independent variables (Z, UZ, UZZ, and so on, are included in three-dimensional problems). Any Fortran real arithmetic expression may be used for the coefficients and right side of the DE.

After a SOLUTION or TRIPLE segment has been executed, the ELLPACK generated Fortran functions U(X,Y), UX(X,Y), UXX(X,Y), etc., are available to the user. In FORTRAN segments, these can be used to evaluate the approximate solution and its first two derivatives.

SUBPROGRAMS segment. The SUBPROGRAMS segment allows user-supplied Fortran subprograms to be incorporated into an ELLPACK program. A single SUBPROGRAMS segment may be used; it must immediately precede the END. It consists of one or more complete user-supplied Fortran SUBROUTINE or FUNCTION subprograms.

The next program fragment illustrates one of the uses of this segment; here the coefficients and right side of the DE are Fortran functions:

```
EQUATION.   C(X,Y,1)*UXX + C(X,Y,2)*UYY = C(X,Y,3)
     . . .
SUBPROGRAMS.
        FUNCTION C(X,Y,ICOF)
        IF(ICOF .EQ. 1) C = - 1. + 0.3*SIN(X + Y)
        IF(ICOF .EQ. 2) C = - 3. + (X**2 + Y**2)/10.
        IF(ICOF .EQ. 3) C = ATAN2(Y,X) + 2.
        RETURN
        END
END.
```

A differential equation in self-adjoint form,

$$(p(x,y)u_x)_x + (q(x,y)u_y)_y - r(x,y)u = f(x,y), \tag{3.1}$$

is specified by

```
EQ. (P(X,Y)*UX)X + (Q(X,Y)*UY)Y - R(X,Y)*U = F(X,Y)
```

either with the functions P, Q, R, and F defined in the SUBPROGRAMS segment, or replaced with explicit arithmetic expressions such as X**2 + 1.

BOUNDARY segment. Boundary conditions in ELLPACK have the form

$$a(x,y)u + b(x,y)u_x + c(x,y)u_y = g(x,y). \tag{3.2}$$

For each piece of the boundary, a boundary condition of the form (3.2) is given, and the piece itself is specified parametrically. Successive pieces must have a common endpoint; they are listed so that the boundary is traversed counterclockwise, with the interior of the domain to the left. Straight-line pieces can be given with LINE and their endpoints. If no boundary condition is given on a piece, that for the preceding piece is used, as in the example of §1.

The boundary of the half-annulus $1 \leqslant r \leqslant 2$, $0 \leqslant \theta \leqslant \pi$ (polar coordinates), and the boundary conditions

$$\begin{aligned} u_n &= \sin\theta && \text{on } r = 2, && 0 < \theta < \pi, \\ u &= 0 && \text{on } 1 \leqslant r \leqslant 2, && \theta = 0 \text{ and } \theta = \pi, \\ u &= \theta^2 - \pi\theta && \text{on } r = 1, && 0 < \theta < \pi, \end{aligned}$$

can be specified in ELLPACK by

```
BOUNDARY.   COS(T)*UX + SIN(T)*UY = SIN(T) &
  ON X = 2.*COS(T), Y = 2.*SIN(T) FOR T = 0. TO PI
      U = 0. ON LINE -2.,0. TO -1.,0.
      U = (PI - T)**2 - (PI - T) &
  ON X = -COS(T),    Y = SIN(T)      FOR T = 0. TO PI
      U = 0. ON X = R, Y = 0.        FOR R = 1. TO 2.
```

Here the first and third statements are continued to the next line by the ampersands. The second piece is given as a LINE; the fourth piece could have been given similarly.

A rectangle with sides parallel to the coordinate axes can be specified simply by giving its sides. Thus, if the DE defined in the EQUATION segment of the preceding example is in polar coordinates with X and Y denoting r and θ, respectively, then one can use

```
BO. UX  = SIN(Y)          ON X = 2. $ U = 0. ON Y = PI
    U   = Y**2 - PI*Y     ON X = 1. $ U = 0. ON Y = 0.
```

where the $ separates statements written on the same line.

Multiply connected domains can be defined with the HOLE and ARC segments (see Rice-Boisvert [84] for details).

GRID segment. A rectangular mesh[5] over the domain is specified by the GRID segment; the mesh (or 'grid') lines are parallel to coordinate axes. To get a uniform mesh, the user simply gives the number of mesh lines (including boundary lines) in both directions and the coordinates of the first and last lines. If the domain is a rectangle, then only the number of horizontal and vertical mesh lines need to be specified.

Nonuniform spacing is obtained by listing the coordinates of the mesh lines. In all cases, the domain of the problem must be between the first and last mesh lines in both directions.

The following GRID specifications are typical:

```
GRID. 9 X POINTS               $ 11 Y POINTS
GRID. 9 X POINTS  .2 TO  .8    $ 21 Y POINTS 0. TO PI/2.
GRID. 4 X POINTS -.5,  .25,   .75,  1.
      7 Y POINTS 0.,   .125,  .25,  .5,  1.,  2.,  4.
```

The first of these grids can be used only for a rectangular domain, the other two can be used for any domain. In the second grid, the spacing is .6/8 in the x-direction and $\pi/40$ in the y-direction. Numerical values, such as 9 and .25, which appear in ELLPACK segments, can be replaced with Fortran expressions which evaluate to appropriate constants.

At most one EQUATION, BOUNDARY, and SUBPROGRAMS segment may appear in an ELLPACK program. Several GRID segments may appear, provided the finest grid is specified in the OPTIONS segment by MAX statements; see §6.

[5] Except for the discretization module P2C0 TRIANGLES (see §4), ELLPACK constructs a rectangular mesh.

4. DISCRETIZATION modules. The DISCRETIZATION segment specifies the discrete approximation scheme used to solve the DE. Because so many special methods have been developed for treating rectangular and box-shaped domains, it is hardly surprising that the same is true of the ELLPACK discretization modules. Most of these modules require INDEXING and SOLUTION segments to get a solution. A TRIPLE segment specifies a module which discretizes and also solves a problem; see §8.

Finite difference modules. We begin with approximation schemes based on finite differences.

5 POINT STAR uses the 5-point divided difference approximation to

$$Au_{xx} + Cu_{yy} + Du_x + Eu_y + Fu = G, \tag{4.1}$$

where the coefficients and right side depend on x and y. This module can be used on general domains with general boundary conditions (3.2) on a nonuniform mesh. If the equation is given in self-adjoint form (as in the EQ displayed after (3.1)) and the domain is a rectangle, then the standard five-point approximation is used (Chapter 3, (4.2)).

7 POINT 3D is the corresponding approximation for three-dimensional problems on boxes. It is limited to DE's in self-adjoint form.

The 9-point HODIE approximations (Chapter 3, §11) are restricted to rectangular domains and uniform meshes. The module HODIE solves (1.1);[6] the other HODIE modules require that $B \equiv 0$. The module HODIE ACF requires that $D = E \equiv 0$, and HODIE HELMHOLTZ requires that $A = C \equiv 1$ and $D = E \equiv 0$ (see the TRIPLE segments for other HODIE modules).

Finite element modules. There are three types of finite element modules: collocation, Rayleigh-Ritz, and Galerkin (for the last, see TRIPLE).

COLLOCATION, HERMITE COLLOCATION, and INTERIOR COLLOCATION use the bicubic Hermite collocation approximation (Chapter 6, §6). The first is used for general boundary conditions or for a general domain; the second and third are restricted to problems on rectangles. The third is the most efficient because it eliminates boundary unknowns (coefficients of the bicubic Hermite approximation) and hence uses fewer unknowns and matrices with smaller bandwidth; it is restricted to problems with either Dirichlet or Neumann boundary conditions. These modules can treat equations with a cross-derivative term (B in (1.1) nonzero). They require that the boundary be specified in a *clockwise* direction.

SPLINE GALERKIN uses Rayleigh-Ritz approximation (Chapter 6, §2) to self-adjoint DE's on rectangles; it uses a tensor product basis of B-splines of degree k with user-specified global continuity as given by the

[6] HODIE also solves the three-dimensional analogue of (1.1) on boxes.

parameters DEGREE and NDERV. The default values are DEGREE = 3 and NDERV = 2, which gives bicubic splines (degree 3 and continuous second derivatives). For example, to get continuous bilinear or bicubic Hermite approximation one can use, respectively,

```
DISCRETIZATION. SPLINE GALERKIN (DEGREE=1,NDERV=0)
DISCRETIZATION. SPLINE GALERKIN (NDERV=1)
```

As illustrated, parameters appear between parentheses when they are used.

5. INDEXING and SOLUTION modules. As explained in Chapter 4, the ordering of the equations and the unknowns affects the cost and accuracy of solving a linear system $A\mathbf{U} = \mathbf{b}$. The coefficients and right sides of the equations are computed by a DISCRETIZATION module in some sequential order, and the ordering of equations and unknowns is left unchanged by the module AS IS.

HERMITE COLLORDER can be applied to the HERMITE COLLOCATION discretization of a DE on a rectangle to obtain a matrix with nonzero diagonal entries.

MINIMUM DEGREE applies a minimum degree algorithm (Chapter 4, §3) to $A + A^T$, where A is the coefficient matrix of the linear system as generated by the DISCRETIZATION module. It attempts to obtain an ordering which minimizes the fill-in during band or profile Gauss elimination.

NESTED DISSECTION does the dissection outlined in Chapter 4, §3. The parameter NDTYPE = k, with $k = 5, 9, 7,$ or 27 gives the type of finite difference stencil: 5- or 9-point in the plane and 7- or 27-point in space.

RED BLACK rearranges the equations and unknowns into a red black ordering (Chapter 4, §3).

REVERSE CUTHILL MCKEE applies the reverse Cuthill-McKee algorithm (Chapter 4, §3) prior to solving the system with a profile or band elimination method.

SOLUTION modules. There are several modules which solve linear systems; some use direct and some use iterative methods.

Direct methods. BAND GE solves the system by band Gauss elimination. Scaled pivoting is used, that is, the equations are normalized by making the largest coefficient of each row unity before solving the system with pivoting (see Chapter 6, §6). BAND GE NO PIVOTING solves the system without pivoting.

ENVELOPE uses the envelope method (Chapter 4, §2). It determines whether or not the matrix is symmetric and produces the LDL^T or LDU factorization; then it solves the vector equation.

LINPACK BAND is a band solver which pivots but it does not scale.

LINPACK SPD BAND solves symmetric positive definite banded systems without pivoting.

SPARSE GE PIVOTING, (or GE NO PIVOTING, LU PIVOTING, LU COMPRESSED, LU UNCOMPRESSED, LDLT) uses sparse matrix techniques[7] and Gauss elimination to form the LU or LDL^T factorization and then solve the vector equation; there are two modes of storage available.

Iterative methods. ELLPACK includes seven modules from ITPACK[8] which solves a linear system by one of seven iterative methods. These modules accept parameters which limit the number of iterations, specify the desired accuracy, and so on; alternatively, the user can use the default values. We mention only four parameters: ITMAX $= m$ specifies the maximum number of iterations; ZETA $= e$ specifies the desired accuracy; OMEGA $= \omega$ gives the value of the SOR overrelaxation parameter; and UINIT = .TRUE. [.FALSE.] indicates that the initial estimate is [is not] user-supplied. The default values for these parameters are 100, $\max\{500*\text{macheps}, 10^{-6}\}$, 0, and .FALSE., respectively. For example,

```
SOLUTION.   SOR ( ITMAX = 150, ZETA = 0.0001 )
```

invokes the SOR iterative scheme, which stops automatically when the estimated error is 0.0001 or after 150 iterations. Here the initial guess $U \equiv 0$ is used; the user can specify a difference guess with the TRIPLE SET segment (see §8).

JACOBI CG, SYMMETRIC SOR CG, and REDUCED SYSTEM CG apply conjugate gradient acceleration (Chapter 5, §6) to the Jacobi, SSOR, and the reduced system one-step methods (Chapter 4, §7, Chapter 5, §§2 and 5).

JACOBI SI, SYMMETRIC SOR SI, and REDUCED SYSTEM SI apply the Chebyshev semi-iterative method (Chapter 5, §4) to these schemes.

Some of the parameters used in the ITPACK routines are obtained adaptively by the procedures outlined in Chapter 5, §7.

6. OUTPUT, OPTIONS, and PROCEDURE modules. To get printed and graphical output from ELLPACK, one specifies what is desired in an OUTPUT segment. For example, execution of

```
OUT.  TABLE(U)  $  TABLE(UXX,9,9)  $  PLOT(U)  $  MAX(U)
```

gives tables of the computed solution at mesh-points of the grid and of the second derivative with respect to x of the solution on a 9-by-9 mesh. The last two statements produce a contour plot[9] of the solution and the discrete l_1, l_2, and l_∞ norms of the solution. When the user supplies TRUE(X,Y), MAX(ERROR) lists these norms of the error; they are stored

[7] These routines are part of the Sparse Matrix Package developed at Yale University.

[8] This is available from IMSL as a separate package of subroutines.

[9] ELLPACK uses the contour plotting routine of W. V. Snyder, Algorithm 531, TOMS 4 (1978), 290; a subroutine DRAW must be supplied which invokes the plot routine of the local computer system.

in the variables R1NRL1, R1NRL2, and R1NRMI, and can be used in user-supplied FORTRAN segments (see §2 and §10 for examples).

OPTIONS segment. The execution time of each module and the amount of memory used are printed when TIME and MEMORY are listed in an OPTIONS segment. For example:

```
OPTIONS.              TIME $ MEMORY
```

When more than one GRID is used, the maximum number of horizontal and vertical mesh lines must be given in MAX statements in an OPTIONS segment; for example:

```
OPTIONS.   MAX X POINTS = 51 $ MAX Y POINTS = 66
```

allows grids with at most 51 vertical and 66 horizontal mesh lines.

If the user gives the boundary with the clockwise orientation (e.g., when collocation is used), then

```
OPTIONS.           CLOCKWISE = .TRUE.
```

must appear before the BOUNDARY segment.

PROCEDURE modules. The PROCEDURE modules perform tasks which are not included in the usual discretization and solution process.

After an INDEXING segment, DISPLAY MATRIX PATTERN shows which entries in the $N \times N$ matrix are nonzero. Because of the amount of output, N is limited to about 100.

EIGENVALUES uses EISPACK routines to compute the eigenvalues of the matrix of the linear system after INDEXING.

Other procedures allow the user to subtract a function v from the solution; instead of solving $L[u]=f$, ELLPACK solves $L[u-v]=f-L[v]$ and adds v to U. For example, v might be a blending function (see Chapter 7, §9) so that the boundary conditions for $u-v$ are homogeneous.

7. Fortran segments. There are four segments in which users can insert Fortran statements[10] into an ELLPACK program. Groups of Fortran statements must be preceded by one of the segment names FORTRAN, DECLARATIONS, GLOBAL, or SUBPROGRAMS. As usual, the user-supplied Fortran statements begin in column 7; continuation is indicated by a character in column 6; and so on. Moreover, the first column of a Fortran statement in an ELLPACK program must be either a blank, a C, a $,[11] or (for a label) a 1.

---

[10] Fortran statements (and arithmetic expressions used in ELLPACK segments) must follow the rules of the dialect of Fortran which is used by the local computer system; e.g., a statement begins in column 7. Throughout we use Fortran real arithmetic expressions. Some installations use a double precision version of ELLPACK in which case Fortran double precision arithmetic expressions should be used.

[11] The ELLPACK preprocessor removes the $ and the characters on the line are moved to the left. This is to allow control cards to be inserted into an ELLPACK program between each Fortran subprogram for those computer systems which require them.

The statements in a FORTRAN segment are inserted into the preprocessor generated ELLPACK control program intermixed with calls to the ELLPACK library routines as specified by other segments and in the order they appear in the ELLPACK program.

If user-defined subprograms are called in a FORTRAN segment, then the subprograms themselves must appear in a SUBPROGRAMS segment which immediately precedes the terminator END.

If the user-supplied Fortran statements require arrays or the specification of variable types and named[12] COMMON blocks, these are given in a DECLARATION or GLOBAL segment. The first is used when the declared variables are used only by the user's subprograms. Declarations in a GLOBAL segment are inserted in all the ELLPACK generated subprograms. Examples illustrating the use of the DECLARATION and GLOBAL segments are given in §9 and §10.

When different methods are compared for accuracy and efficiency, one often solves collections of problems with known solutions. Such collections can be generated by first choosing the domain, the coefficients of the DE and of the boundary equation, and the *solution*. Then one constructs the right sides of the DE and the boundary conditions. When this is done, the errors in the computed solution can be determined. The Fortran function TRUE(X,Y) (or TRUE(X,Y,Z)), which must be user-supplied in the SUBPROGRAMS segment, is used by ELLPACK when output pertaining to the error is requested; see §2 for an example.

8. TRIPLE segments. Some discretization modules also generate the solution, for example, routines which use FFT, tensor products, or marching methods. For these the INDEXING and SOLUTION segments are not used.

CMM HIGHER ORDER uses the capacitance matrix method (Chapter 8, §9) and deferred correction to solve the Poisson equation on a general domain with Dirichlet or Neumann boundary conditions. For the same boundary conditions, CMM EXPLICIT and CMM IMPLICIT solve the Helmholtz equation

$$-u_{xx}-u_{yy}+\lambda u=f(x,y).$$

The first should be used when the capacitance matrix fits into core and, otherwise, the second.

DYAKANOV CG solves the standard 5-point approximation to separable self-adjoint equations with general boundary conditions on a rectangle. A conjugate gradient iteration is applied (Chapter 5, §6), with a

[12] The ELLPACK system uses blank COMMON; if the user needs a COMMON block, it must be a named COMMON block.

preconditioning matrix which is a scaled separable approximation to the coefficient matrix of the linear system; then the generalized marching method is applied (Chapter 4, §5).

DYAKANOV CG 4 is similar to DYAKANOV CG, but the problem is solved on two meshes and fourth-order accuracy is obtained with Richardson extrapolation (Chapter 3, §§7 and 10).

FFT 9 POINT uses a HODIE 9-point approximation to

$$Au_{xx} + Cu_{yy} + f(\mathbf{x})u = g(\mathbf{x}),$$

with A, C constants, on a rectangle with a uniform mesh and Dirichlet boundary conditions. It uses FFT and cyclic reduction (Chapter 4, §6) to obtain the solution. An $O(h^p)$ approximation is used, where p is specified by the parameter IORDER $= p$, $p = 2$, 4, or 6. The choice $p = 6$ is restricted to the Poisson equation on a square mesh. The number of mesh lines (X POINTS or Y POINTS) is restricted to $1+2^k$, $k = 3, \cdots, 7$.

FISHPAK HELMHOLTZ solves the constant coefficient Helmholtz equation with variable right side on a rectangle for Dirichlet, Neumann, or periodic boundary conditions by FFT. The standard 5-point approximation is used on a uniform mesh.

HODIE 27 POINT 3D solves the $O(h^6)$ HODIE approximation (Chapter 3, §11) to the Poisson equation on a box with Dirichlet boundary conditions. A cubic mesh is used and the solution is obtained by tensor product evaluation (Chapter 4, §6).

HODIE FFT uses FFT (Chapter 4, §6) to solve the $O(h^6)$ HODIE approximation to the Poisson equation on a rectangle.

MARCHING ALGORITHM solves separable self-adjoint problems with general boundary conditions on a rectangle. The 5-point approximation on a uniform mesh is solved by the marching method (Chapter 4, §5).

MULTIGRID MG00 uses the multigrid method (Chapter 5, §10) to solve (4.1) in which D and E are zero.

P2C0 TRIANGLES[13] uses a triangulation of the GRID. There is no restriction on the DE, the domain, or the boundary conditions. It uses a Galerkin approximation with continuous piecewise quadratic triangular elements (Chapter 7, §4). The domain is subdivided adaptively in an attempt to obtain a uniform error distribution. The user specifies the number of triangles with the parameter NTRI, e.g.:

```
TRIPLE.   P2C0 TRIANGLES (NTRI = 50)
```

The user can assign values to the nodal values with SET U = F, where F is the name of a user-supplied function with arguments (X,Y); this can be

[13] This is a specialization of a much more general routine called TWODEPEP, written by E. G. Sewell and available as a separate package from IMSL.

used to construct the first approximation for an iterative method. For example

```
TRIPLE.    SET ( U = QUAD )
  . . .
SOLUTION. JACOBI CG (ITMAX = 150, UINIT = .TRUE.)
  . . .
SUBPROGRAMS.
        FUNCTION QUAD(X,Y)
        QUAD = X*(1. - X)*Y*(1. - Y)
        RETURN
        END
END.
```

The user can also set the values of U with a blending function or a Hermite piecewise bicubic polynomial which interpolates at point of a rectangular boundary (see Rice-Boisvert [84]).

9. Sample ELLPACK programs. Here we give some sample ELLPACK programs and parts of programs, to illustrate how ELLPACK can be used to solve some of the problems discussed in Chapters 3–8, and to show how ELLPACK can be used to compare solution methods.

Slit square. We begin with the Model Problem $\nabla^2 u = -1$ on the slit square considered in Chapter 3, §7. Here $\Omega = (0,1) \times (-1,1)$, and the boundary conditions are $u_x(0,y) = 0$, $0 < y < 1$, and $u = 0$ on the rest of the boundary. An ELLPACK program which solves this problem is:

```
EQUATION.          UXX + UYY = -1.
BOUNDARY. U   = 0. ON LINE 0.,0. TO 0.,-1. &
              TO 1.,-1. TO 1.,1. TO 0., 1.
          UX  = 0. ON LINE 0.,1. TO 0., 0.
GRID.              11 X POINTS $ 21 Y POINTS
DISCRETIZATION.    5 POINT STAR
INDEXING.          AS IS
SOLUTION.          LINPACK BAND
OUTPUT.            TABLE(U)
END.
```

Effect of order of accuracy. To find the maximum error for $O(h^2)$, $O(h^4)$, and $O(h^6)$ HODIE approximations to the Poisson equation using FFT on five grids, the following ELLPACK program can be used:

```
EQUATION.        UXX + UYY = F(X,Y)
BOUNDARY.   U = 0. ON Y = 0. $ ON X = 1.
                   ON Y = 1. $ ON X = 0.
OPTIONS.    MAX X POINTS = 129 $ MAX Y POINTS = 129
FORTRAN.
        DO 100 NGRID = 1, 5
           NPNTS = 2**NGRID + 1
GRID.      NPNTS X POINTS $ NPNTS Y POINTS
FORTRAN.
           DO 50 IORD = 2, 6, 2
```

```
TRIPLE.       FFT 9 POINT (IORDER = IORD)
OUTPUT.       MAX( ERROR )
FORTRAN.
   50      CONTINUE
  100 CONTINUE
SUBPROGRAMS.
      FUNCTION TRUE(X,Y)
  . . .
      FUNCTION F(X,Y)
  . . .
END.
```

Comparing methods.[14] Tables 2 and 3 of §§5 and 6 in Chapter 5 give results for Chebyshev and conjugate gradient acceleration of one-step iterative schemes for solving the Model Problem. The results for the second of these tables are obtained with the following program.

```
OPT. TIME $ MAX X POINTS = 65 $ MAX Y POINTS = 65
EQUATION. UXX + UYY = -1.
BOUNDARY. U = 0. ON LINE 0.,0. TO 1.,0. TO 1.,1. &
                                TO 0.,1. TO 0.,0.
FORTRAN.
      DO 10 I = 3, 6
         IGRID = 1 + 2**I
GRID.    IGRID X POINTS $ IGRID Y POINTS
DISCRETIZATION.  5 POINT STAR
INDEXING.        AS IS
SOLUTION.  JACOBI CG           (IMAX=100,ZETA=0.0001)
SOLUTION.  SYMMETRIC SOR CG    (IMAX=100,ZETA=0.0001)
INDEXING.        RED BLACK
SOLUTION.  REDUCED SYSTEM CG   (IMAX=100,ZETA=0.0001)
   10 CONTINUE
END.
```

Here only a single discretization is required. Both the JACOBI CG and SYMMETRIC SOR CG use the AS IS (or 'natural') ordering of equations and unknowns; the REDUCED SYSTEM CG requires the RED BLACK ordering.

Finite elements. The next program determines the error in Rayleigh-Ritz approximation to the problem of Example 13 in Chapter 3, §11:

$$-(e^{xy}u_x)_x-(e^{xy}u_y)_y+2(x^2+y^2)e^{xy}u=0,$$

on the unit square with $u=e^{xy}$ on its boundary; the exact solution is $u=e^{xy}$. Three types of elements are compared: bilinear, bicubic Hermite, and cubic splines.

---

[14] For a 'population' of elliptic problems, see J. R. Rice, E. N. Houstis, and W. R. Dyksen, Math. Comp. 36 (1981), 475–484.

```
EQUATION.  (P(X,Y)*UX)X+(P(X,Y)*UY)Y+Q(X,Y)*U = 0.
BOUNDARY.  U = TRUE(X,Y)  ON Y = 0.  $ ON X = 1.
                          ON Y = 1.  $ ON X = 0.
GRID.         5 X POINTS $ 5 Y POINTS
FORTRAN.
         DO 100 ICASE = 1, 3
            IF( ICASE .EQ. 1 ) THEN
               IDGR  = 1
               ISMTH = 0
            ELSE IF( ICASE .EQ. 2 ) THEN
               IDGR  = 3
               ISMTH = 1
            ELSE IF( ICASE .EQ. 3 ) THEN
               IDGR  = 3
               ISMTH = 2
            END IF
DISCRET.  SPLINE GALERKIN (DEGREE=IDGR,NDERV=ISMTH)
INDEXING. AS IS
SOLUTION. LINPACK SPD BAND
OUTPUT.   MAX( ERROR )
FORTRAN.
  100 CONTINUE
SUBPROGRAMS.
         FUNCTION P(X,Y)
         P = - EXP(X*Y)
         RETURN
         END
         FUNCTION Q(X,Y)
         Q = 2.*(X**2 + Y**2)*EXP(X*Y)
         RETURN
         END
         FUNCTION TRUE(X,Y)
         TRUE = EXP(X*Y)
         RETURN
         END
END.
```

Singularities. The program given below determines experimentally the effect of the corner singularities on the error for three examples in Chapter 3, §7.

The program solves the five-point approximation to $-\nabla^2 u = 0$ on the unit square and $u = g(\mathbf{y})$ on the boundary. The boundary values are set equal to TRUE(X,Y) which is controlled by the parameter ISING, passed by COMMON /COMSIN/. Depending on ISING, the function TRUE(X,Y) is one of

$$u = \operatorname{Re}[z^2 \log z], \qquad u = \operatorname{Re}[z^{2/3}], \qquad u = \operatorname{Re}[z^{1/2}],$$

which has the appropriate asymptotic behavior.

FFT is used to obtain the solutions quickly on a very fine mesh. The norms of the errors are saved and tabulated at the end of the run.

```
EQUATION.    UXX + UYY = 0.
BOUNDARY.    U = TRUE(X,Y)  ON X = 0. $ ON Y = 0.
                            ON X = 1. $ ON Y = 1.
OPTIONS.     MAX X POINTS = 129 $ MAX Y POINTS = 129
DECLARATIONS.
        COMMON /COMSIN/ SAVEE(6,3), H(6), ISING
FORTRAN.
        DO 200 IGRID = 1, 6
           IPNTS     = 1 + 2**IGRID
           H(IPNTS) = 1./FLOAT(IPNTS-1)
           DO 100 ISING = 1, 3
GRID.          IPNTS X POINTS $ IPNTS Y POINTS
TRIPLE.        FFT 9 POINT (IORDER=2)
OUTPUT.        NORM(ERROR)
FORTRAN.
               SAVEE(IGRID,ISING) = R1NRMI
FORTRAN.
  100      CONTINUE
  200   CONTINUE
        DO 400 ISING = 1, 3
           PRINT 300,
     A        ISING, (H(I),(SAVEE(I,ISING),I=1,2)
  300      FORMAT(//22H   SUMMARY, SINGULARITY, I5//
     A        31H      H       MAX ERROR     L2 ERROR/
     B        (F9.6,2E12.4))
  400   CONTINUE
SUBPROGRAMS.
        FUNCTION TRUE(X,Y)
        COMMON / COMSIN / SAVEE(6,3),H(6),ISING
        DATA TWOTHR / 0.66666666666666 /
        TRUE = 0.
        R    = SQRT( X**2 + Y**2 )
        IF( R .GT. 0. ) THEN
           THETA = ATAN2(Y,X)
           IF( ISING .EQ. 1 )
     A        TRUE = R**2( ALOG(R)*COS(2.*THETA)
     B                   - SIN(2.*THETA)*THETA )
           IF( ISING .EQ. 2 )
     A        TRUE = R**TWOTHR*SIN(TWOTHR*THETA)
           IF( ISING .EQ. 3 )
     A        TRUE = SQRT(R)*SIN(0.5*THETA)
        ENDIF
        END
END.
```

Here the DECLARATIONS segment is used to define the arrays in the user's program and to pass the parameter ISING; they are not needed by any ELLPACK generated subprogram.

10. Additional capabilities. Although originally designed to solve linear second-order elliptic problems, more general problems can be solved. Systems of equations can be solved by parameterizing the coefficients and right side of the DE; we illustrate this with a program to solve the fourth-order freely supported plate problem. We use the Plateau problem to illustrate solution of nonlinear problems.

Biharmonic equation. The biharmonic problem for the loaded simply supported flat plate, $\nabla^4 u = f(x,y)$, on Ω, $u = \nabla^2 u = 0$ on $\partial\Omega$, factors into a pair of Poisson problems each with homogeneous Dirichlet boundary conditions: $\nabla^2 v = f(x,y)$, $\nabla^2 u = v(x,y)$. An ELLPACK program to approximate u is outlined below:

```
EQUATION.        UXX + UYY = F(X,Y,IPASS)
BOUNDARY.        U = 0 ON . . .
GRID.            . . .
GLOBAL.
         COMMON / COMBIH / IPASS
FORTRAN.
         DO 100 IPASS = 1, 2
DISCRETIZATION.        5 POINT STAR
INDEXING.              AS IS
SOLUTION.              BAND GE
FORTRAN.
   100 CONTINUE
OUTPUT.          . . .
SUBPROGRAMS.
         FUNCTION F(X,Y,IPASS)
         IF(IPASS .EQ. 1) F = . . .
         IF(IPASS .EQ. 2) F = U(X,Y)
         RETURN
         END
END.
```

When IPASS is 1, the approximate solution of $\nabla^2 v = f(x,y)$ is obtained. After the SOLUTION module has been executed, the ELLPACK function U(X,Y) evaluates this approximation. Then, when IPASS is equal to 2, the right side of the Poisson equation is set equal to U(X,Y) $\approx v(x,y)$.

Here the named COMMON containing IPASS is declared in a GLOBAL segment, because this COMMON block must also appear in the subprogram (generated by the preprocessor) which evaluates the right side of the DE.

Plateau problem. The next program obtains an approximate solution to the Plateau problem by the Newton-Kantorovich method.[15] The

[15] See Chapter 6, §10, for the formulas and numerical results. The program we used was the first ELLPACK program to solve the Plateau problem. A simpler program is given by J. R. Rice in Birkhoff-Schoenstadt [84]; we give his version here, which uses features which were unavailable when we did our experiments.

subprograms which evaluate the boundary function $g(\mathbf{y})$ are not shown.

```
EQ.  (1.+UY(X,Y)**2)*UXX +(1.+UX(X,Y)**2)*UYY           &
          - 2.*UX(X,Y)*UY(X,Y)*UXY                      &
  +2.*(UX(X,Y)*UYY(X,Y)-UY(X,Y)*UXY(X,Y))*UX            &
  +2.*(UY(X,Y)*UXX(X,Y)-UX(X,Y)*UXY(X,Y))*UY            &
  =2.*(UX(X,Y)*UYY(X,Y)-UY(X,Y)*UXY(X,Y))*UX(X,Y)       &
  +2.*(UY(X,Y)*UXX(X,Y)-UX(X,Y)*UXY(X,Y))*UY(X,Y)
BO.  U = G1(X) ON Y = 0.  $ U = G2(Y) ON X = 1.
     U = G3(X) ON Y = 1.  $ U = G4(Y) ON X = 0.
GRID.           5 X POINTS $ 5 Y POINTS
TRIPLE.          SET U = 0.
FORTRAN.
        DO 100 IT = 1, 8
DISCRETIZATION.    INTERIOR COLLOCATION
INDEXING.          AS IS
SOLUTION.          BAND GE
OUTPUT.            TABLE(U)
FORTRAN.
   100 CONTINUE
SUBPROGRAMS.
          FUNCTION G1(X,Y)
            . . .
END.
```

The coefficients of the DE include UX and UY, and the right side uses also UXX, UXY, and UYY. These ELLPACK generated Fortran functions are available only after the execution of a SOLUTION or TRIPLE segment. The TRIPLE SET, which appears above after the GRID segment, initializes U and its derivatives to zero, and thus, in the first pass through the DO-loop, the Laplace equation is discretized. After each execution of BAND GE the values and derivatives of the current approximation are available, and these are used during the next discretization.

Bibliography

Frequently used references:

[FW] = Forsythe and Wasow [1960].
[HY] = Hageman and Young [1981].
[K] = Kellogg [1929].
[KK] = Kantorovich and Krylov [1958].
[V] = Varga [1962].
[Y] = Young [1971].

Adams, R. A., [1975]: *Sobolev Spaces*, Academic Press.

Ahlberg, J. H., E. N. Nilson, and J. L. Walsh, [1967]: *The Theory of Splines and Their Applications*, Academic Press.

Ahlfors, L. V., [1979]: *Complex Analysis*, 3rd ed., McGraw-Hill.

Anselone, P., (ed.), [1964]: *Nonlinear Integral Equations*, University of Wisconsin Press.

Arthurs, A. M., [1970]: *Complementary Variational Principles*, Oxford University Press.

Atkinson, K., [1976]: *A Survey of Numerical Methods for the Solution of Fredholm Integral Equations of the Second Kind,* Society for Industrial and Applied Mathematics.

Aziz, A. K., (ed.), [1969]: *Lecture Series in Differential Equations*, vol. II, Van Nostrand Mathematical Study 19, Van Nostrand.

Aziz, A. K., (ed.), [1972]: *The Mathematical Foundations of the Finite Element Method, with Applications to Partial Differential Equations*, Academic Press.

Bank, R. E., [1977]: "Marching algorithms for elliptic boundary value problems II. The variable coefficient case", SIAM J. Numer. Anal. 14, 950–970.

Bank, R. E., and D. J. Rose, [1975]: "An $O(n^2)$ method for solving constant coefficient boundary value problems in two dimensions", SIAM J. Numer. Anal. 12, 529–540.

Bank, R. E., and D. J. Rose, [1976]: "Extrapolated fast direct algorithms for elliptic boundary value problems", pp. 201–250 of J. E. Traub (ed.), *Algorithms and Complexity*, Academic Press.

Barker, V. A., (ed.), [1977]: *Sparse Matrix Techniques*, Lecture Notes in Mathematics 572, Springer.

Beckenbach, E. F., (ed.), [1952]: *Construction and Application of Conformal Maps*, NBS Applied Mathematics Series 18, National Bureau of Standards.

BERGMAN, S., AND J. G. HERRIOT, [1961]: "Applications of the method of the kernel function for solving boundary-value problems", Numer. Math. 3, 209–225.

BERGMAN, S., AND M. SCHIFFER, [1953]: *Kernel Functions and Elliptic Differential Equations in Mathematical Physics*, Academic Press.

BERS, L., F. JOHN, AND M. SCHECHTER, [1964]: *Partial Differential Equations*, Interscience.

BIRKHOFF, G., [1971]: *The Numerical Solution of Elliptic Equations*, CBMS-NSF Regional Conference Series in Applied Mathematics 1, Society for Industrial and Applied Mathematics.

BIRKHOFF, G., AND T. C. BARTEE, [1970]: *Modern Applied Algebra*, McGraw-Hill.

BIRKHOFF, G., C. DE BOOR, B. SWARTZ, AND B. WENDROFF, [1966]: "Rayleigh-Ritz approximation by piecewise cubic polynomials", SIAM J. Numer. Anal. 3, 188–203.

BIRKHOFF, G., AND S. GULATI, [1974]: "Optimal few-point discretizations of linear source problems", SIAM J. Numer. Anal. 11, 700–728.

BIRKHOFF, G., AND S. MAC LANE, [1977]: *A Survey of Modern Algebra*, 4th ed., Macmillan.

BIRKHOFF, G., AND G-C. ROTA, [1978]: *Ordinary Differential Equations*, 3rd ed., Wiley.

BIRKHOFF, G., AND A. SCHOENSTADT (eds.), [1984]: *Elliptic Problem Solvers* II, Academic Press.

BIRKHOFF, G., M. H. SCHULTZ, AND R. S. VARGA, [1968]: "Piecewise Hermite interpolation in one and two variables with applications to partial differential equations", Numer. Math. 11, 232–256.

BIRKHOFF, G., AND R. S. VARGA, [1959]: "Implicit alternating direction methods", Trans. Am. Math. Soc. 92, 13–24.

BIRKHOFF, G., AND R. S. VARGA (eds.), [1970]: *Numerical Solution of Field Problems in Continuum Physics*, SIAM-AMS Proceedings II, American Mathematical Society.

BIRKHOFF, G., R. S. VARGA, AND D. M. YOUNG, [1962]: "Alternating direction implicit methods", Advances in Computers 3, 189–273.

BIRKHOFF, G., D. M. YOUNG, AND E. H. ZARANTONELLO, [1953]: "Numerical meth-ods in conformal mapping", Proc. Symposia in Applied Mathematics IV, American Mathematical Society, 117–140.

BRAMBLE, J. H., AND S. R. HILBERT, [1970]: "Error estimates for spline interpolants", SIAM J. Numer. Anal. 7, 112–124.

BRAMBLE, J. H., AND S. R. HILBERT, [1971]: "Bounds for a class of linear functionals with applications to Hermite interpolation", Numer. Math. 16, 362–369.

BRAMBLE, J. H., AND B. E. HUBBARD, [1964]: "Approximation of derivatives by difference methods in elliptic boundary value problems", Contributions to Differential Equations 3, 399–410.

BRAMBLE, J. H., AND B. E. HUBBARD, [1964a]: "New monotone type approximations for elliptic problems", Math. Comp. 18, 349–367.

BRAMBLE, J. H., AND B. E. HUBBARD, [1964b]: "On a finite difference analogue of an elliptic boundary problem which is neither diagonally dominant nor of non-negative type", J. Math. and Phys. 43, 117–132.

BRAMBLE, J. H., B. E. HUBBARD, AND V. THOMEE, [1969]: "Convergence estimates for essentially positive type discrete Dirichlet problems", Math. Comp. 23, 695–709.

BRAMBLE, J. H., AND A. H. SCHATZ, [1970]: "On the numerical solution of elliptic boundary value problems by least squares approximation of the data", in Hubbard [71, pp. 107–131].

BRANDT, A., [1977]: "Multi-level adaptive solution to boundary value problems", Math. Comp. 31, 333–391.

BUNCH, J. R., AND D. J. ROSE (eds.), [1976]: *Sparse Matrix Computations*, Academic Press.

BUTZER, P. L., [1971]: *Fourier Analysis and Approximation*, Birkhäuser.

BUZBEE, B. L., G. H. GOLUB, AND C. W. NIELSON, [1970]: "The method of odd/even reduction and factorization with application to Poisson's equations", SIAM J. Numer. Anal. 7, 617–656.

BUZBEE, B. L., F. W. DORR, J. A. GEORGE, AND G. H. GOLUB, [1971]: "The direct solutions of the discrete Poisson equation on irregular regions", SIAM J. Numer. Anal. 8, 722–736.

CIARLET, P. G., [1978]: *The Finite Element Method for Elliptic Problems*, North-Holland.

COFFMAN, C. V., AND G. FIX (eds.), [1980]: *Constructive Approximation to Mathematical Models*, Academic Press.

COLLATZ, L., [1960]: *Numerical Treatment of Differential Equations*, 3rd ed., Springer.

COLLATZ, L., [1978]: "Application of multivariate approximation to the solution of boundary value problems", in *Multivariate Approximation* (Sympos. Univ. Durham, Durham, 1977), 12–29, Academic Press.

CONCUS, P., AND G. H. GOLUB, [1973]: "Use of fast direct methods for the efficient numerical solution of nonseparable elliptic equations", SIAM J. Numer. Anal. 10, 1103–1120.

CONTE, S. D., AND C. DE BOOR, [1980]: *Elementary Numerical Analysis*, 3rd ed., McGraw-Hill.

COTTLE, R. W., AND C. E. LEMKE (eds.), [1976]: *Nonlinear Programming*, SIAM-AMS Proceedings IX, American Mathematical Society.

COURANT, R., K. FRIEDRICHS, AND H. LEWY, [1928]: Math. Ann. 100, 32–74; translated by Phyllis Fox: "On the partial difference equations of mathematical physics", IBM J. Research 11 (1967), 215–238.

COURANT, R., [1943]: "Variational methods for the solution of problems of equilibrium and vibrations", Bull. American Mathematical Society 49, 1–23.

COURANT, R., [1950]: *Dirichlet's Principle*, Interscience.

COURANT, R., AND D. HILBERT, [1953, 1962]: *Methods of Mathematical Physics*, vols. I, II, Interscience.

COWELL, W., [1984]: *Sources and Development of Mathematical Software*, Prentice-Hall.

CURRY, H. B., AND I. J. SCHOENBERG, [1966]: "On Pólya frequency functions. IV: The fundamental spline functions and their limits", J. d'Analyse Math. 17, 71–107.

DANIELS, R. W., [1978]: *An Introduction to Numerical Methods and Optimization Techniques*, North-Holland.

DANTZIG, B. G, AND B. C. EAVES (eds.), [1974]: *Studies in Optimization*, Mathematical Association of America.

DAVIS, P. J., [1963]: *Interpolation and Approximation*, Blaisdell.

DAVIS, P. J., AND P. RABINOWITZ, [1961]: "Advances in orthonormalizing computation", Advances in Computers 2, 55–133.

DE BOOR, C., (ed.), [1974]: *Mathematical Aspects of Finite Elements . . .*, Academic Press.

DE BOOR, C., [1978]: *A Practical Guide to Splines*, Applied Mathhematical Sciences Series 27, Springer.

DE BOOR, C., AND G. FIX, [1973]: "Spline approximation by quasi-interpolants", J. Approx. Theory 8, 19–45.

DE BOOR, C., AND G. H. GOLUB (eds.), [1978]: *Recent Advances in Numerical Analysis*, Academic Press.

DELVES, L. M., AND J. WALSH (eds), [1974]: *Numerical Solution of Integral Equations*, Clarendon Press.

DENNIS, J. E., AND J. J. MORÉ, [1977]: "Quasi-Newton methods, motivation and theory", SIAM Rev. 19, 46–89.

DE VEUBEKE, B. F., [1971]: *Upper and Lower Bounds in Matrix Structural Analysis*, Pergamon.

DORR, F. W., [1970]: "The direct solution of the discrete Poisson equation on a rectangle", SIAM Rev. 12, 248–263.

DOUGLAS, J., JR., AND H. RACHFORD, [1956]: "On the numerical solution of heat conduction problems in two and three space variables", Trans. American Mathematical Society 82, 421–439.

DUFF, I. S., [1977]: "A survey of sparse matrix research", Proc. IEEE, 65, 500–525.

DUFF, I. S., AND J. K. REID, [1979]: "Some design features of a sparse matrix code", TOMS 5, 18–35.

DUFF, I. S., AND G. W. STEWART (eds.), [1979]: *Sparse Matrix Proceedings*, 1978, Society for Industrial and Applied Mathematics.

DUFFIN, R. J., [1956]: "Discrete potential theory", Duke Math. J. 23, 233–251.

DUPONT, T., R. P. KENDALL, AND H. H. RACHFORD, Jr., [1968]: "An approximate factorization procedure for self adjoint elliptic difference equations", SIAM J. Numer. Anal. 5, 559–573.

DUPONT, T., AND R. SCOTT, [1980]: "Polynomial approximation of functions in Sobolev spaces", Math. Comp. 34, 441–463.

ENGELI, M., TH. GINSBURG, H. RUTISHAUSER, AND E. STIEFEL, [1959]: "Refined iterative methods . . . self-adjoint boundary value problems", Mitt. Inst. ang. Math. ETH Zurich, No. 8, Birkhäuser.

EVANS, D. J., [1967]: "The use of preconditioning in iterative methods for solving linear equations with symmetric positive definite matrices", J. Inst. Math. Appl. 4, 295–314.

FAIRWEATHER, G., J. F. RIZZO, D. J. SHIPPY, AND Y. S. WU, [1979]: "On the numerical solution of two-dimensional potential problems by an improved boundary integral equation method", J. Comp. Phys. 31, 96–112.

FIKE, C. T., [1968]: *Computer Evaluation of Mathematical Functions*, Prentice-Hall.

FIX, G. J., AND G. STRANG, [1969]: "Fourier analysis of the finite element method in Ritz-Galerkin theory", Studies in Applied Math. 48, 265–273.

FLETCHER, R., (ed.), [1969]: *Optimization*, Academic Press.

FORNBERG, B., [1980]: "A numerical method for conformal mappings", SIAM J. Sci. Stat. Comp. 1, 386–400.

FORSYTHE, G. E., AND C. B. MOLER, [1967]: *Computer Solutions of Linear Algebraic Systems*, Prentice-Hall.

FORSYTHE, G. E., AND W. R. WASOW, [1960]: *Finite Difference Methods for Partial Differential Equations*, Wiley.

FOX, L., [1948]: "A short account of relaxation methods", Quart. J. Mech. Appl. Math. 1, 253–280.

FOX, L., [1950]: "The numerical solution of elliptic differential equations when the boundary conditions involve a derivative", Phil. Trans. Roy. Soc. London A242, 345–378.

FOX, L., [1962]: *Numerical Solution of Ordinary and Partial Differential Equations*, Addison-Wesley.

FOX, L., [1971]: "Some experiments with singularities in linear elliptic partial differential equations", Proc. Roy. Soc. London, A323, 179–190.

FOX, L., P. HENRICI, AND C. MOLER, [1967]: "Approximations and bounds for eigenvalues of elliptic operators", SIAM J. Numer. Anal. 4, 89–102.

FRANK, Ph., AND R. VON MISES, [1930, 1935]: *Die Differential- und Integralgleichungen der Mechanik und Physik*, 2 vols., Vieweg, Braunschweig.

FRANKEL, S. P., [1950]: "Convergence rates of iterative treatments of partial differential equations", Math. Tables Aids Comp. 4, 65–75.

GAIER, D, [1964]: *Konstruktive Methododen der konformen Abbildung*, Springer.

GALLIGANI, I., AND E. MAGENES (eds.), [1977]: *Mathematical Aspects of the Finite Element Method*, Lecture Notes in Mathematics 606, Springer.

GARABEDIAN, H. L., (ed.), [1965]: *Approximation of Functions*, Elsevier.

GARABEDIAN, P. R., [1956]: "Estimation of the relaxation factor for small mesh size," Math. Tables Aids Comput. 10 183–185.

GARABEDIAN, P. R., [1964]: *Partial Differential Equations*, Wiley.

GEORGE, A., [1973]: "Nested dissection of a regular finite element mesh", SIAM J. Numer. Anal. 10, 345–363.

GEORGE, A., AND J. LIU, [1981]: *Computer Solution of Large Sparse Positive Definite Problems*, Prentice-Hall.

GELINAS, R. J., S. K. DOSS, AND K. MILLER, [1981]: "The moving finite element method," J. Comput. Phys. 40, 202–249.

GERSCHGORIN, G., [1930]: "Fehlerabschätzung für das Differenzen-verfahren zur Lösung partieller Differentialgleichungen", ZaMM 10, 373–382.

GILBARG, D., AND N. S. TRUDINGER, [1977]: *Elliptic Partial Differential Equations of Second Order,* Springer.

GLASSTONE, S., AND M. EDLUND, [1952]: *Elements of Nuclear Reactor Theory*, Van Nostrand.

GOLUB, G. H., AND R. S. VARGA, [1961]: "Chebyshev semi-iterative methods, successive overrelaxation iterative methods, and second order Richardson iterative methods", Numer. Math. 3, 147–156 (Part I), 157–168 (Part II).

GOULD, S. H., [1964]: *Variational Methods in Mathematical Physics*, Macmillan.

GREGORY, J. A., D. FISHELOV, B. SCHIFF, AND J. R. WHITEMAN, [1978]: "Local mesh refinement with finite elements for elliptic problems", J. Comp. Phys. 29, 133–140.

HAGEMAN, L. A., AND D. M. YOUNG, [1981]: *Applied Iterative Methods*, Academic Press.

HAGEMAN, L. A., AND R. S. Varga, [1964]: "Block iterative methods for cyclically reduced matrix equations", Numer. Math. 6 106–119.

HART, JOHN F., et al., [1968]: *Computer Approximations*, Wiley.

HENRICI, P., [1981]: *Essentials of Numerical Analysis, with Pocket Calculator Demonstrations*, Wiley.

HESTENES, M., [1980]: *Conjugate Direction Methods in Optimization*, Springer.

HOCKNEY, R. W., [1965]: "A fast direct solution of Poisson's equation using Fourier analysis", J. Assoc. Comp. Mach. 12, 95–113.

HOCKNEY, R. W., [1970]: "The potential calculation and some applications", Methods in Comp. Physics 9, 135–211.

HOFMANN, P., [1967]: "Asymptotic expansions of the discretization error of boundary value problems of the Laplace equation in rectangular domains", Numer. Math. 9, 302–322.

HOUSTIS, E. N., R. E. LYNCH, T. S. PAPATHEODOROU, AND J. R. RICE, [1978]: "Evaluation of numerical methods for elliptic partial differential equations", J. Comp. Physics 27, 323–350.

HUBBARD, B., (ed.), [1971]: *Numerical Solution of Partial Differential Equations* II, Academic Press.

HULL, T. E., (ed.), [1970]: *Studies in Optimization* 1, Society for Industrial and Applied Mathematics.

ISAACSON, E., AND H. B. KELLER, [1966]: *Analysis of Numerical Methods*, Wiley.

JACKSON, D., [1930]: *The Theory of Approximation*, American Mathematical Society.

JAKOB, Max, [1950, 1957]: *Heat Transfer*, vols. I, II, Wiley.

JASWON, M. A., AND G. T. SYMME, [1977]: *Integral Equations Methods in Potential Theory and Electrostatics*, Academic Press.

JEANS, James, [1941]: *The Mathematical Theory of Electricity and Magnetism*, 5th ed., Cambridge University Press.

JOYCE, D. C., [1971]: "Survey of extrapolation processes in numerical analysis", SIAM Rev. 13, 435–490.

KAHAN, W., [1958]: "Gauss-Seidel methods of solving large systems of linear equations", Ph.D. Thesis, Univ. of Toronto.

KANTOROVICH, L. V., AND V. I. KRYLOV, [1958]: *Approximate Methods of Higher Analysis*, Noordhoff-Interscience.

KARLIN, S., [1976]: *Studies in Spline Functions and Approximation Theory*, Academic Press.

KELLOGG, O. D., [1929]: *Foundations of Potential Theory*, Springer; Dover reprint, 1953.

KINCAID, D. R., AND D. M. YOUNG, [1979]: "Survey of iterative methods", in J. Belzer, A. G. Holtzman, and A. Kent (eds.), *Encyclopedia of Computer Science and Technology* 13, pp. 354–391.

KOBER, H., [1952]: *Dictionary of Conformal Representations*, Dover.

LAASONEN, P., [1957]: "On the degree of convergence of discrete approximation for the solutions of the Dirichlet problem", Ann. Acad. Sci. Fenn. Ser. A246, 1–19.

LAASONEN, P., [1958]: "On the solution of Poisson's difference equation", J. Assoc. Comp. Mach. 5, 370–382.

LAMB, H., [1932]: *Hydrodynamics*, 6th ed., Cambridge University Press.

LANCZOS, C., [1952]: "Solution of systems of linear equations by minimized iterations", J. Res. Nat. Bur. Standards 49, 33–53.

LANCZOS, C., [1956]: *Applied Analysis*, Prentice-Hall.

LANGER, R. E., (ed.), [1959]: *On Numerical Approximation*, University of Wisconsin Press.

LANGER, R. E., (ed.), [1960]: *Frontiers of Numerical Mathematics*, University of Wisconsin Press.

LANGER, R. E., (ed.), [1960a]: *Boundary Problems in Differential Equations*, University of Wisconsin Press.

LOOTSMA, F. A., (ed.), [1972]: *Numerical Methods for Non-linear Optimization*, Academic Press.

LORENTZ, G. G., K. JETTER, and S. D. RIEMENSCHNEIDER, [1983]: *Birkhoff Interpolation*, Addison-Wesley.

LYNCH, R. E., AND J. R. RICE, [1968]: "Convergence rates of ADI methods with smooth initial error", Math. Comp. 22, 311–335.

LYNCH, R. E., AND J. R. RICE, [1978]: "High accuracy finite difference approximation to solutions of elliptic partial differential equations", Proc. Nat. Acad. Sci. 75, 2541–2544.

LYNCH, R. E., J. R. RICE, AND D. H. THOMAS, [1964]: "Tensor product analysis of partial difference equations", Bull. American Mathematical Society 70, 378–384.

LYNCH, R. E., J. R. RICE, AND D. H. THOMAS, [1964a]: "Direct solution of partial difference equations by tensor product methods", Numer. Math. 6, 185–199.

LYNCH, R. E., J. R. RICE, AND D. H. THOMAS, [1965]: "Tensor product analysis of alternating direction implicit methods", J. SIAM 13, 995–1006.

MANSFIELD, L., [1974]: "On the variational approach to defining splines on L-shaped regions", J. Approx. Theory 12, 99–112.

MEINARDUS, G., [1967]: *Approximation of Functions, Theory and Numerical Methods*, Springer.

MENIKOFF, R., AND C. ZEMACH, [1980]: "Methods for numerical conformal mapping", J. Comp. Phys. 36, 366–410.

MIKHLIN, S. G., [1964]: *Variational Methods in Mathematical Physics*, Macmillan.

MIKHLIN, S. G., [1971]: *The Numerical Performance of Variational Methods*, Noordhoff.

MILLER, J. J. H., (ed.), [1973, 1976, 1977]: *Topics in Numerical Analysis*, I, II, III, Academic Press.

MILLER, R., AND J. THATCHER (eds.), [1972]: *Complexity of Computer Computations*, Plenum Press.

MORREY, C. B., [1966]: *Multiple Integrals in the Calculus of Variations*, Springer.

MORREY, C. B., [1969]: "Differentiability theorems for weak solutions of nonlinear elliptic differential equations", Bull. American Mathematical Society 4, 684–705.

MORSE, P. M., AND H. FESHBACH, [1953]: *Methods of Theoretical Physics*, vols. I, II, McGraw-Hill.

MUSHKELISHVILI, N. I., [1954]: *Some Fundamental Problems of the Theory of Elasticity*, 4th ed., Noordhoff.

MUSKAT, M., [1937]: *Flow of Homogeneous Fluids through Porous Media*, McGraw-Hill.

NASH, J. C., [1979]: *Compact Numerical Methods for Computers, Linear Algebra and Function Minimization*, Wiley.

NICOLAIDES, R. A., [1979]: "On some theoretical and practical aspects of multigrid methods", Math. Comp. 33, 933–952.

NICOLESCO, M., [1936]: *Les Fonctions Polyharmoniques*, Hermann.

NITSCHE, J., AND J. C. C. NITSCHE, [1960]: "Error estimates for the numerical solution of elliptic differential equations", Arch. Rational Mech. Anal. 5, 293–306 and 307–314.

ORTEGA, J., AND W. RHEINBOLDT, [1970]: *Iterative Solution of Nonlinear Equations in Several Variables*, Academic Press.

PAPATHEODOROU, T., [1979]: "Fast tensor product Poisson solvers for the Dirichlet problem on a rectangle", Ann. Sci. Math. Québec 3, 135–141.

PARTER, S. V., (ed.), [1979]: *Numerical Methods for Partial Differential Equations*, Academic Press.

PEACEMAN, D. W., AND H. H. RACHFORD, JR., [1955]: "The numerical solution of parabolic and elliptic differential equations", J. SIAM 3, 28–41.

PEREYRA, V., [1966]: "On improving an approximate solution of a functional equation by deferred corrections", Numer. Math. 8, 376–391.

PEREYRA, V., [1967]: "Accelerating the convergence of discretization algorithms", SIAM J. Numer. Anal. 4, 508–533.

PEREYRA, V., [1968]: "Iterated deferred corrections for nonlinear boundary value problems", Numer. Math. 11, 111–125.

PHILIPS, H. B., AND N. WIENER, [1923]: "Nets and the Dirichlet problem", J. Math. and Phys. 2, 105–124.

PÓLYA, G., [1954]: "Estimates for eigenvalues", in *Studies in Mathematics and Mechanics Presented to Richard von Mises*, Academic Press, 200–207.

PÓLYA, G., AND G. SZEGÖ, [1951]: *Isoperimetric Inequalities in Mathematical Physics*, Princeton University Press.

POWELL, M. J. D., [1970]: "A survey of numerical methods for unconstrained optimization", SIAM Rev. 12, 79–97.

PROTTER, M. H., AND H. F. WEINBERGER, [1967]: *Maximum Principles in Differential Equations*, Prentice-Hall, Chap. 3.

RALSTON, A., AND H. S. WILF (eds.), [1960, 1967]: *Numerical Methods for Digital Computers*, vols. I, II, Wiley.

READ, R. C., (ed.), [1972]: *Graph Theory and Computing*, Academic Press.

REID, J. K., (ed.), [1971]: *Large Sparse Sets of Linear Equations*, Academic Press.

RHEINBOLDT, W., [1974]: *Methods for Solving Systems of Nonlinear Equations*, CBMS-NSF Regional Conference Series in Applied Mathematics 14, Society for Industrial and Applied Mathematics.

RICE, J. R., [1964, 1969]: *Approximation of Functions*, vols. I, II, Addison-Wesley.

RICE, J. R., (ed.), [1977]: *Mathematical Software* III, Academic Press.

RICE, J. R., AND R. F. BOISVERT, [1984]: *Solving Elliptic Problems Using ELLPACK,* Springer.

RICHARDSON, L. F., [1910]: "The approximate arithmetical solution by finite differences of physical problems involving differential equations, with an application to the stresses in a masonry dam", Philos. Trans. Roy. Soc. London A 210, 307–357.

RICHARDSON, L. F., [1927]: "The deferred approach to the limit I: Single lattice", and, with J. A. GAUNT, "II: Interpenetrating lattices", Philos. Trans. Roy. Soc. London A 226, 299–361.

ROACHE, Patrick J. [1978]: "Marching methods for elliptic problems", Numerical Heat Transfer 1, 1–25 and 163–181.

ROSE, D. J., AND R. A. WILLOUGHLY (eds.), [1972]: *Sparse Matrices and their Applications*, Plenum Press.

ROSSER, J. B., [1975]: "Nine point difference solutions for Poisson's equation", Comp. and Math. with Appls. 1, 351–360.

SAIGAL, R., AND M. J. TODD, [1978]: "Efficient acceleration techniques for fixed point algorithms", SIAM J. Numer. Anal. 15, 997–1007.

SARD, A., [1963]: *Linear Approximation*, Mathematical Surveys 9, American Mathematical Society.

SCHOENBERG, I., (ed.), [1969]: *Approximation with Special Emphasis on Spline Functions*, University of Wisconsin Press.

SCHULTZ, M. H., [1973]: *Spline Analysis*, Prentice-Hall.

SCHULTZ, M. H., (ed.), [1981]: *Elliptic Problem Solvers*, Academic Press.

SHORTLEY, G. H., AND R. WELLER, [1938]: "The numerical solution of Laplace's equation", J. Appl. Phys. 9, 334–344.

SOUTHWELL, R. V., [1940]: *Relaxation Methods in Engineering Science*, Oxford University Press.

SOUTHWELL, R. V., [1946]: *Relaxation Methods in Theoretical Physics*, Oxford University Press.

STAKGOLD, I., [1979]: *Green's Functions and Boundary Value Problems*, Dekker.

STEFFENSEN, J. F., [1950]: *Interpolation*, 2nd ed., Chelsea.

STEPHAN, E., AND W. WENDLAND, [1980]: *Mathematische Grundlagen der finiten Element-Methoden*, Peter Lang.

STEWART, G. W., [1973]: *Introduction to Matrix Computation*, Academic Press.

STIEFEL, E. L., [1954]: "Recent developments in relaxation techniques", Proc. Internat. Congress Math., Amsterdam, 1, Noordhoff, 384–391.

STIEFEL, E., [1956]: "On solving Fredholm integral equations", J. SIAM 4, 63–85.

STONE, H. L., [1968]: "Iterative solution of implicit approximations of multidimensional partial differential equations", SIAM J. Numer. Anal. 5, 530–558.

STRANG, G. H., AND G. FIX, [1973]: *Analysis of the Finite Element Method*, Prentice-Hall.

SWARZTRAUBER, P. N., [1974]: "A direct method for the discrete solution of separable elliptic equations", SIAM J. Numer. Anal. 11, 1136–1150.

SWARZTRAUBER, P. N., [1977]: "The methods of cyclic reduction, Fourier analysis and the FACR algorithm for the discrete solution of Poisson's equation on a rectangle", SIAM Rev. 19, 490–501.

SWEET, R. A., [1977]: "Cyclic reduction algorithm for solving Poisson's equation on a grid of arbitrary size", SIAM J. Numer. Anal. 14, 706–720.

SYNGE, J. L., [1957]: *The Hypercircle in Mathematical Physics*, Cambridge University Press.

SYNGE, J. L., AND B. A. GRIFFITH, [1949]: *Principles of Mechanics*, 2nd ed., McGraw-Hill.

TEMPLE, G., AND W. G. BICKLEY, [1933]: *Rayleigh's Principle and its Applications to Engineering*, Oxford University Press; Dover reprint, 1956.

THEODORSEN, T., [1931]: "Theory of wing sections of arbitrary shape", NACA Tech. Rep. 411.

THOMÉE, V, [1969]: "Discrete interior Schauder estimates for elliptic difference operators", SIAM J. Numer. Anal. 5, 626–645.

TIMAN, A. E., [1963]: *Theory of Approximation of Functions of a Real Variable*, Pergamon-Macmillan.

TIMOSHENKO, S., AND S. WOINOWSKY-KRIEGER, [1959]: *Theory of Plates and Shells*, McGraw-Hill.

TIMOSHENKO, S., AND J. N. GOODIER, [1951]: *Theory of Elasticity*, McGraw-Hill.

TODD, J., (ed.), [1955]: *Experiments in the Computation of Conformal Maps*, NBS Applied Mathematics Series 42, National Bureau of Standards.

TODD, J., (ed.), [1962]: *Survey of Numerical Analysis*, McGraw-Hill.

TRAUB, J. E., (ed.), [1973]: *Complexity of Sequential and Parallel Numerical Algorithms*, Academic Press.

TYCHONOV, A. N., AND A. A. SAMARSKII, [1964]: *Partial Differential Equations of Mathematical Physics*, vols. I, II, Holden-Day.

VAINBERG, M. M., [1973]: *Variational Method and Method of Monotone Operators in the Theory of Nonlinear Equations*, Wiley.

VARGA, R. S., [1957]: "A comparison of the successive overrelaxation method and semi-iterative methods using Chebyshev polynomials", J. SIAM 5, 39–46.

VARGA, R. S., [1959]: "Ordering of successive overrelaxation schemes", Pacific J. Math. 9, 925–936.

VARGA, R. S., [1962]: *Matrix Iterative Analysis*, Prentice-Hall.

VARGA, R. S., [1971]: *Functional Analysis and Approximation Theory in Numerical Analysis*, CBMS-NSF Regional Conference Series in Applied Mathematics 3, Society for Industrial and Applied Mathematics.

WACHSPRESS, E. L., [1966]: *Iterative Solutions of Elliptic Systems*, Prentice-Hall.

WACHSPRESS, E. L., AND G. J. Habetler, [1960]: "An alternating direction implicit iteration technique", J. SIAM 8, 403–424.

WALSH, J. L., [1969]: *Interpolation and Approximation*, 5th ed., American Mathemaical Society Colloquium Publ. XX.

WARSCHAWSKI, S. E., [1945]: "On Theodorsen's method of conformal mapping of nearly circular regions", Quart. Appl. Math. 3, 12–28.

WARSCHAWSKI, S., [1956]: "Recent results in numerical methods in conformal mapping", Proc. Symposia in Applied Mathematics VI, American Mathematical Society, 219–250.

WASOW, W. R., [1957]: "The accuracy of difference approximations to plane Dirichlet problems with piecewise analytic boundary values", Quart. Appl. Math. 15, 53–63.

WEINBERGER, H. K., [1974]: *Variational Methods for Eigenvalue Approximation*, CBMS-NSF Regional Conference Series in Applied Mathematics 15, Society for Industrial and Applied Mathematics.

WILKINSON, J. H., [1965]: *The Algebraic Eigenvalue Problem*, Oxford University Press, Oxford.

YOUNG, D. M., [1950]: "Iterative methods for solving partial difference equations of elliptic type", Ph.D. Thesis, Harvard University.

YOUNG, D. M., [1954]: "Iterative methods for solving partial difference equations of elliptic type", Trans. American Mathematical Society 76, 92–111.

YOUNG, D. M., [1971]: *Iterative Solution of Large Linear Systems*, Academic Press.

YOUNG, D. M., [1977]: "On the accelerated SSOR method for solving large linear systems", Advances in Math. 23, 215–271.

ZIENKIEWICZ, O. M., [1977]: *The Finite Element Method*, 3rd ed., McGraw-Hill.

Index